THE FOURTH FAMILY OF QUARKS AND LEPTONS

First International Symposium

ANNALS OF THE NEW YORK ACADEMY OF SCIENCES
Volume 518

THE FOURTH FAMILY OF QUARKS AND LEPTONS

First International Symposium

Edited by David B. Cline and Amarjit Soni

The New York Academy of Sciences
New York, New York
1987

Library of Congress Cataloging-in-Publication Data

The Fourth family of quarks and leptons/edited by David B. Cline and Amarjit Soni.
p. cm.—(Annals of the New York Academy of sciences; v. 518
"The First International Symposium on the Fourth Family of Quarks and Leptons was held in Santa Monica, California, from February 26 to 28, 1987"—Pref.
Bibliography: p.
Includes index.
1. Quarks—Congresses. 2. Leptons (Nuclear physics)—Congresses.
I. Cline, D. (David), 1933– . II. Soni, Amarjit.
III. International Symposium on the Fourth Family of Quarks and Leptons (1st : 1987 : Santa Monica, Calif.) IV. New York Academy of Sciences. V. Series.
Q11.N5 vol. 518
[QC793.5.Q252]
500 s—dc19
[539.7'21]

SP
Printed in the United States of America
ISBN 0-89766-435-3 (cloth)
ISBN 0-89766-436-1 (paper)
ISSN 0077-8923

ANNALS OF THE NEW YORK ACADEMY OF SCIENCES

Volume 518
December 31, 1987

THE FOURTH FAMILY OF QUARKS AND LEPTONS[a]

First International Symposium

Editors

DAVID B. CLINE AND AMARJIT SONI

CONTENTS

[a]These papers are the result of a conference entitled International Symposium on the Fourth Family of Quarks and Leptons held on February 26–28, 1987 in Santa Monica, California.

Financial assistance was received from:

- U.S. DEPARTMENT OF ENERGY (Grant Number DE-A503-S1ER40024, Task J)

Preface

DAVID B. CLINE AND AMARJIT SONI
Department of Physics
University of California
Los Angeles, California 90024

The First International Symposium on the Fourth Family of Quarks and Leptons was held in Santa Monica, California, from February 26 to 28, 1987. The spirit of the meeting was best described by Cecilia Jarlskog: "Not only is this a meeting of the Fourth Family but it is also a meeting of our family . . . ."

The meeting had a heavy schedule: almost 30 invited talks and about 20 contributed papers were delivered in less than three working days. While the main topic was certainly of a speculative character, several talks covered the status of various ongoing experiments and that of upcoming new machines. Some new experimental results were also reported, perhaps the most significant of which was the Argus result of $B_d - \overline{B}_d$ mixing.

It is of course very unlikely that the question of the existence or nonexistence of the fourth family will be settled in the near future. We have therefore promised to hold this symposium once every two years. By the time of the next meeting (February 1989) we hope that the crucial question of the mass of the top quark will be fully settled and the quest for the fourth family will be well on its way.

Low-Thrust Hadron Events with Isolated μ or e from PETRA

JAMES G. BRANSON

Massachusetts Institute of Technology
Cambridge, Massachusetts 02139

INTRODUCTION

Since the beginning of 1979, when PETRA first ran succeeding Doris as the highest energy electron positron collider in the world, the PETRA experiments have been searching for new kinds of particles and putting limits on their masses. Probably the most often repeated search, as the PETRA energy increased from 13 GeV to 46.78 GeV, was that for a new quark flavor. New quark flavors were searched for in a scan for the 1S state of toponium, by looking for a step in R, and, most sensitively, by looking for a step in the rate of spherical hadronic events containing muons. This is probably the one search technique that is sensitive to a charge $-1/3$ quark as well as to the charge $2/3$ top quark.

In the upper 400 MeV of PETRA's energy range, some evidence was found for such spherical events containing muons, first by Mark-J, which also found that the muons were quite isolated from hadronic energy, then by JADE, which used the same cuts as Mark-J to add supporting evidence. The excess of events found is particularly striking because the expected background rate need not be computed from Monte Carlo simulations, but can be directly measured from lower energy running using the same detectors, the same analysis, and the same cuts. Previous measurements of muons in hadron events have been used to determine the semileptonic branching ratios in b and c quark decay and the fragmentation functions for those quarks.[1]

MEASUREMENTS WITH MUONS

Early in 1984, PETRA was nearing the end of a long high-energy scan. With great difficulty the machine group had raised the maximum center of mass energy above 46 GeV, moving in 30-MeV steps to search for toponium and for other new particles. As the highest energies were approached, the background conditions in the experiments became somewhat worse. It was clear that, with the luminosity dwindling, it was pointless to try to push the energy up further.

By the end of the scan, each group was able to rule out the existence of the 1S state of toponium in the scanned region. For example, Mark-J found that at the 95% confidence level, $B_h \cdot \Gamma_{ee} < 3.0$ keV,[2] while 3.5 to 7.0 keV is expected for toponium. No significant increase in R was seen.

Up to 46.3 GeV everything else appeared to be normal also. However, in the energy bin above 46.3 GeV, Mark-J saw a slight increase (2 sigma) in the rate of low-thrust events[3] and a more significant increase in the rate of low-thrust events containing muons. The expected background rate in this sensitive indicator of new physics was quite small so that even a few extra events could be significant. In order to get further information out of the small number of events found, Mark-J compared low-thrust events from an open top quark pair production Monte Carlo to those low-thrust events from the standard background processes that are weak decays of *c* and *b* quarks in hadron events having three jets. They found that the muons from the top quark decay should be isolated, while the muons from the background processes nearly always occur inside a hadron jet. Trying several variables to compare the isolation of the excess events above 46.3 GeV to those below, Mark-J found that the high-energy events had isolated muons like in the top Monte Carlo. The variables tried were the energy in a 30 degree cone around the muons' direction, the thrust using the muon as the thrust axis, the energy flow vs. angle from the muon direction, and the angle between the muon direction and the thrust axis. All of these gave similar results and showed no significant deviation from the top Monte Carlo. At this point Mark-J presented its data to the DESY community and requested an additional 3 pb^{-1} running at 46.57 GeV and the DESY directorate agreed to 1 pb^{-1}.

By the end of the additional running, for center of mass energies larger than 46.3 GeV, Mark-J had 8 events with thrust less than 0.8, the standard cut used for searching for open production of top. Just 1.9 events were expected based on previous data. The probability of observing 8 or more when 1.9 are expected is 8×10^{-4}. When dealing with the statistics of small numbers one has to be quite careful that physicists did not unwittingly stack the odds in favor of an anomaly. Were the cuts chosen to maximize the signal? How many variables were checked before finding one anomaly? Is the small background estimated correctly? In this case, the thrust cut in the hadron events containing muons was set up before seeing any signal. Results from the same cuts were published in other energy ranges to rule out open production of top. There are only a few new particle searches that were checked as consistently as this one. In addition the 46.3 GeV cut was set up before seeing the data. In fact our lowest energy event comes at 46.39 GeV.

On the other hand, adding an isolation criterion was done after seeing the data and noting that the muons were isolated. Therefore statistics using this cut could be questioned somewhat as far as the Mark-J data are concerned. Data from the other experiments were not seen in setting up this cut and are therefore independent.

FIGURE 1 shows the plot of events in the thrust cos δ plane, where δ is the angle between the thrust axis and the muon's direction. The region with thrust less than 0.8 and cos δ less than 0.7 is used to select low-thrust hadron events with isolated muons. Two energy regions are shown in the figure, the scan below 46.3 GeV and above 46.3 GeV. Higher statistics data taken around 36 GeV agree well with the lower energy scan data, as do Monte Carlo simulations with five quark flavors produced. Above 46.3 GeV we select 7 events, while 0.8 is expected based on previous data. The probability of finding 7 or more with 0.8 expected is just 2×10^{-5}.

The higher than normal background during the high-energy running did not affect the measurement of the muons or of the thrust of the events. This was checked by overlaying beam gate triggers with actual events and finding that the thrust and muon identification did not change. Indeed the muon identification in these events with isolated muons can be checked much better than for the normal kind of hadron event containing muons because in the normal event, the muon is in a jet and not only do

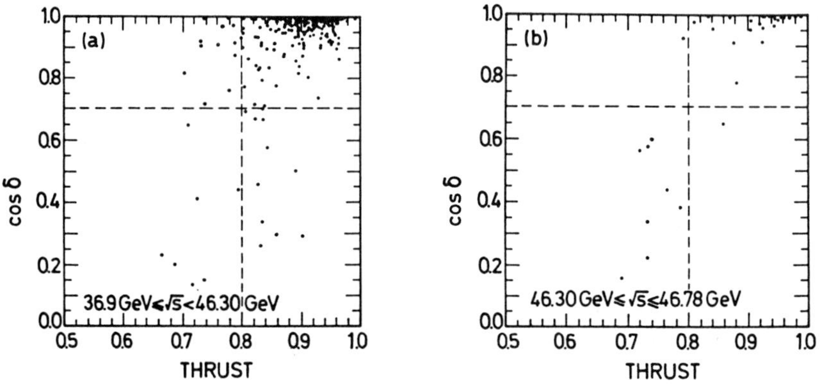

FIGURE 1. Cos δ vs. thrust for Mark-J events containing muons.

decay and punchthrough become more probable, but matching the muon chamber track with data from the inner detector is more difficult. In all of the eight events found above 46.3 GeV, the muon is clearly identified and gives consistent data near the vertex, after one interaction length of shower counters, after three interaction lengths, and outside the six interaction lengths of the hadron absorber where its trajectory is measured for momentum fitting.

TABLE 1 gives the detailed parameters of the events with thrust less than 0.8 and center of mass energy greater than 46.3 GeV. The lowest energy event is at 46.39 GeV, well above the lower limit of the energy bin. The fraction of visible energy is normal for either hadron events or hadron events including muons, since the resolution is nearly 20%. The broad jet oblateness O_B is large if the event is planar. The cut O_B greater than 0.3 was used to select planar events. These events tend to be quite planar. The average value of O_B agrees with what we would expect from the dominant background and is nearly two standard deviations higher than what our Monte Carlo predicts for a t quark or a b' quark. The muon momentum distribution is as expected for a heavy quark where most of the muons come from a cascade decay, but some high-energy muons can come from the primary decay of the heavy quark. The transverse momentum relative to the thrust axis is quite high as we expect for isolated muons. The muons' momentum out of the event plane is also quite high unlike the expected

TABLE 1. Parameters of Mark-J Events

	$\sqrt{s}$	$E_{vis}/\sqrt{s}$	Thrust	O_B	P_μ	$P_{\perp\mu}$	$P_{out\mu}$	cos δ	e or γ
A	46.39	0.74	0.79	0.25	2.8	2.6	1.1	0.38	14 GeV γ
B	46.57	0.89	0.74	0.49	2.4	1.9	0.4	0.60	12 GeV e
C	46.57	0.63	0.72	0.26	3.0	2.5	2.2	0.56	10 GeV e or γ
D	46.69	0.83	0.69	0.36	4.2	4.2	3.6	0.16	
E	46.78	0.89	0.73	0.38	2.2	1.8	1.0	0.58	
					4.5	4.2	1.7	0.34	
F	46.57	0.80	0.73	0.33	16.1	15.7	7.7	0.22	
G	46.57	1.08	0.77	0.41	2.7	2.4	0.9	0.44	
H	46.57	1.05	0.79	0.18	2.2	0.8	0.8	0.93	

background. That is, although the events are planar like the expected background, the muon is not in the plane as expected. The cosine of the angle from the thrust axis is small indicating large angles with the thrust axis and isolated muons. Although Mark-J's electron and photon identification is not ideal, high-energy electromagnetic showers can be clearly seen in three of the events. This is unusual for hadron events in general and also for those containing muons, although it could be due to photon radiation in the initial state.

The other three experiments at PETRA did not find supporting evidence for the effect found by Mark-J and therefore the PETRA machine was modified to go back to lower energy and higher luminosity. The situation would have been left there but for the further analysis of JADE. During the high-energy running JADE had installed a new vertex chamber and was devoting most of its effort to that. However, later analysis[4] showed that JADE also had a signal at energies above 46.3 GeV.

FIGURE 2 shows the cross sections measured by JADE and Mark-J for production of hadron events with thrust less than 0.8 and cos δ less than 0.7. Both groups find a substantially larger cross section in the energy bin above 46.3 GeV. The error bars in

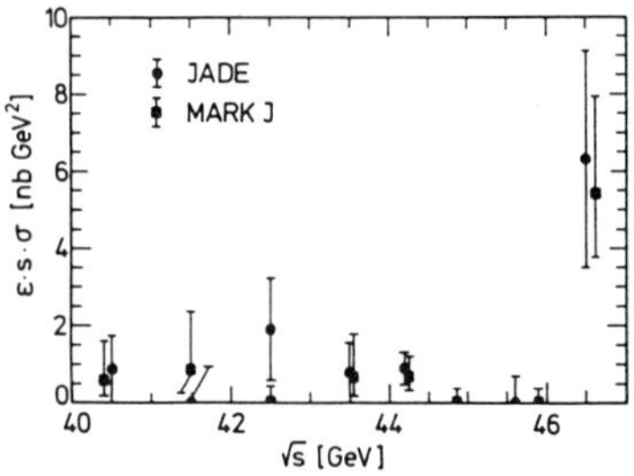

FIGURE 2. Cross section for events with thrust < 0.8 and cos δ < 0.7 vs. energy from JADE and Mark-J.

the figure are based simply on the number of events in the bin, not on the expected number.

FIGURE 3 shows the JADE data in the thrust vs. cos δ plane. Below 46.3 GeV, the JADE data look quite similar to those from Mark-J as shown in FIGURE 1. Above 46.3 GeV JADE has five events in the selection region. The distribution of events in the plane is notably different at high energy as it was for Mark-J. Based on the lower energy data JADE expects 0.56 event. They therefore have an excess of about the same statistical significance as did Mark-J. This is indeed a completely independent confirmation, because the cuts are identical and because JADE did not do this analysis until long after Mark-J had presented their final results.

TABLE 2 shows the details of the five events from JADE. Again the actual center of mass energies at which the events are seen starts well above 46.3 GeV. The muon momentum distribution is again as we would expect from t or b' decay, this time with two very high-momentum muons, which are unusual from conventional sources. Three events have now clearly identified electrons or photons with high momentum. In the JADE events, these are at large angle to the beam direction. Four of the five events are quite planar when the muon is excluded from the analysis again like the Mark-J events. In general, the similarity between the events seen by Mark-J and those seen by JADE

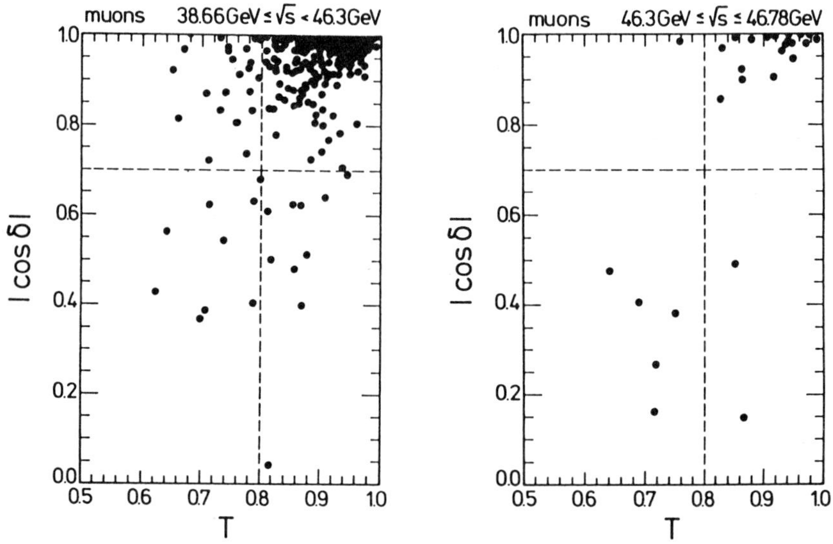

FIGURE 3. Cos δ vs. thrust for JADE events containing muons.

is quite amazing. Even assuming both were detecting the same new phenomenon, one would expect some fluctuations in the events seen. It should be noted that the five events listed below pass all of JADE's cuts used to select inclusive muon events for new particle searches. More restrictive cuts, used to study the decays of charm and bottom quarks, eliminate three of the events mainly because they go into less accurate regions of the detector.

The other two experiments have also searched for low-thrust hadron events containing muons using the same cuts as did Mark-J. They have found no large excess of events. The results of all four experiments[5] are combined and compared in TABLE 3. To estimate the sensitivities of each experiment to this type of event that is from an unknown source, we have simply used the sensitivities to the standard inclusive muons from b and c decay. This should be approximately accurate since in general the muon identification cuts are not complicated and depend mainly on solid angle coverage.

TABLE 2. Parameters of JADE Events

	$\sqrt{s}$	Thrust	P_μ	δ	e or γ
1	46.45	0.72	$15.2^{+3.7}_{-2.5}$	74.3	10.7 GeV e or γ
2	46.54	0.72	$2.5^{+0.1}_{-0.1}$	80.8	
3	46.57	0.69	$2.0^{+0.1}_{-0.1}$	65.8	
4	46.57	0.64	$21.8^{+19.2}_{-6.8}$	61.3	13.4 GeV γ
			$3.2^{+0.3}_{-0.3}$		
5	46.57	0.75	$4.1^{+0.2}_{-0.2}$	67.7	12.3 GeV e and 9.6 GeV γ

The table shows that TASSO had substantially less sensitivity than did the other three experiments, and in fact that the excess cross section, which is the cross section in addition to the expected background rate, observed by TASSO is nearly equal to the overall average. Therefore we are only faced with the question of whether the result from CELLO is consistent with that from Mark-J and JADE. To address this we have calculated the errors on the observed excess cross section by assuming that the average over experiments was the correct number. Therefore Mark-J and CELLO, which have the same sensitivity, have the same error on the cross section, although Mark-J observes more events. Thus we can test whether all the experiments are consistent with the average of 2.0 pb. Certainly Mark-J, JADE, and TASSO are quite consistent. CELLO is nearly two standard deviations off, but with four measurements this is not an unlikely distribution of events. We therefore conclude that the four measurements are consistent as far as can be tested and that combining the results of the four experiments is justified and is the best way of judging whether there is a significant signal.

The signal then is 14 events found with 2.4 pb expected from standard sources measured at lower energy. This size of fluctuation has a probability of 3×10^{-7} which,

TABLE 3. Summary of Measurements from PETRA

	Mark-J	JADE	CELLO	TASSO	Total
Luminosity (pb^{-1})	2.8	1.7	2.8	1.1	
Acceptance	70%	62%	72%	38%	
Luminosity × Acceptance	2.0	1.1	2.0	0.4	5.5 pb^{-1}
Number of μ's	31	32	28	9	96 events
N_{evt} ($T < 0.8$, $\cos\delta < 0.7$)	7	5	1	1	14 events
N_{evt} expected	0.8	0.56	0.65	0.3	2.4 events
Observed excess σ	3.1 ± 1.0	4.0 ± 1.4	0.2 ± 1.0	1.8 ± 2.3	2.0 ± 0.6 pb

even given the bias of physicists toward finding something, makes it quite unlikely that this is a fluctuation.

MEASUREMENTS WITH ELECTRONS

There is no corresponding signal found in low-thrust hadron events containing isolated electrons. Here the total luminosity times acceptance is 1.9 pb^{-1} and the number of events found by 3 experiments is zero with 1.2 events expected from standard sources and an additional 3.9 events expected to agree with the signal observed in the muon channel. The probability of observing zero when 5.5 is expected is just 6×10^{-3}. Thus the electron data are in some disagreement with the muon data, although the electron statistics are still poor. It should be emphasized, however, that electron identification has much more finely tuned cuts and depends more on the type of event that is being observed. Remember that the JADE and Mark-J events with muons probably also have electrons in them that fail some identification criterion, like the cuts removing photon conversions into pairs.

SUMMARY AND CONCLUSION

In a search for new quark flavors, an excess of low-thrust events containing muons has been found for PETRA energies greater than 46.3 GeV. The muons in these events are quite isolated from hadronic energy, as would be expected from pair production of massive quarks. Combining data from all 4 PETRA experiments, there are 14 events with thrust less than 0.8 and the cosine of the angle between the muons' direction and the thrust axis less than 0.7. Based directly on measurements taken at lower energy, the expected number of events is quite accurately predicted to be 2.4. The probability of a statistical fluctuation of this size or larger occurring in a given event selection is 3×10^{-7}. Although these events are consistent with the production of a new quark flavor, the evidence is as yet insufficient to prove that this is the source. No excess of events is seen in the selection of low-thrust events with isolated electrons where the sensitivity is substantially lower.

Thus, although there is a quite significant excess of low-thrust hadron events containing muons at the highest PETRA energies, the source of these events cannot be determined with the statistics available, but the most conventional would be a new charge $-1/3$ quark with a mass around 23 GeV. However, the observed events tend to be a bit more planar than expected from this source and the lack of a corresponding electron signal tends to discourage this explanation. There is more than enough evidence to warrant further study of this energy region.

REFERENCES

1. ADEVA, B. *et al.* 1983. Phys. Rev. Lett. **51:** 443.
2. ADEVA, B. *et al.* 1986. Phys. Rev. Lett. **53:** 134.
3. ADEVA, B. *et al.* 1986. Phys. Rev. D **34:** 681.
4. M. KUHLEN. 1986. DESY 86-052.
5. TASSO collaboration and CELLO collaboration. Private communication.

Missing Transverse Energy Events and a Search for Additional Neutrino Families and a Fourth-Generation Lepton

DAVID B. CLINE

Department of Physics
University of California
Los Angeles, California 90024

MISSING ENERGY EVENTS IN UA1

FIGURE 1 summarizes the cuts to and the final sample of events obtained in the UA1 experiments from 1983 to 1985. An example of a missing energy event that is due to a $W \rightarrow \tau + \nu$ is shown in FIGURE 2. A non-τ-like event (monojet) is shown in FIGURE 3. The objective of the missing transverse energy event selection is to define an event sample for which the background due to fluctuations in the detector response is small, thereby isolating events containing one or more energetic neutrinos (or other noninteracting particles). The missing transverse energy events were selected using the following criteria:[1,2]

(1) Missing transverse energy, $E_T^{\text{miss}} > 15$ GeV and $N_\sigma > 4$, where N_σ is defined as $E_T^{\text{miss}}/(0.7\sqrt{\Sigma E_T})$. The term ΣE_T is the scalar sum of transverse energy observed in all calorimeter cells where transverse energies are measured in gigaelectronvolts; N_σ is thus a measure of the experimental significance of the missing transverse energy.
(2) One or more jets observed in the calorimeters with $E_T > 12$ GeV and $|\eta| < 2.5$.
(3) One or more tracks observed in the drift chambers with $p_T > 1$ GeV/C within a cone of size $\Delta R = 0.4$ centered on the calorimeter jet axis ($\Delta R^2 \equiv \Delta\phi^2 + \Delta\eta^2$, where ϕ is the azimuthal angle and η is the pseudorapidity).
(4) Isolation of the missing transverse energy, by rejecting events with a jet observed in the calorimeter with $E_T > 8$ GeV and $|\eta| > 2.5$ (or a jet observed in the central detector drift chambers with $p_T > 5$ GeV/C) within ± 30 degrees in ϕ of the missing transverse energy direction.
(5) Veto of coplanar jet activity, by rejecting events with a calorimeter jet of $E_T >$ 8 GeV and $|\eta| < 2.5$ (or a central detector jet with $p_T > 5$ GeV/C) within 150–210 degrees in ϕ of the direction of the highest E_T jet.
(6) Veto of events with an electron or muon candidate in order to remove contributions from $W \rightarrow e\nu$ or $\mu\nu$ or heavy flavor semileptonic decays. For this purpose we have developed an algorithm for removing electrons or muons having tracks with $p_T > 10$ GeV/C. In addition, events were scanned and removed if they had muon candidates with $p_T > 3$ GeV/C. The cuts used to define electrons and muons for veto purposes in this analysis are less stringent,

especially on isolation requirements, than those used to select low background samples of W events.

For additional information see references 1 and 2. A total of 56 events satisfied the cuts described above and in FIGURE 1. We may now make use of measured properties

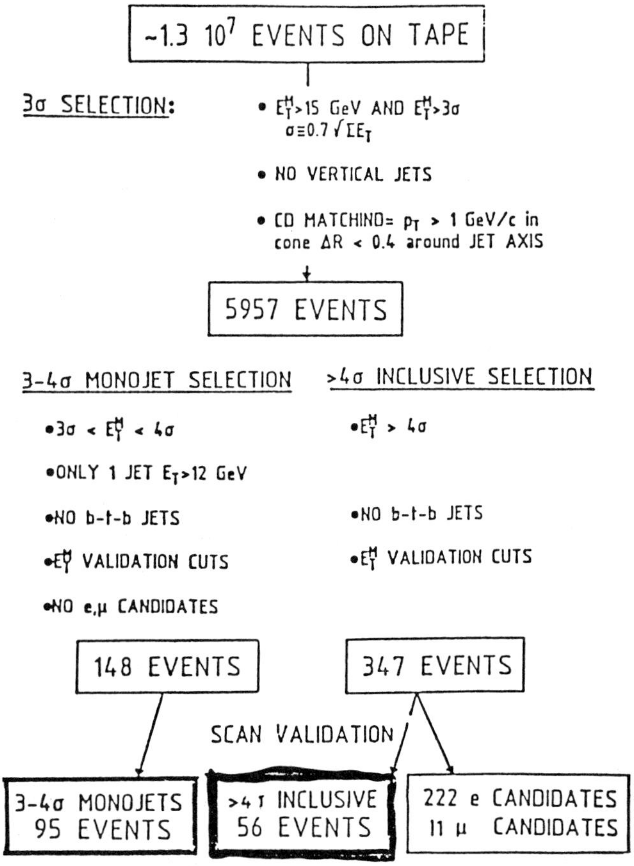

FIGURE 1. UA1 missing E_T events selection.

of semihadronic τ-decays, namely the low multiplicity, low mass jet, to separate out the production and decay of a $W^{\pm}$ into tau-lepton plus neutrino, through the process:

$$p\bar{p} \rightarrow W + X;\ W \rightarrow \tau\upsilon;\ \tau \rightarrow \text{hadrons} + \upsilon.$$

The signature is a high transverse momentum narrow hadronic jet with associated missing transverse energy. For this purpose, we define three variables particularly

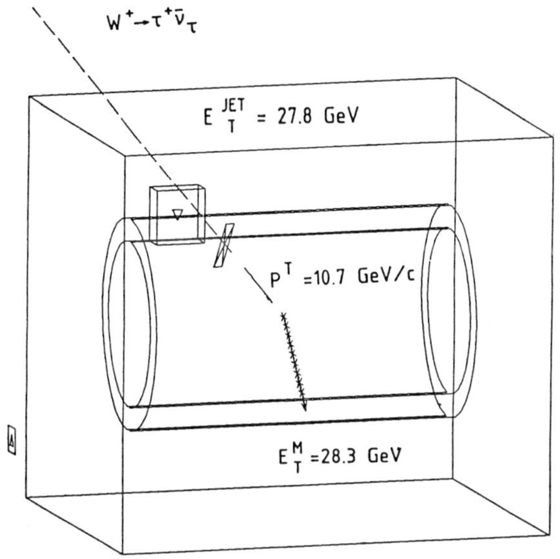

FIGURE 2.

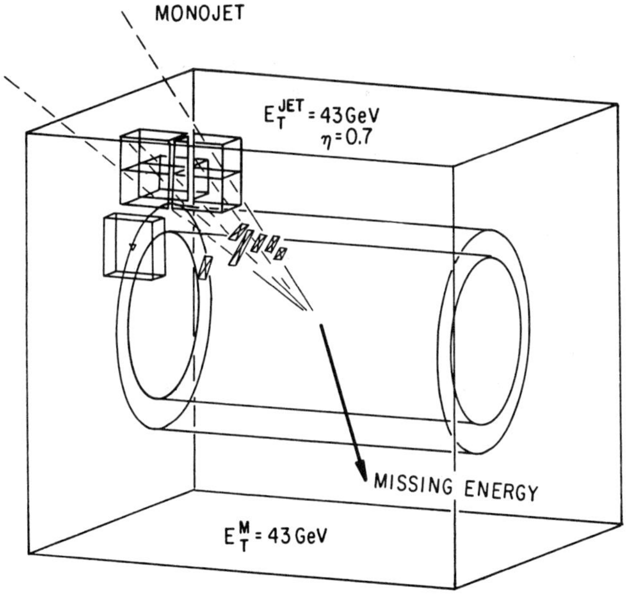

FIGURE 3.

suitable for separating high p_T τ-jets from ordinary jets:

(1) Fraction of jet energy (F), as measured in the calorimetry, contained within a cone of size $\Delta R = 1.0$.
(2) Angular separation ($r = \Delta R$) between the leading track of the jet, as measured in the central detector drift chambers, and the jet axis as measured in the calorimeters.
(3) Charged particle multiplicity (n) of tracks with $p_T > 1$ GeV/C within a cone of $\Delta R = 0.4$ around the calorimeter jet axis.

Using the τ Monte Carlo described later, we have produced the distributions of F, r, and n for the jet coming from the τ decay. We calculate a τ-log-likelihood, $L_\tau = \ln(f_F f_r f_n)$, for the highest E_T jet in an event, where f_F, f_r, and f_n are the values of the τ Monte Carlo distributions for the measured values of F, r, and n for this jet. The distributions of L_τ for the data, τ Monte Carlo, and jet data are shown in FIGURE 3; the range of L_τ is -16 to $+4$. The probability that a jet comes from a τ decay is greater with increasing L_τ. The cut $L_\tau > 0$ keeps 78% of the $\tau \rightarrow$ hadrons + v decays and rejects 89% of the ordinary jet events. Therefore, we define our final τ sample as those events with $L_\tau > 0$ (32 events). The remaining events are used in the search for:[1,2]

$$W \rightarrow L + \nu_L$$

and

$$Z \rightarrow \nu\bar{\nu}.$$

TECHNIQUE TO SEARCH FOR THE FOURTH-GENERATION LEPTON IN W DECAY

The search for a fourth generation of leptons and quarks is of great importance for the Standard Model. In the past the lepton of each generation was detected first (electron before the neutron, μ before strangeness and charm, the τ before the b or t quarks). It is possible this will happen again due to the unique signature of the heavy lepton process. Previously, it was pointed out that the $W^\pm$ particle decays provide a unique mass range in which one can search for new leptons.[1,2,3] Furthermore, we expect copious production of $W^\pm$ particles at the CERN and Fermilab $\bar{p}p$ colliders in the next few years. We follow the initial ideas of early papers and carry out a study of the experimental method needed to separate the signal from the production and decay of the new lepton into one or two jet final states from the backgrounds.[3,4,5] We have applied this analysis to the UA1 detector at CERN; however, the technique is general enough to be applied to the other detectors at CERN and Fermilab.

We consider the decays of the $W^\pm$ particles

$$W^+ \rightarrow L^+ + \nu_L$$

$$W^- \rightarrow L^- + \bar{\nu}_L$$

where L is a new sequential lepton with mass in the range 23 to 70 GeV/C^2 and ν_L, $\bar{\nu}_L$

are a fourth generation of neutrinos and antineutrinos that are assumed to be massless. The hadronic decays of the $L^{\pm}$ are expected to be[3,4,5]

$$L^+ \rightarrow u + \bar{d} + \bar{\nu}_L$$

$$\rightarrow c + \bar{s} + \bar{\nu}_L$$

$$L^- \rightarrow \bar{u} + d + \nu_L.$$

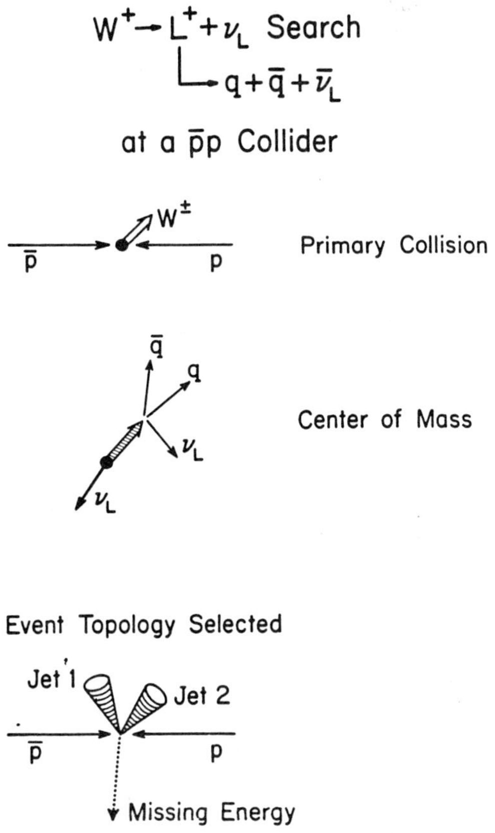

FIGURE 4.

Two techniques have been proposed to detect the L in the W decay debris.

(1) Observation of an excess of single lepton events with missing energy in a region where $W \rightarrow l + \bar{\nu}_l$ is suppressed and reduction of the corresponding background from $W \rightarrow \tau + \bar{\nu}_\tau \rightarrow l + \bar{\nu}_l + \nu_\tau + \bar{\nu}_\tau$ with a microvertex detector.

(2) Observation of one or two jet events with missing energy from the L decay. So far only technique 2 has been applied in the UA1 experiment. The experimental technique proposed here makes use of a background suppression by detecting two jets from the heavy lepton production and decay that are on the same side of the $\bar{p}p$ line of interaction (i.e., within 180° of each other). As we will show this selection and certain requirements on the jets from the decay serve to reduce the expected backgrounds. FIGURE 4 illustrates the technique proposed here.

In this analysis we make use of a modified Isajet Monte Carlo. The full matrix element of the decay, including the polarization of the L, has been included with both right and wrong helicity states generated. In reference 5 it was shown that the W decay gives a kinematic window up to ~(65–70) GeV/C^2 to search for the new lepton. In this paper we report on calculations up to a lepton mass of 55 GeV/C^2. Higher mass calculations will be reported in a subsequent report.

LIMITS ON M_L

For this analysis only events with ln $\tau < 0$ are used.[2] A careful calculation of the background and a subtraction from the data yields the curves shown in FIGURE 5 for

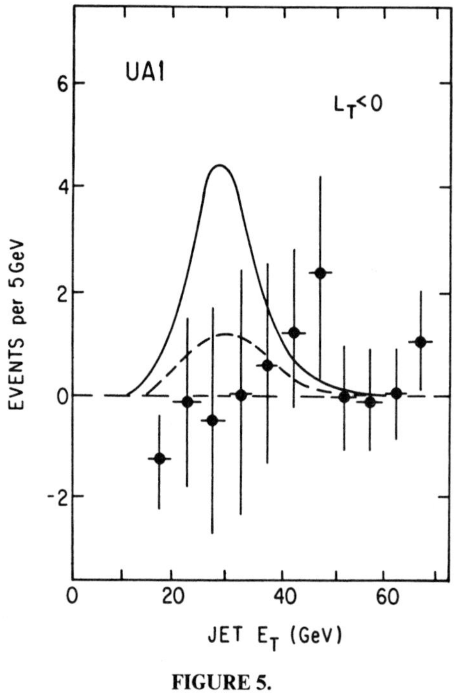

FIGURE 5.

TABLE 1. Techniques to Search for Additional Neutrino Families

Process	Dependence on δN_ν	r
$e^+e^- \rightarrow \nu\nu\gamma$	$\sigma = \alpha G F^2 s\left(1.75 + \frac{\delta N_\nu}{4}\right)$	0.13
Γ_Z/Γ_W	$R = 2.7x \frac{3.3 \pm 0.2}{[QCD]}\left[\frac{2.83 + 0.18\delta N_\nu}{2.82}\right]$	≤0.06
$\bar{p}p \rightarrow Z + g \cdots \rightarrow \nu\bar{\nu}$	$\sigma \sim (A + B\delta N_\nu)^a$	
	$A \sim 10$	0.1
	$B \sim 1$	to
	[UA1 Data]	~0.2
Γ_Z measurement (direct)	$\Gamma_Z = 2.83 + 0.18\delta N_\nu$	0.06

[a]Improved calorimetry can change the ratio of B/A to ~1

$$r = \frac{[d\sigma/d[\delta N_\nu]]N_\nu = 3}{[\sigma]N_\nu = 3}.$$

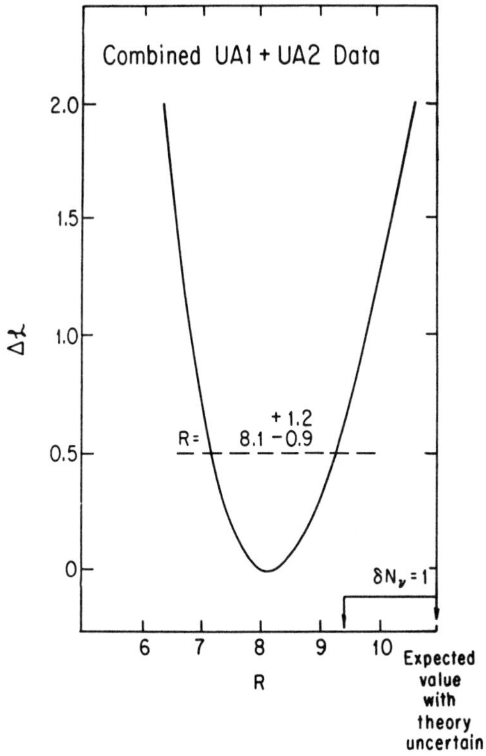

FIGURE 6.

$M_L = 35$ GeV/C^2 and 55 GeV/C^2. The subsequent limit on the L mass is

$$M_L > 41 \text{ GeV } 90\% \text{ C.L.}$$

including systematic errors.[2]

TECHNIQUES TO COUNT NEUTRINO SPECIES

Several techniques have been invented to search for additional neutrino families including

$$e^+e^- \rightarrow \nu\bar{\nu}\gamma$$

$$\bar{p}p \rightarrow Z_{\rightarrow\nu\bar{\nu}} + \text{gluon}$$

(Γ_z/Γ_w) technique to measure the Z width.

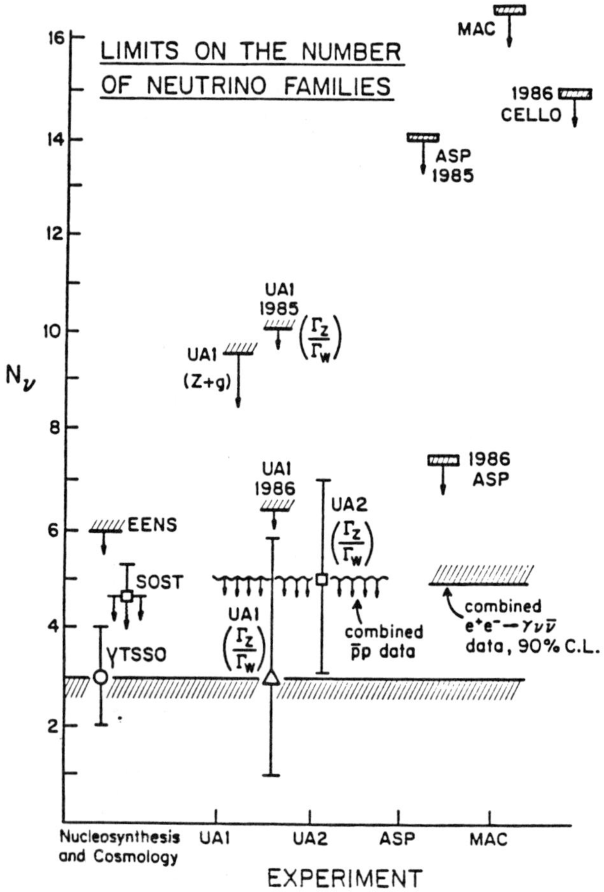

FIGURE 7.

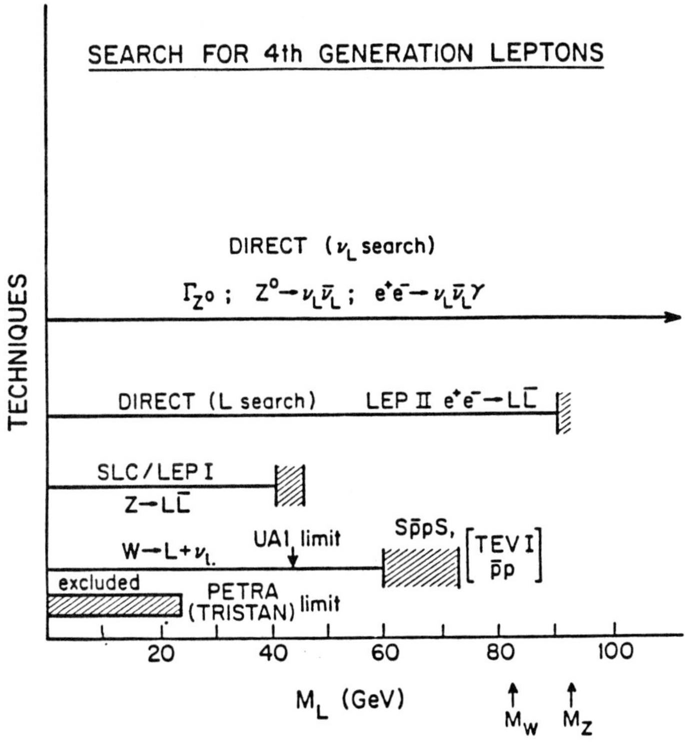

FIGURE 8.

In TABLE 1 we report the sensitivity of these techniques to additional neutrino families. These range from a 6% effect to a ~25% effect per unit family increase.

LIMITS ON δN_ν

The ratio of widths technique yields the best limits for present $\bar{p}p$ collider data.[6] FIGURE 6 shows the most recent fit to the UA1 and UA2 data values of R where

$$R = N(W \to l_\nu)/N(Z \to e^+e^-)$$

when fully corrected for acceptance and efficiency.[7] This results in a $\delta N_\nu = (N_\nu - 3)$ value that depends on the t-quark mass but reaches a value of

$$\delta N_\nu < 5$$

for $M_t > 50$ GeV. When combined with the results from e^+e^- collisions this points to the overall value of less than or equal to four neutrino families. This is in good agreement with cosmological limits[9] (see FIGURE 7).

FUTURE SEARCH FOR L AND δN_ν

We may anticipate the different mass ranges to be searched for the fourth-generation lepton L as shown in FIGURE 8. The present UA1 limit appears to limit the search region for e^+e^- colliders, whereas W decay can extend the range to 65–70 GeV. Beyond this, LEP II is needed. Finally, if $M_L > 100$ GeV, it will be necessary to discover this lepton indirectly by the observation of

$$\delta N_\nu = 1$$

or wait until a new e^+e^- high-energy collider is constructed. As shown in Figure 8, all techniques reach some intrinsic limit.[10] We wish to thank our colleagues in UA1 for helpful discussions.

SUMMARY

We reported on the most recent analysis of the missing transverse energy events in the UA1 experiment and discussed the search for a fourth-generation heavy lepton from these data. We also reported the latest results on the number of neutrino species from Z° decay using the ratio-of-widths technique. A summary of the world's data on δN_ν was also given.

REFERENCES

1. ABAJAN, C. *et al.* Events with large missing transverse energy at the CERN collider: W → $\tau\nu$ decay and test of τ-μ-e universality at $Q^2 = M_W^2$. CERN-EP/86-81. Submitted for publication in Phys. Lett.
2. ABAJAN, C. *et al.* Events with large missing transverse energy at the CERN collider: Search for the decays of $W^\pm$ into heavy leptons and of the Z° into non-interacting particles. CERN EP/86-82. Submitted for publication in Phys. Lett.
3. CLINE, D. & C. RUBBIA. 1983. Phys. Lett. **127B:** 277; CLINE, D. 1983. Recontre de Moriond. (Published in the 1983 Proceedings.)
4. BARGER, V. *et al.* 1983. Phys. Lett. **133B:** 449.
5. CLINE, D. & M. MOHAMMADI. 1986. The search for the 4th generation leptons from $W^\pm$ decay: A method to extract the signal. Wisconsin preprint WISC EX-86-269.
6. CLINE, D. & J. ROHLF. 1983. UA1 TN 83-40; HIKASA, K. 1984. P.R. **D29:** 1939; DESHPANDE, N. *et al.* 1985. P.P.L. **54:** 1757; HALZEN, F. & K. MURSADA. 1983. Phys. Rev. Lett. **51:** 857.
7. GEER, S. Private communication.
8. LAVINE, T. L. 1986. Limits on neutrino, selectrons and Winos from single-photon searches. UWISC Preprint MAC-748.
9. STEIGMAN, G., D. N. SCHRAMM & J. GUNN. 1977. Phys. Lett. **866:** 502; YANG, J., M. S. TURNER, G. STEIGMAN, D. N. SCHRAMM & K. A. OLIVE. 1984. Astrophys. J. **281:** 493; BOESGAARD, A. M. & G. STEIGMAN. 1985. Ann. Rev. Astron. Astrophys. **23:** 319.
10. See CLINE, D. *In* Proc. 1986 Aachen Antiproton-Proton Collider Conf. To be published.

Search for Neutral Higgs Bosons in $e^+e^- \rightarrow H^\circ Z^\circ$ at LEP 200[a,b]

SAU LAN WU

Department of Physics
University of Wisconsin
Madison, Wisconsin 53706

INTRODUCTION

In this report, we concentrate on looking for the Higgs boson[1] through the process:

$$e^+e^- \rightarrow H^\circ Z^\circ$$

via the diagram shown in FIGURE 1. Standard Model couplings with three generations of quarks and leptons are used throughout the rest of this report. The center-of-mass energy (E_{cm}) is chosen to be 200 GeV and the integrated luminosity used for this study is 500 pb^{-1}. Since, for an average luminosity of 10^{31} cm^{-2} sec^{-1}, we can get about 1 pb^{-1} per day, it takes about two to three years to accumulate 500 pb^{-1}. The energy 200 GeV is chosen so that the physics includes phenomena above the W^+W^- and $Z^\circ Z^\circ$ thresholds.

We consider three different classes of event topologies for different Z° decay modes. First,

$$e^+e^- \rightarrow H^\circ Z^\circ$$
$$Z^\circ \rightarrow \nu\bar{\nu}$$
$$H^\circ \rightarrow b\bar{b} \text{ jets.}$$

The branching ratio of $Z^\circ \rightarrow \nu\bar{\nu}$ is 18%. In this channel, the Higgs boson, which decays into two $b\bar{b}$ jets, recoils against nothing (FIGURE 2a). Second,

$$e^+e^- \rightarrow H^\circ Z^\circ$$
$$Z^\circ \rightarrow e^+e^-, \mu^+\mu^-, \tau^+\tau^-$$
$$H^\circ \rightarrow b\bar{b} \text{ jets.}$$

The branching ratio of $Z^\circ \rightarrow e^+e^-$, $\mu^+\mu^-$, or $\tau^+\tau^-$ is 3% each. In this case, the signals

[a]This work was supported by the U.S. Department of Energy under Contract DE-AC02-76ER00881 and by U.S. National Science Foundation Grant INT-8313994 for travel.

[b]These results were also presented at the ECFA Workshop—LEP 200, Aachen, West Germany, September 29–October 1, 1987.

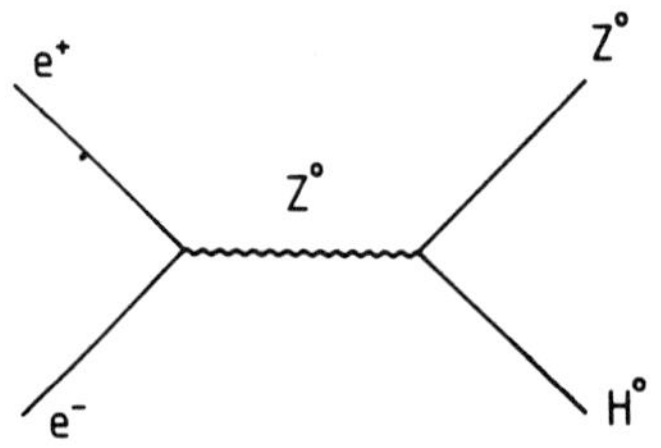

FIGURE 1. Lowest order diagram for $H°Z°$ production.

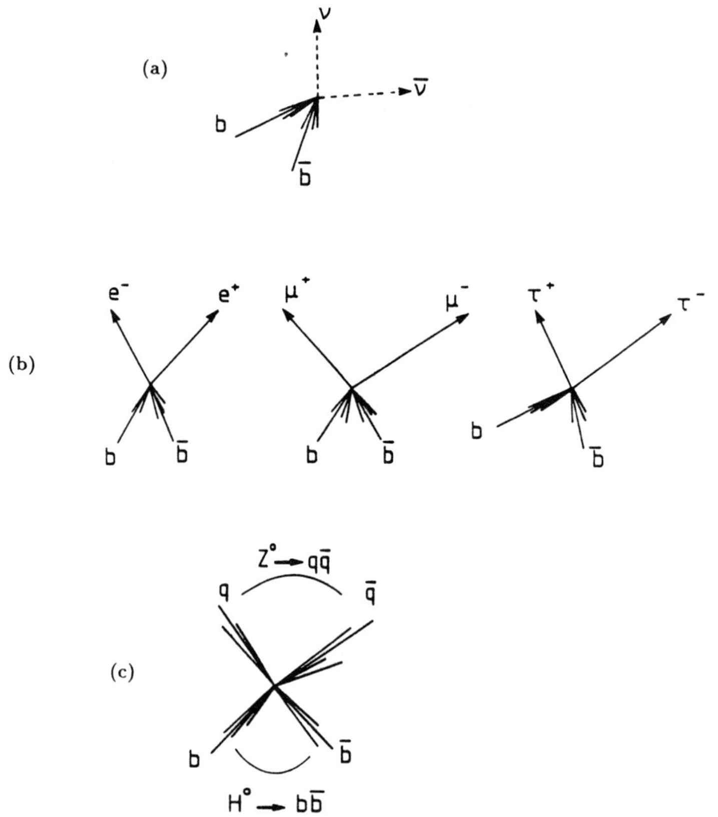

FIGURE 2. Final state configurations for the process $e^+e^- \rightarrow H°Z°$, where the $Z°$ decays to (a) two neutrinos, (b) two charged leptons, and (c) two quarks.

are a pair of leptons and two jets (FIGURE 2b). Finally,

$$e^+e^- \longrightarrow H^\circ Z^\circ$$

$$\hookrightarrow q\bar{q}\ \text{jets}$$

$$\hookrightarrow b\bar{b}\ \text{jets.}$$

The branching ratio of $Z^\circ \longrightarrow$ jets is about 73%. In this case, the signals are four jets (FIGURE 2c).

Here we perform the search for the Higgs boson from $e^+e^- \longrightarrow H^\circ Z^\circ$ at $E_{cm} = 200$ GeV using a realistic detector simulation, including a track-finding algorithm and pattern recognition, and detailed background studies. Sophisticated analysis methods have been used to demonstrate what levels of signals and what Higgs masses we can see at LEP II with $E_{cm} = 200$ GeV.

THE PROCESS $e^+e^- \longrightarrow H^\circ Z^\circ$

$$\hookrightarrow \nu\bar{\nu}$$

$$\hookrightarrow b\bar{b}\ \textbf{jets}$$

As shown in FIGURE 2a, the signal for this process is two jets recoiling against "nothing." Hence the event has large missing energy and large missing transverse momentum. The total number of events for 500 pb^{-1} integrated luminosity at $E_{cm} =$ 200 GeV is given in TABLE 1 for different Higgs masses.

The backgrounds considered for this process are

$$e^+e^- \longrightarrow Z^\circ Z^\circ$$

$$\hookrightarrow \nu\bar{\nu}$$

$$\hookrightarrow q\bar{q}\ \text{jets} \qquad (1)$$

$$e^+e^- \longrightarrow W+W-$$

$$\hookrightarrow \text{two jets}$$

$$\hookrightarrow \nu\tau \longrightarrow \nu\bar{\nu} + \text{soft charged particles} \qquad (2)$$

$$e^+e^- \longrightarrow q\bar{q},\ q\bar{q}g,\ q\bar{q}gg,\ q\bar{q}q\bar{q},\ \text{etc.} \quad (\text{QCD}). \qquad (3)$$

In the case of photon radiation in the initial state where the photon goes along the beam pipe undetected, we would have $e^+e^- \longrightarrow (\gamma)q\bar{q}$ with two jets recoiling against "nothing." However, these events have small missing transverse momenta.

A large number of selection criteria are necessary to extract the $H^\circ Z^\circ$ signals due to the large background cross sections of quantum chromodynamics (QCD), W^+W^-, and $Z^\circ Z^\circ$. The selection criteria are given in Wu.[2]

In $e^+e^- \longrightarrow H^\circ Z^\circ$, where $Z^\circ \longrightarrow \nu\bar{\nu}$, the invariant mass of the observed hadronic

TABLE 1. $E_{cm} = 200$ GeV; Integrated Luminosity $= 500$ pb^{-1}

M_{H° GeV	Expected Number of Events for $e^+e^- \rightarrow H^\circ Z^\circ\ (Z^\circ \rightarrow \nu\bar{\nu})$
40	107
60	83
80	55
100	21
M_{H° GeV	Expected Number of Events for $e^+e^- \rightarrow H^\circ Z^\circ\ (Z^\circ \rightarrow e^+e^-, \mu^+\mu^-)$
40	36
60	28
80	18
100	7
M_{H° GeV	Expected Number of Events for $e^+e^- \rightarrow H^\circ Z^\circ \rightarrow 4$ jets
40	430
60	340
80	230

system should peak approximately at the H° mass. However, due to the fact that there are undetected particles in the decay $H^\circ \rightarrow b\bar{b}$ (neutrinos or other particles escaping the detector), the invariant mass distribution has a sharp edge near M_{H° and a falling distribution for decreasing invariant mass as shown in FIGURE 3a.

To correct for the undetected particles we make use of the fact that we know the mass of the $\nu\bar{\nu}$ system.

Let $\mathbf{P}_i$ and E_i be the momentum and energy of the detected particles, respectively. Let $\mathbf{P}_{obs}$ be the total momentum of the observed particles and E_{obs} be the total energy of the observed particles,

$$\mathbf{P}_{obs} = \sum_i \mathbf{P}_i$$

$$E_{obs} = \sum_i E_i.$$

Instead of $M_{H^\circ} = \sqrt{E^2_{obs} - \mathbf{P}^2_{obs}}$, we use the more accurate relation

$$M_{H^\circ} = \frac{E_{cm}E_{obs} - \sqrt{E^2_{cm}E^2_{obs} - (E^2_{obs} - \mathbf{P}^2_{obs})(E^2_{cm} - M^2_{Z^\circ})}}{\sqrt{E^2_{obs} - \mathbf{P}^2_{obs}}}.$$

This formula is equation 4 derived in Hilgart *et al.*[3]

FIGURE 3b shows the invariant mass distribution for $M_{H^\circ} = 60$ GeV on the basis of this improved formula. This distribution centers at the M_{H° with a full width at half maximum (FWHM) of about 5 GeV.

Note that FIGURE 3b demonstrates the power of a e^+e^- colliding beam machine. The knowledge of the total center-of-mass energy and total momentum responsible for the reaction enables us to reconstruct the process without complete measurements of all the particles.

The final invariant mass distributions for integrated luminosity of 500 pb^{-1} are shown in FIGURE 4a–4c for $M_{H^\circ} = 40$ GeV, 60 GeV, and 80 GeV, respectively. These distributions show the Higgs signals significantly standing out of the W^+W^-, $Z^\circ Z^\circ$, and QCD backgrounds.

In conclusion, at $E_{cm} = 200$ GeV and for 500 pb^{-1} integrated luminosity, one can get significant signals up to a Higgs mass of about 80 GeV from the process $e^+e^- \rightarrow H^\circ Z^\circ$, where $Z^\circ \rightarrow \nu\bar{\nu}$ and $H^\circ \rightarrow b\bar{b}$ jets. It will be difficult to extend M_{H° to 90 GeV

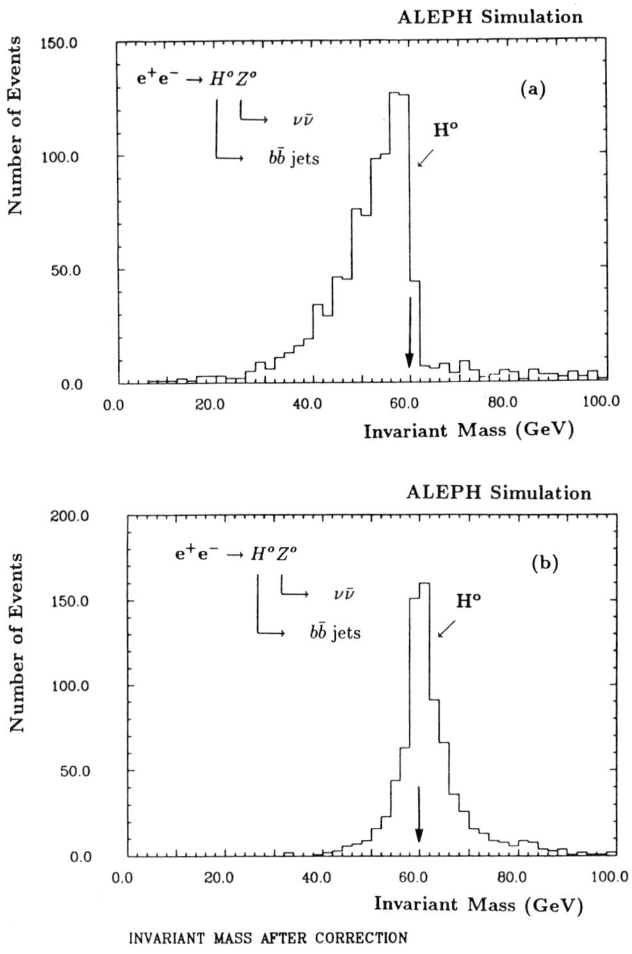

FIGURE 3. Invariant mass distribution from the process $e^+e^- \rightarrow \overline{H}^\circ Z^\circ \rightarrow b\bar{b}\nu\bar{\nu}$. (a) The invariant mass (GeV) of the observed hadronic system. (b) The invariant mass of the observed hadronic system calculated to correct for missing particles as described in the text. $M_{H^\circ} = 60$ GeV.

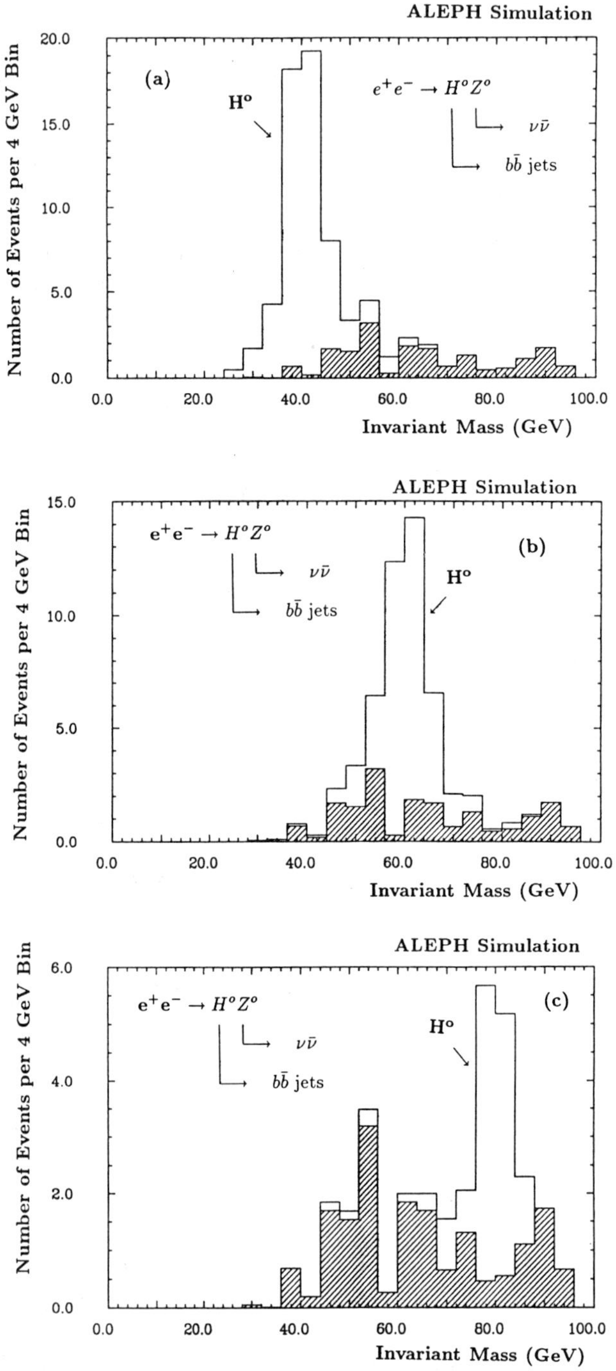

FIGURE 4. Invariant mass distribution after final analysis for the process $e^+e^- \longrightarrow H^\circ Z^\circ$ ($H^\circ \longrightarrow b\bar{b}$ jets; $Z^\circ \longrightarrow \nu\bar{\nu}$) for (a) $M_{H^\circ} = 40$ GeV; (b) $M_{H^\circ} = 60$ GeV; (c) $M_{H^\circ} = 80$ GeV. The shaded areas are background due to QCD, W^+W^-, and $Z^\circ Z^\circ$ production. The distribution is normalized to 500 pb^{-1} and $E_{cm} = 200$ GeV.

due to the $Z^\circ Z^\circ$ background. For M_{H° above 90 GeV the rates are lower and it would require a more sophisticated analysis.

THE PROCESS $e^+e^- \rightarrow H^\circ Z^\circ$

$$e^+e^- \rightarrow H^\circ Z^\circ$$

- $Z^\circ \rightarrow e^+e^-, \mu^+\mu^-, \tau^+\tau^-$
- $H^\circ \rightarrow b\bar{b}$ **jets**

As shown in FIGURE 2b, the signal for this process is two jets and two leptons. The number of events for different Higgs masses for 500 pb^{-1} integrated luminosity and $E_{cm} = 200$ GeV is given in TABLE 1.

The major backgrounds for this process are:

$$e^+e^- \rightarrow Z^\circ Z^\circ$$

- $\rightarrow e^+e^-, \mu^+\mu^-, \tau^+\tau^-$
- $\rightarrow q\bar{q}$ jets

$$e^+e^- \rightarrow W^+W^-$$

- $\rightarrow e\nu$ or $\mu\nu$
- $\rightarrow q\bar{q}$ jets with π misidentified as e or μ

$$e^+e^- \rightarrow t\bar{t}$$

- $\rightarrow e$ or μ + hadrons
- $\rightarrow e$ or μ + hadrons.

The analysis for this case is simpler because the background is much less severe. It is carried out in two steps:

(1) Calculate the invariant mass $M_{e^+e^-}$ (or $M_{\mu^+\mu^-}$) of the e^+e^- (or $\mu^+\mu^-$) pairs. For this study, we ignore the case where $Z^\circ \rightarrow \tau^+\tau^-$. For invariant masses between 88 and 99 GeV and charged particle multiplicity of the event greater than 4, calculate the recoil mass against the e^+e^- (or $\mu^+\mu^-$) pairs. The distributions of the recoil mass against e^+e^- and $\mu^+\mu^-$ pairs for $M_{H^\circ} = 40$, 60, and 80 GeV are shown in FIGURES 5a, 6a, and 7a, respectively, normalized to an integrated luminosity of 500 pb^{-1}.

(2) Calculate the invariant mass of the hadronic system (excluding the e^+e^- and $\mu^+\mu^-$ pairs). Using the method given in the second section above (equation 4 of reference 3), the distributions of the calculated invariant mass (i.e., the reconstructed M_{H°) for $M_{H^\circ} = 40$, 60, and 80 GeV are shown in FIGURES 5b, 6b, and 7b, respectively, normalized to an integrated luminosity of 500 pb^{-1}.

In conclusion, at $E_{cm} = 200$ GeV and for 500 pb^{-1} integrated luminosity, one can

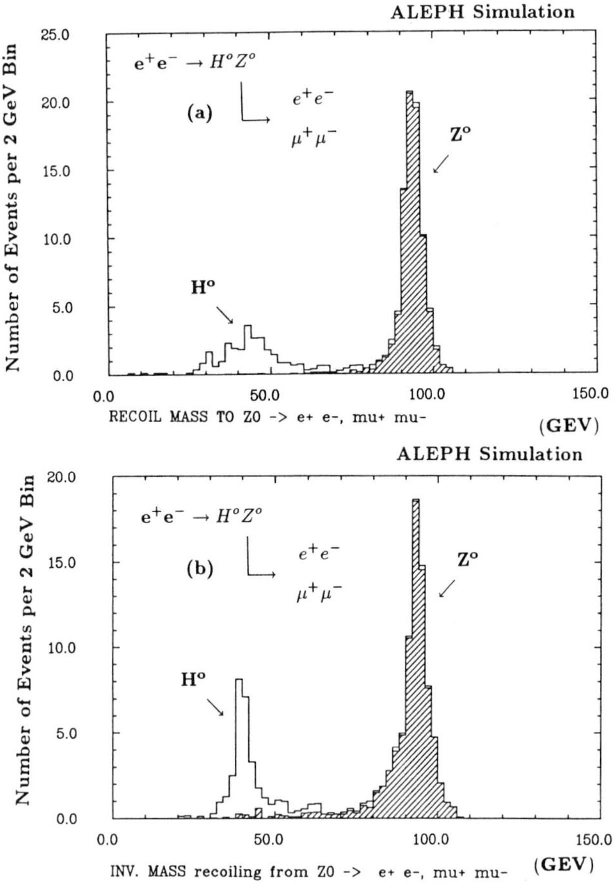

FIGURE 5. For the process $e^+e^- \to H^\circ Z^\circ \to b\bar{b}l^+l^-$ (e,μ). (a) Recoil mass to $Z^\circ \to l^+l^-$. (b) Reconstructed invariant mass, correcting for missing particles as described in the text for $M_{H^\circ} =$ 40 GeV. The shaded area indicates the sum of the backgrounds from QCD, W^+W^-, and $Z^\circ Z^\circ$ processes. It is clear that the background from $e^+e^- \to Z^\circ Z^\circ$ dominates.

get significant signals up to Higgs mass about 80 GeV from $e^+e^- \to H^\circ Z^\circ$, where $Z^\circ \to e^+e^-$ and $\mu^+\mu^-$. Again it is difficult to extend M_{H° to 90 GeV due to the $Z^\circ Z^\circ$ background. For M_{H° greater than 90 GeV, the rates seem too low to be useful.

THE PROCESS $e^+e^- \to H^\circ Z^\circ$

$Z^\circ \to$ $q\bar{q}$ **jets**

$H^\circ \to$ $b\bar{b}$ **jets**

As shown in FIGURE 2c, the signal for this process is four jets, and the numbers of events for 500 pb^{-1} integrated luminosity and $E_{cm} = 200$ GeV are given in TABLE 1 for different Higgs masses. The backgrounds considered for this process are

$$e^+e^- \longrightarrow W^+W^-, Z^\circ Z^\circ \longrightarrow 4 \text{ jets}$$

$$e^+e^- \longrightarrow q\bar{q}, q\bar{q}g, q\bar{q}gg, q\bar{q}q\bar{q}, \text{etc.} \longrightarrow 4 \text{ jets}.$$

The following analysis strategies are used.

(1) In order to reconstruct $Z^\circ H^\circ$ signals we proceed to calculate jet–jet invariant masses in the following steps: (a) Apply cluster analysis[4] to the detected particles of an event. Accept events with jet multiplicity = 4. (b) To obtain jet

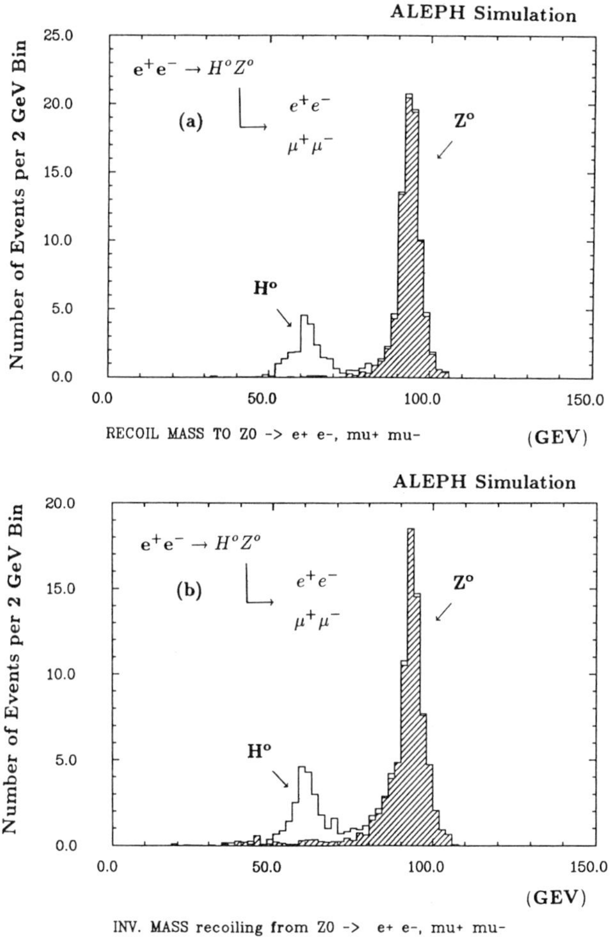

FIGURE 6. (a), (b) Same as FIGURE 5 for $M_{H^\circ} = 60$ GeV.

energies, we apply the 4-jet reconstruction method of Hilgart *et al.* and Wu.[3,5] The idea is to make use of energy-momentum conservation (see equation 5 of reference 3). Using these new jet energies and the original jet directions, calculate the jet–jet invariant masses as shown in FIGURE 8. The backgrounds due to W^+W^- and $Z°Z°$ are severe. Applying the same jet analysis to $e^+e^- \rightarrow$

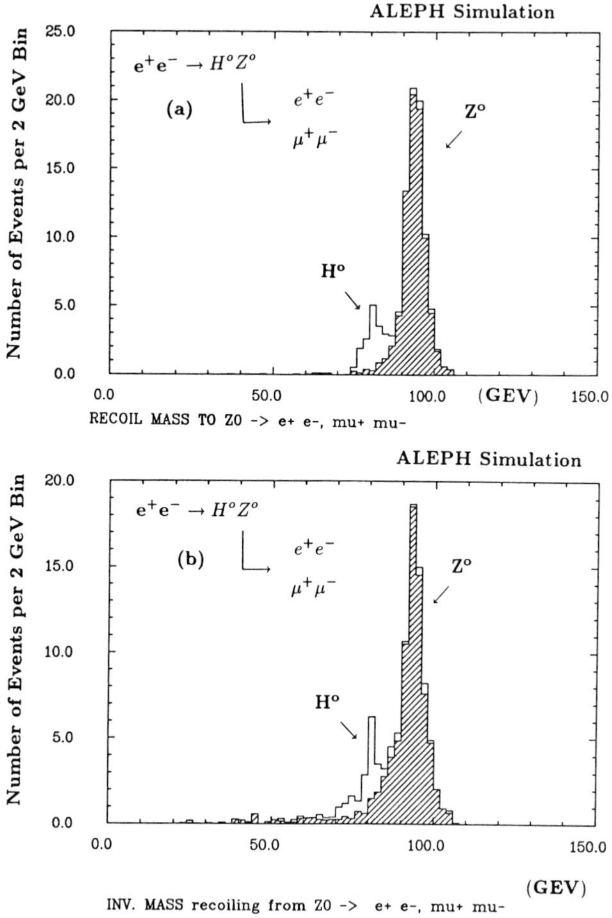

FIGURE 7. (a), (b) Same as FIGURE 5 for $M_{H°} = 80$ GeV.

W^+W^- and $Z°Z°$ events, we obtain the jet–jet invariant mass spectrum as shown in FIGURE 9. Superimposed are the $e^+e^- \rightarrow H°Z°$ signals, and it is clear that the background is extremely large.

(2) To reduce the $e^+e^- \rightarrow W^+W^-$ and $Z°Z°$ backgrounds, we apply a new fitting procedure with the following two hypotheses as described in Hilgart *et al.*:[3] (a) A fit procedure with the $e^+e^- \rightarrow W^+W^-$, $Z°Z°$ hypotheses. (b) A fit procedure

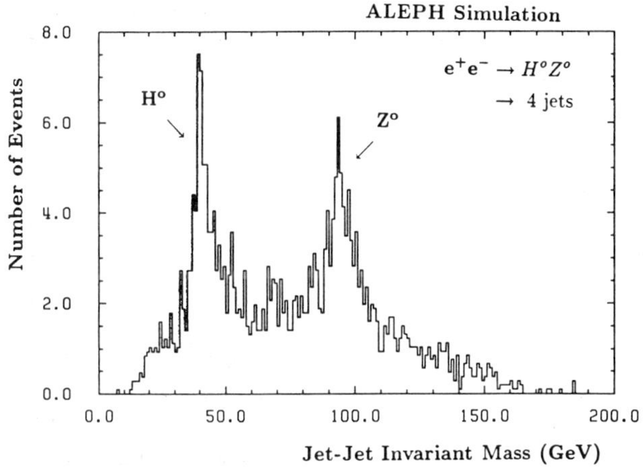

FIGURE 8. All jet–jet invariant mass combinations (6 per event) for $e^+e^- \rightarrow H°Z° \rightarrow$ 4 jets after applying the 4-jet reconstruction method of Hilgart *et al.* and Wu.[3,5]

with the $e^+e^- \rightarrow H°Z°$ hypothesis. These procedures are described in section 3 of reference 3. In this analysis, no hadron calorimeter information is used. The charged particles are reconstructed by the ALEPH TPC pattern recognition programs and the electromagnetic showers are simulated with realistic detector energy resolution. To suppress W^+W^- and $Z°Z°$ backgrounds, we accept events that satisfy the following two conditions: (i) Events that either do not converge in the $e^+e^- \rightarrow W^+W^-$ and $Z°Z°$ hypothesis fits, or converge in the

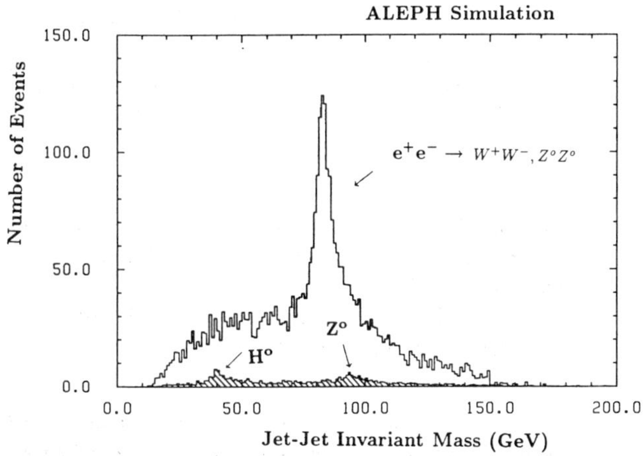

FIGURE 9. All jet–jet invariant mass combinations (6 entries per event) for $e^+e^- \rightarrow W^+W^-$, $Z°Z° \rightarrow$ 4 jets after applying the 4-jet reconstruction method in Hilgart *et al.* and Wu.[3,5] A $W^{\pm}$ peak is seen with a hint of a $Z°$ shoulder. The shaded histogram near the bottom is FIGURE 8 superimposed with the correct relative normalization. The $H°$ signal is buried in the W^+W^- background.

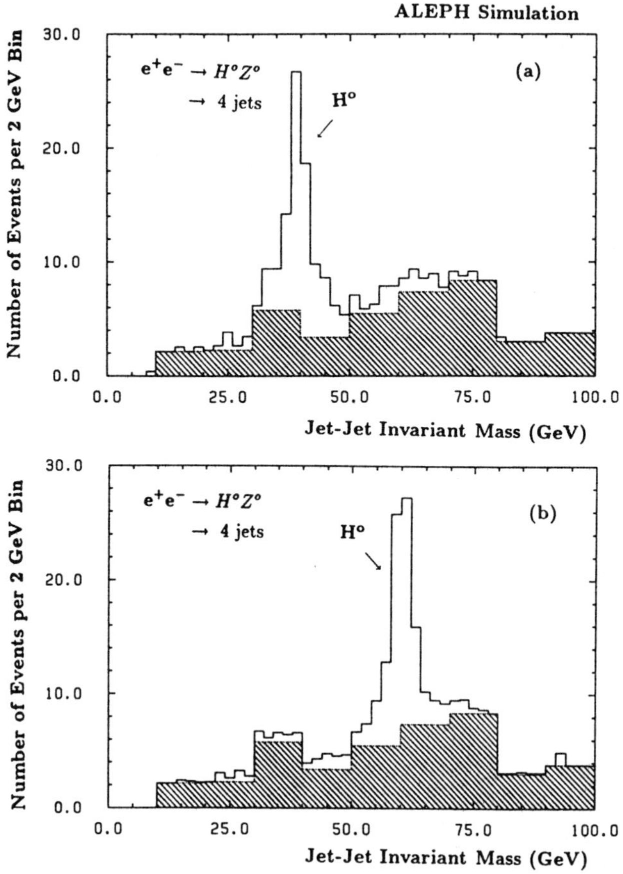

FIGURE 10. The jet–jet invariant mass distribution after analysis from the process $e^+e^- \rightarrow H^\circ Z^\circ$ ($H^\circ \rightarrow b\bar{b}$ jets and $Z^\circ \rightarrow q\bar{q}$ jets) for (a) $M_{H^\circ} = 40$ GeV, and (b) $M_{H^\circ} = 60$ GeV. The shaded areas are background due to QCD, W^+W^-, and $Z^\circ Z^\circ$ production. The distribution is normalized to 500 pb^{-1} and $E_{cm} = 200$ GeV.

$e^+e^- \rightarrow W^+W^-$ and $Z^\circ Z^\circ$ hypothesis fits, but with jet–jet invariant mass < 75 GeV (of the pair with best χ^2). (ii) Events that fit the $H^\circ Z^\circ$ hypothesis with probability greater than 1%.

(3) We proceed to reduce the background due to the QCD processes $e^+e^- \rightarrow q\bar{q}$, $q\bar{q}g$, $q\bar{q}gg$, etc. The cuts are given in Wu.[2]

The final results are given in FIGURE 10a and b for Higgs masses of 40 GeV and 60 GeV, respectively. In this jet–jet invariant mass distribution the H° mass peak stands out clearly above the W^+W^-, $Z^\circ Z^\circ$, and QCD backgrounds. These figures are normalized to 500 pb^{-1} integrated luminosity. Again, we conclude that at $E_{cm} = 200$ GeV, 500 pb^{-1} integrated luminosity, one can get significant signals of Higgs masses up to about 70 GeV from the process $e^+e^- \longrightarrow H^\circ Z^\circ \longrightarrow$ 4 jets. It is difficult to extend the Higgs mass to 80 GeV (due to $W^\pm$) or 90 GeV (due to Z°). Above 90 GeV, the rate reduces and a more sophisticated analysis method will be required to extract the Higgs signals.

REFERENCES

1. HIGGS, P. W. 1964. Phys. Lett. **12:** 132; HIGGS, P. W. 1964. Phys. Rev. Lett. **13:** 508; HIGGS, P. W. 1966. Phys. Rev. **145:** 1156; ENGLERT, F. & R. BROUT. 1964. Phys. Rev. Lett. **13:** 321; GURALNIK, G. S., C. R. HAGEN & T. W. B. KIBBLE. 1964. Phys. Rev. Lett. **13:** 585; KIBBLE, T. W. B. 1967. Phys. Rev. **155:** 1554.
2. WU, S. L. 1986. *In* Proc. ECFA Workshop—LEP 200. (Aachen, West Germany, Sept. 29–Oct. 1, 1986). Also CERN-EP/87-40.
3. HILGART, J., M. MERMIKIDES, S. RITZ, S. L. WU & G. ZOBERNIG. Jet analysis in Higgs search. To be published in Z. Phys. C-Particles and Fields.
4. ANDERSSON, B., G. GUSTAFSON & T. SJÖSTRAND. 1980. Z. Phys. C-Particles and Fields **6:** 235; SJÖSTRAND, T. 1980 and 1982. Lund Preprint LU TP 80-3, LU TP 82-3; ANDERSSON, B., G. GUSTAFSON & T. SJÖSTRAND. 1982. Nucl. Phys. **B197:** 45.
5. WU, S. L. 1981. Z. Phys. C-Particles and Fields **9:** 329; BARKLOW, T. L. 1983. A search for the production of charged Higgs and technipions with large hadronic branching ratios in e^+e^- annihilation at 34.5 GeV center-of-mass energy. Ph.D. Thesis, University of Wisconsin, Madison.

The Status of SLC Commissioning[a]

W. B. ATWOOD

Stanford Linear Accelerator Center
Stanford University
Stanford, California 94305

INTRODUCTION

It has been recognized for some time that the cost of conventional e^+e^- colliders using storage-ring technology would become prohibitive as the maximum energy of the machines increased. At Stanford we had a single, aging electron LINAC. As a test bed for the collider concept we have built a "folded" linear collider by revitalizing and energy upgrading our existing facility. The Stanford Linear Collider (SLC) is "folded" in the sense that our one LINAC is used to accelerate both the electron and positron bunch, which are then separately directed around two arcs, and thus bring them into a head-on collision with each other. The SLAC LINAC had a major upgrade in energy and greatly improved steering and focusing, allowing for much higher charges per radio frequency (R.F.) bucket in order to produce Z° particles at a high rate.

The construction and installation of technical components was essentially completed in January 1987. Since then the rate of progress in bringing on the SLC has been wonderful. TABLE 1 gives a list of the milestones achieved since then.

TABLE 1. Achievements in the Commissioning of the SLC

Milestone	Date
Energy = 50 GeV	Jan. 19, 1987
Positrons to damping ring	Feb. 8, 1987
Electron beam to intersection point	Feb. 22, 1987

FIGURE 1 shows a schematic of the SLC. The new construction includes the two arc tunnels, the final focus section and collider hall, two damping rings (at the injector end of the LINAC), and a positron source at the two-thirds point along the LINAC. The purpose of the two arcs is obvious. The final focus section on either side of the collision point is designed to produce micron-size beam spots and steer these tiny cross-section beams into head-on collisions. The damping rings at the beginning of the LINAC were necessary to reduce the emittance (phase space) of the electron and positron bunches and they accomplish this via synchrotron radiation cooling. The positron source produces the bunches of positrons that are to be collided with electrons. Positrons are made through pair production in a high Z target from an incident electron beam,

[a]This work was supported by the U.S. Department of Energy under Contract DE-AC03-76SF00515.

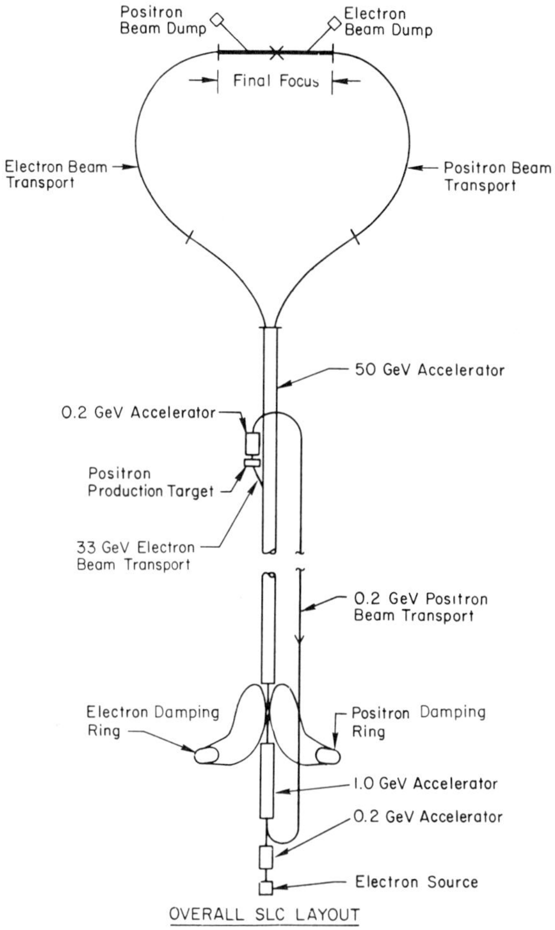

FIGURE 1. Schematic layout of the SLC. The new construction consisted of the two arc tunnels and the collider hall experimental area, the positron source area, and two damping rings just downstream of the injector.

accelerated up to 200 MeV, and shipped back to the injector for cooling and final acceleration.

Not listed above under new construction are the numerous upgrades to the LINAC. Most of the klystrons have been replaced with a new high-power model capable of producing in excess of 65 MW for a pulse duration of 3.5 μsec. These tubes when "SLEDed" each give the beam a 240-MeV kick. When all the klystron sockets on the machine are filled with new tubes, a maximum energy in excess of 55 MeV will be attainable (i.e., much more than is required to make Z° particles).

Other necessary LINAC improvements included the addition of many quadrupole magnets, beam position monitors, and steering magnets. These additions keep the small emittance beams produced by the damping rings from "blowing up" in their trip down the LINAC due to wakefield effects and results in what is perhaps the most surprising improvement in the LINAC. The "old" SLAC LINAC produced beams of $\sim 5 \times 10^{11}$ electrons distributed over some 4600 R.F. buckets in 1.6 μsec. Hence the charge per R.F. bucket was $\sim 1 \times 10^{8}$. We now routinely accelerate 1×10^{10} electrons in a single R.F. bucket, and will eventually push this up to more than 5×10^{10}. This is more than a two orders of magnitude increase in the charge density!

A summary of the present operating conditions, as well as the design goals, are shown in TABLE 2. The spot sizes, luminosity, and rate of $Z°$ production in the Present Value column are calculated quantities based on the present LINAC performance.

The SLC is essentially complete, but a few components still need to be changed due to inadequate design, malfunction, or as part of finishing the general upgrade. In the damping rings, the original kicker design has too slow a rise time to allow for two-bunch operation in these rings. Replacement kickers were installed in the fall of 1987. Several components in the positron source developed problems during initial operation. Although they didn't fail completely, they are compromising the positron yield by a factor of 2. Again, replacement components of improved design and manufacture are on the way. Finally, of the 244 klystrons associated with the LINAC, some 44 are still old, low-power tubes, and as a sufficient pool of spare high-power tubes is developed, these will be changed.

The electron beam and its associated components have been exercised all the way to the collider hall. The positron beam has been returned to the injector from the source and will soon be stored in the south damping ring. (*Note:* Since the conference the positrons have been stored, accelerated in the LINAC, and brought to the collider hall, and finally on March 27, 1987, both electrons and positrons at 47 GeV were transported there on the same machine pulse.)

THE FRONT END

FIGURE 2 shows us schematically the path electrons follow from the "gun" to the beginning of the main LINAC (starting with Sector 2). The present gun is a thermionic device that produces high-current electron pulses. These bursts of particles are compressed in length to about 2 mm and boosted to 200 MeV for injection into Sector 1. This sector provides the boost to 1.15 GeV for injection into the north damping ring where the electrons are cooled by radiating synchrotron light. Extraction follows after one machine pulse period and the electrons are conveyed along the "ring

TABLE 2. The Present and Design Values of SLC Operating Parameters

Parameter	Present Value	Design Value
e^- bunch population	1.0×10^{10}	7.2×10^{10}
e^+ bunch population	1.0×10^{10}	7.2×10^{10}
Pulse rate	60 pps	180 pps
Horizontal beam size	$3 - 5\ \mu m$	$1.7\ \mu m$
Vertical beam size	$2 - 3\ \mu m$	$1.7\ \mu m$
Luminosity	$3 - 8 \times 10^{27}\ cm^{-2}\ sec^{-1}$	$6 \times 10^{30}\ cm^{-2}\ sec^{-1}$
$Z°$'s per day	6–17	12,400

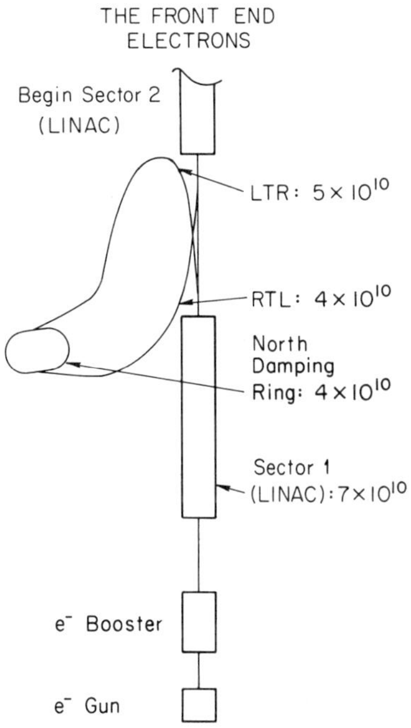

FIGURE 2. Electron source for the SLC. Electrons are produced by a thermionic gun and bunched in the injector section. Next they are accelerated up to 1.15 GeV in Sector 1 and are transported to and stored in the north damping ring for cooling. Finally, they are ejected from the damping ring and conveyed to the beginning of the main LINAC starting at Sector 2.

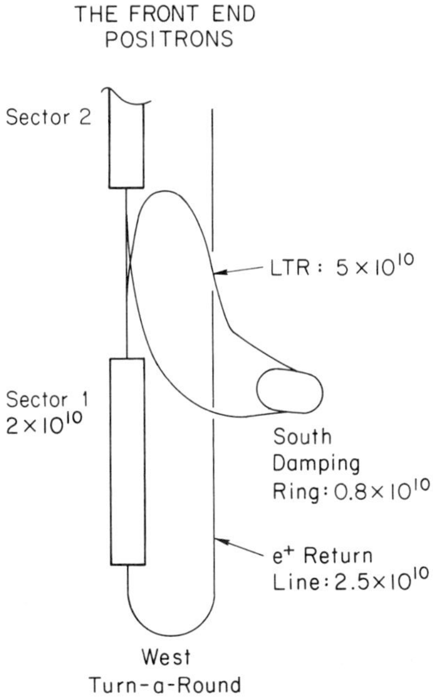

FIGURE 3. Positron front end of the SLC. Positrons are produced at a target far downstream and are shipped back to the front end for cooling and acceleration. Upon arriving, the positrons are first accelerated up to 1.15 GeV in Sector 1 and are then transported and stored in the south damping ring for cooling. They are then ejected from the damping ring and transported to the beginning of the main LINAC starting at Sector 2.

to LINAC (RTL)" transport to the entrance of Sector 2. In the damping ring the natural length of the electron bunch is about 6 mm and to restore it to the necessary 1–2-mm bunch length, an energy position correlation is introduced by a special R.F. section in the RTL. Compression is then accomplished by the remaining leg of the RTL that is anisochronous. On FIGURE 2 are noted the currently achieved electron bunch populations. Finally, this electron injection system has been the operational injector for all downstream programs for the past year; it is debugged and reliable.

FIGURE 3 complements the last figure and shows the path the positrons follow, beginning with the last leg of the return line from the positron source far downstream. The positron bunch populations achieved so far are recorded on this figure. As of the date of this conference, positrons are "knocking at the door" of the south damping ring, but have not been stored.

THE POSITRON SOURCE AND THE LINAC

The positron source area is located two thirds of the way down the LINAC. Electron bunches at about 33 GeV are extracted by a fast kicker magnet and

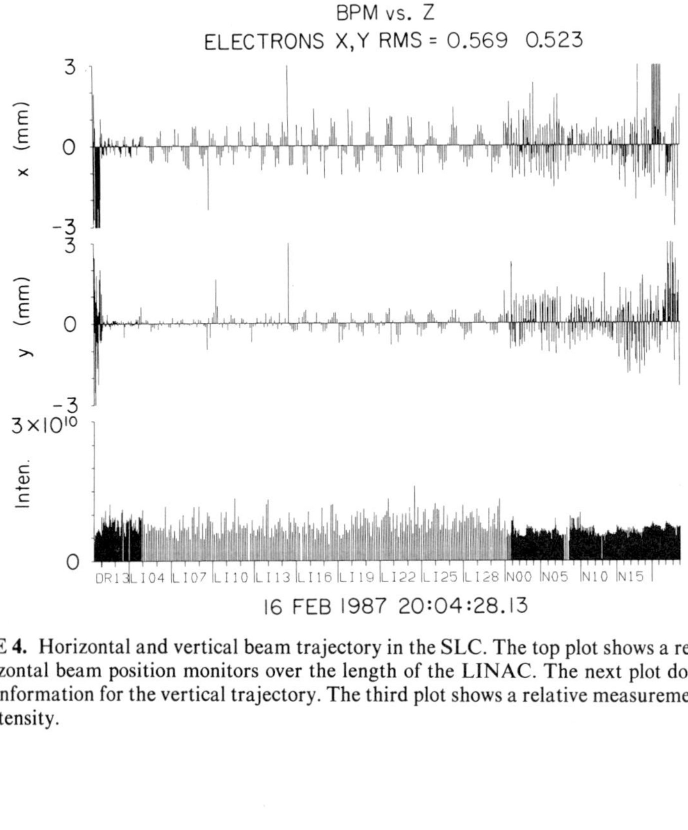

FIGURE 4. Horizontal and vertical beam trajectory in the SLC. The top plot shows a readout of the horizontal beam position monitors over the length of the LINAC. The next plot down gives similar information for the vertical trajectory. The third plot shows a relative measurement of the beam intensity.

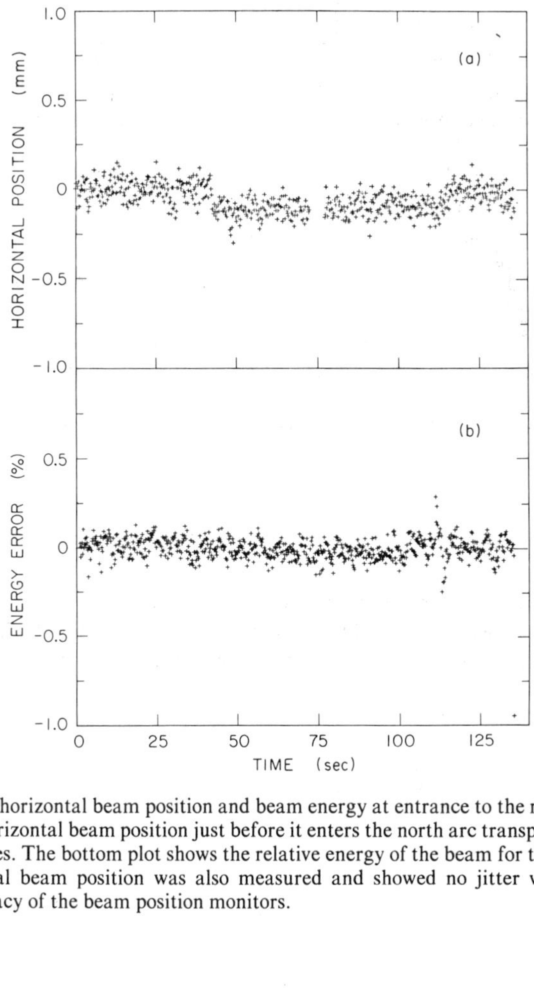

FIGURE 5. The horizontal beam position and beam energy at entrance to the north arc. The top plot shows the horizontal beam position just before it enters the north arc transport line measured over a few minutes. The bottom plot shows the relative energy of the beam for the same period of time. The vertical beam position was also measured and showed no jitter within the 30-μm measuring accuracy of the beam position monitors.

Lambertson septum magnets from the LINAC and are transported to a high Z target that is about six radiations long. Positrons from the resulting electromagnetic cascade showers are collected by a high field pulsed flux concentrator and a solenoidal magnetic field. Immediately downstream of the target begins a high gradient "capture" accelerator section whose purpose it is to make the positrons ultrarelativistic before velocity dispersion has appreciably lengthened the bunch. Further acceleration to about 200 MeV is obtained with three additional sections of LINAC of ordinary SLAC design. The positron bunches are then turned through 180 degrees and begin the trip back to the injector.

Of critical importance to the success of the SLC is the quality and stability of the accelerated beams. To hold wakefield effects to acceptably small values, the particle bunches must be steered to the center of the R.F. structure with high accuracy. FIGURE

TABLE 3. A Timetable for the Turn-on of the SLC

Milestone	Projected Date
Store e^+ in south damping ring	Imminent
Positrons through south arc	Beginning of March
Coacceleration of e^+ and e^-	End of March
First beam–beam collisions	April
Mark II detector rolls on beam line	May

4 shows an on-line display of the horizontal and vertical orbits down the LINAC. Note that full scale is only ±3 mm.

Stability in energy and trajectory at the end of the LINAC is also critical, and FIGURE 5 shows the energy jitter and horizontal position over a period of a few minutes as measured at the end of the LINAC. These stability studies have shown that well less than 10% of the pulses delivered by the LINAC fall outside "SLC specs" for collisions.

ARCs, FINAL FOCUS, AND THE FUTURE

The key issue in the arcs and final focus areas is survey. Great effort was exerted here, employing a space satellite referenced grid, and it has paid off! The first attempt to thread the electron beam through some 19 columnators and many more magnetic

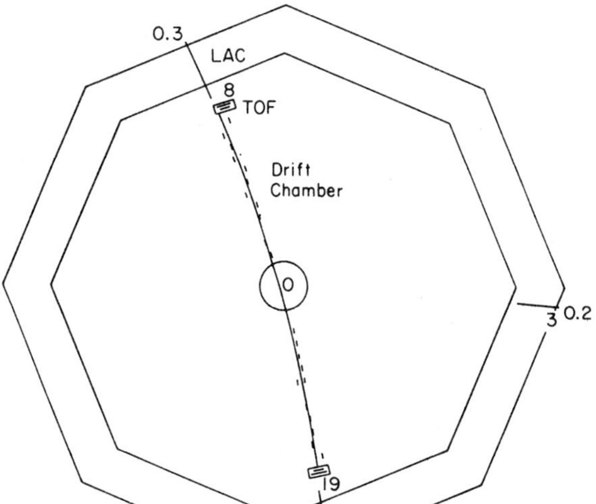

FIGURE 6. A cosmic ray event registered in the Mark II detector. This event is recorded by the Mark II detector as it sits beside the beam line in the collider hall at SLC. All systems are seen to be active by the readouts shown on the display and the track is curved, indicating that the magnet is on.

elements in the final focus achieved success using precalculated magnet settings on the first try!

The pace of commissioning the SLC is brisk and spirits are high around SLAC. TABLE 3 presents some of this author's guesses as to when further milestones will be past.

THE DETECTORS

The Mark II detector is sitting beside the interaction point waiting to be rolled onto the beam line. This detector is currently running on cosmic rays, and FIGURE 6 shows a typical cosmic ray track. Note that signals are registered from all components: liquid argon calorimeter, time-of-flight scintillators, and central-drift chamber.

The SLC's new detector, SLD, has begun to arrive on site. Large shipments of prefabricated steel from Kawasaki are now being assembled on the other side of the beam line from the Mark II. The SLD is definitely no longer a paper detector.

Finding the Top Quark at the SLAC Linear Collider with the Mark II Experiment[a]

FRANK C. PORTER

Lauritsen Laboratory of Physics
California Institute of Technology
Pasadena, California 91125

INTRODUCTION

It may happen that the mass of the expected sixth ("top," or "t") quark is less than half the mass of the Z^0 boson. If so, then the decay $Z^0 \rightarrow t\bar{t}$ will occur. The possibility of finding and studying the top quark in such decays is attractive, because the Z^0 is expected to be produced cleanly and copiously in e^+e^- experiments in the near future. One such experiment is the Mark II, in final stages of preparation to be the first experiment to take data at a new e^+e^- colliding-beam machine capable of reaching the Z^0, the Stanford Linear Collider (SLC). This paper summarizes the essential features relevant to the discovery and identification of a top quark in this experiment.[1] We discuss three aspects: (1) discovery of the top quark; (2) establishing identity of a new object as top quark; and (3) determination of its mass.

DISCOVERY OF THE TOP QUARK

We first set the stage by giving some rates: the cross section for Z^0 production in e^+e^- collisions at a center-of-mass energy equal to the Z^0 mass is ~40 nb. The partial width for $Z^0 \rightarrow t\bar{t}$ varies between ~230 MeV for a top quark mass of $m_t = 25$ GeV and ~40 MeV for $m_t = 45$ GeV. Note that these numbers include the fact that the width near threshold is significantly enhanced by the first-order (α_s) QCD correction to the lowest order Standard Model calculation.[2] Thus, the cross section for $t\bar{t}$ production in $e^+e^- \rightarrow Z^0 \rightarrow t\bar{t}$ is expected to be of order 0.6–3 nb. The design luminosity of the SLC is 6×10^{30} cm^{-2}s^{-1}. At this luminosity, we would expect ~300–1500 $t\bar{t}$ events/day. However, the accelerator will not achieve design luminosity for some time; a more realistic number for early data is ~1% of design. If we wish to find the top quark in a few months of data at this luminosity, we must consider a data set consisting of order 10^4 Z^0 events.

Various possible signatures for the top quark in Z^0 decays have been considered: (a) The event "shape" for $t\bar{t}$ events is expected to be different from the lighter quark backgrounds, that is, $t\bar{t}$ events tend to involve several jets instead of two, and tend not to

[a]This work was supported in part by the U.S. Department of Energy under Contracts DE-AC03-81-ER40050 and DE-AC03-76SF00515.

be as planar as the background processes with gluon emission. However, it is found that precise predictions of the distinguishability of $t\bar{t}$ from background are complicated by model dependence in the calculation of background process event shapes. (b) The top quark decays semileptonically a large fraction (~30%) of the time. Because of its large mass, the lepton from such a decay will tend to have a large transverse momentum (p_T) relative to the direction of the decaying quark. (c) Similarly, the lepton from a semileptonic decay of a heavy quark will tend to be isolated from other particles in the event. (d) The number of jets or "clusters" in a $t\bar{t}$ event is distributed differently than for the light-quark background. (e) Top-quark events tend to involve more semileptonic decays than the average background event. An excess of multilepton events is thus a signature for top. (f) The presence of a top quark accessible in Z^0 decays increases the total hadronic cross section. (g) A special case occurs if the mass of the top quark is approximately half the Z^0 mass. In this case, the vector state of toponium mixes, or interferes, with the Z^0, producing dramatic peaks and valleys in the total

TABLE 1. Monte Carlo Calculation of Background and $t\bar{t}$ Signal Expected Using Isolation and Transverse Momentum Cuts in Semileptonic Events[a]

m_t (GeV^2)	$N_{\text{events produced}}$	$N_e + N_\mu$ $p_T > 3$ GeV	$N_e + N_\mu$ $p_T > 3$ GeV "Isolated" Lepton
Background Signal	10^4 *udscb*	80–90	~3
25	1283	300–350	120–180
30	1058	290–300	120–150
35	795	230–250	90–120
40	512	150–170	~75
42.5	372	110–120	~60
45	240	~80	~40
46	195	~60	~30

[a]The terms N_e and N_μ refer to numbers of electrons and muons, respectively, satisfying the indicated criteria. The background numbers include full detector simulation, while the other numbers use simple cuts on the generated kinematic quantities that adequately approximate the detector.

cross section.[3] Unfortunately, this structure is very sharp, and hence is largely washed out by the finite energy spread of the accelerator beams. However, it remains as an observable possibility with sufficient statistics if this special situation actually occurs.

Several of these possibilities will probably lead to convincing signals. We will restrict further discussion, however, to what appears to be the most promising approach to discovering the top quark in a small data set. TABLE 1 gives the number of events produced for light-quark background and for $t\bar{t}$ production as a function of m_t. All numbers are normalized to a data set that has 10^4 $Z^0 \rightarrow$ light-quark (*udscb*) events. The produced numbers of $t\bar{t}$ events are a cross-section calculation, including the first-order QCD correction. The right two columns are the result of Monte Carlo calculations, based on the LUND and Webber models with different fragmentation assumptions. The ranges indicate the uncertainty introduced by model dependence.

We can see from TABLE 1 that we already achieve a reasonable signal-to-

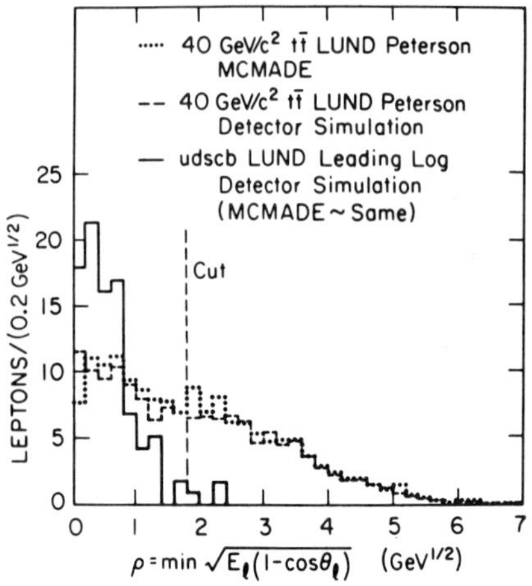

FIGURE 1. Distribution of the quantity ρ for background (*udscb*) and signal (t) events. The normalization is to 10^4 *udscb* events at the Z^0.

background ratio simply by requiring events to have an electron or muon with $p_T > 3$ GeV (transverse momentum is here defined with respect to the event thrust axis). However, we can do even better if we note that the remaining background tends to be from events containing both a semileptonic decay of a lighter quark and a radiated gluon. The effect of the radiated gluon is to sometimes cause the lepton to have a high transverse momentum relative to the thrust axis, but still not relative to the direction of its parent jet. On the other hand, a semileptonic decay of a top quark will typically produce a lepton that has a high momentum transverse to the top-quark direction, and hence produce an "isolated" lepton.

Various cuts have been invented to select events with such isolated leptons. The one that has so far received the most study is as follows: First select a lepton with $p_T > 3$ GeV. Second, find "clusters" of particles using all of the other particles in the event.

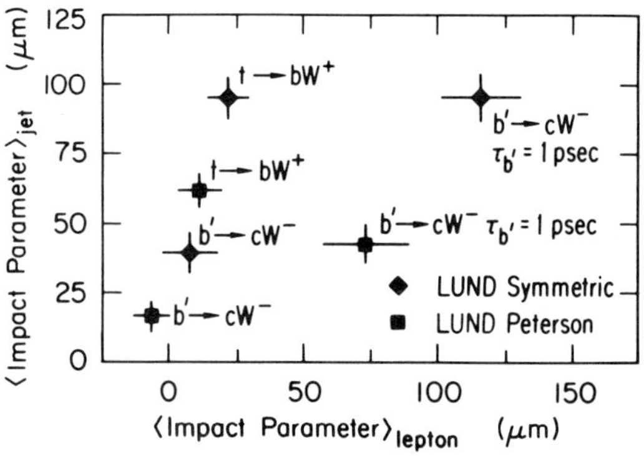

FIGURE 2. Monte Carlo predictions for measured jet-impact parameter vs. lepton-impact parameter for a 25-GeV t or b' quark.

Then calculate the quantity

$$\rho = \min_i \sqrt{E_l(1 - \cos\theta_{li})}$$

where E_l is the energy of the lepton, θ_{li} is the angle between the lepton and cluster i, and the minimum is taken considering all clusters. Note the ρ is related to the minimum mass that the lepton makes with any of the clusters, and hence should be large if the lepton is really not associated with any of the clusters. FIGURE 1 shows a Monte Carlo calculation of the distribution of ρ for signal and background events. A cut selecting $\rho > 1.8$ GeV$^{1/2}$ removes nearly all of the background, but has reasonable efficiency to keep the signal. The effect of this cut for different m_t is summarized in TABLE 1. We see that a very clean top-quark signal can be observed with this size data set for masses nearly up to half the Z^0 mass.

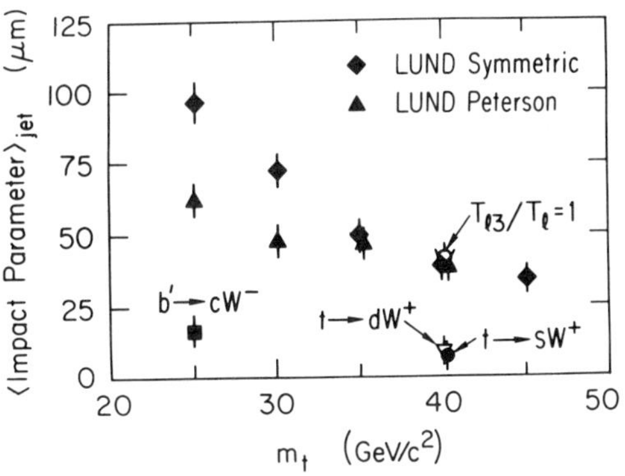

FIGURE 3. Monte Carlo predictions for measured jet-impact parameter as a function of t (or b') quark mass.

ESTABLISHING THAT A SIGNAL IS TOP

Once a signal for a new particle is observed, as in the preceding section, it is still necessary to establish its identity. Perhaps it is top, but other possibilities include a fourth-generation quark or a scalar quark. Most of the effort has concerned the distinction between the top quark and a fourth-generation, charge $-1/3$ quark (b'). Naturally, various potential ways of distinguishing these possibilities exist: (a) The partial width for $Z^0 \rightarrow t\bar{t}$ is about 80% of the corresponding width for a b' quark of the same mass. This is probably not a large enough difference to provide a very convincing distinction, given the systematic uncertainties likely to be present. (b) Rather strong distinctions can be made if a search is made for the "onium" bound state of the new particle and its antiparticle. However, this requires a luminosity near the design luminosity of the SLC. (c) With high statistics, the $t\bar{t}$ forward–backward asymmetry could be measured with lepton tags, and could yield convincing confirmation of the "topness" of the new particle. (d) With only limited statistics available, probably the most promising approach is to find evidence for the decay of a top meson (T or T^*) to a bottom meson (B or B^*). We consider this possibility further.

The idea to show that the $T \rightarrow B$ decay is present is to try to demonstrate that there is a high-p_T lepton (from the semileptonic decay of the top quark) plus a jet that has an impact parameter (i.e., projected distance of closest approach to the primary event vertex) consistent with the decay of a B-meson. As the top quark (and hence the top meson) is expected to have a short lifetime, the high-p_T lepton should have a small impact parameter. The bottom quark (and hence the B-meson) is long-lived ($\gtrsim 1$ ps), so large impact parameters are expected. We may compare this with what we might expect for a b' quark, supposing that it decays principally via $b' \rightarrow cW^-$. Since the b' is a new generation, we really don't know a priori how big the matrix element is that couples it to the charmed quark. If we take the guess that the element is very small, then the lifetime is large, and hence both the high-p_T lepton and the charm jet will have large impact parameters. If, on the other hand, the element is not small, then the lifetime is short, and both impact parameters will be relatively small, given that the c quark is shorter lived than the b quark. Thus, in either situation, we can expect a difference between the t quark and a b' quark. In practice, the situation should be approximately as illustrated in FIGURES 2 and 3. The predictions for the jet impact parameter are fairly model-dependent at low quark mass, because of the influence of assumptions concerning the hardness of the fragmentation of the heavy quark. However, this looks like a promising test to distinguish t from b'.

DETERMINING THE TOP-QUARK MASS

It is of interest to ask how well we can determine the mass of the top quark once we have found it. Again, several ideas have been investigated: (a) The shape of the lepton

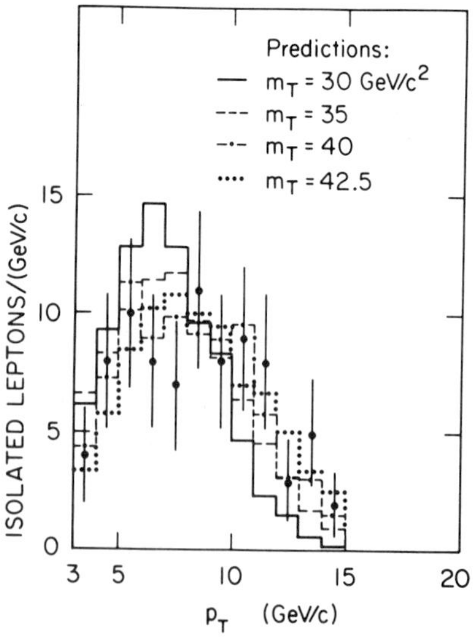

FIGURE 4. The distribution in transverse momentum for isolated leptons for $3 < p_T < 15$ GeV and $m_t = 40$ GeV corresponding to a data sample of 10^4 $Z^0 \rightarrow udscb$ events (points with error bars). Overlaid are histograms giving the predictions for different values of m_t.

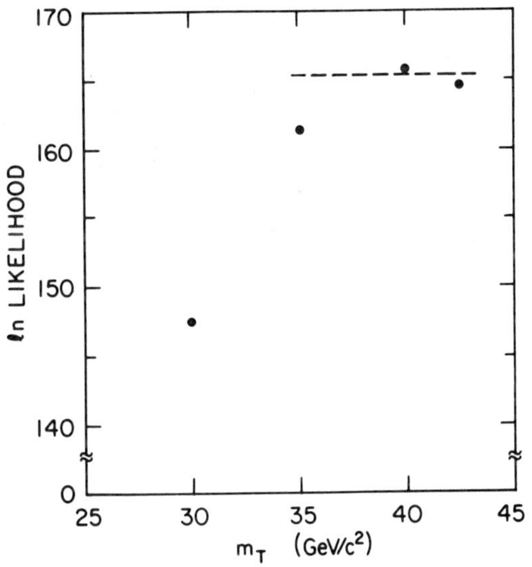

FIGURE 5. Result of a maximum likelihood fit to the "data" in FIGURE 4, with m_t as the varied parameter. The dashed line is drawn at the maximum likelihood minus ½.

transverse momentum distribution in semileptonic top decays depends on m_t. (b) Even the number of high-p_T leptons depends on m_t, both because the number of $t\bar{t}$ events depends fairly strongly on mass, and also because a fixed definition of "high-p_T" will yield different efficiencies for different masses. (c) Various "direct" mass measurements have been investigated, all based on the reconstruction of jet masses. The differences in approach concern whether one or both of the top quarks decay semileptonically. (d) Ultimately, a really "precision" measurement of m_t is possible by finding toponium, where an uncertainty of order 100 MeV may be achieved.

Several possibilities are likely to be fruitful: Let us see how well we can do using the shape of the p_T distribution with a small data set. As an example, FIGURE 4 shows the comparison of the p_T distribution for isolated leptons for m_t = 40 GeV with the predictions for various masses. The distribution is cut off at high-p_T so that a maximum likelihood fit can be made that does not depend on possibly spurious effects in the tail. The result of a maximum likelihood fit to the "data" in FIGURE 4 is shown in FIGURE 5. With a data set corresponding to ~ 10^4 $Z^0 \rightarrow udscb$ events, it seems possible to determine m_t to ~ ±2–3 GeV by this method, improving with higher statistics.

CONCLUSION

The essential conclusion is that it should not be difficult for the Mark II experiment at the SLC to find the top quark with a data set corresponding to several thousand Z^0 events, even if m_t is nearly half the Z^0 mass. Of course, we hope that the top quark is, in fact, light enough to be accessible in Z^0 decays!

REFERENCES

1. The analyses discussed here are principally the work of the "Open Top Physics" subgroup of the Mark II collaboration (including some theorists): BARKLOW, T., B. BARNETT, J. BARTELT, I. BIGI, P. DAUNCEY, G. FELDMAN, R. FREY, D. FUJINO, G. HANSON, C. JUNG, S. KOMAMIYA, W. KOSKA, K. O'SHAUGHNESSY, A. PETERSEN, D. STOKER, S. WAGNER, B. WARD, A. WEIR, and E. WICKLUND. More thorough discussions, on which most of this paper is based, are to be found in HANSON, G. 1986. Mark II/SLC. Physics Working Group Note 3-10. In SLAC Report 306; HANSON, G. 1987. Mark II/SLC. Physics Working Group Note 3-18.
2. KÜHN, J. H., A. REITER & P. M. ZERWAS. 1986. Nucl. Phys. **B272:** 560.
3. GÜSKEN, S., J. H. KÜHN & P. M. ZERWAS. 1985. Nucl. Phys. **B262:** 393; FRANZINI, P. & F. GILMAN. 1985. Phys. Rev. D **32:** 237.

Prospects for Top Search at the Tevatron Collider

JOHN YOH

Fermilab
P.O. Box 500
Batavia, Illinois 60510

INTRODUCTION

We discuss the prospects for the search for top production at the Tevatron collider, at center-of-mass energy of 2 TeV. At least 300 produced T pairs are needed for their discovery above background. Within the next several years, T masses of up to 100 GeV should be accessible. With a TEV I upgrade in luminosity (by a factor of 50), it will be possible to reach T masses above 200 GeV.

DISCUSSION

The CDF detector of the Fermilab Tevatron collider has begun taking data at a center-of-mass (cms) energy of 1.8 TeV. The 1987 run recorded 35 events/nb of data, with the observation of several dozen W's, a few Z's, and jets with Pt approaching 200 GeV. Although the integrated luminosity is only one-twentieth that of the CERN SPS collider, the factor of 3 increase in cms energy results in a production of high mass objects that has already exceeded that of the SPS collider for processes due to gluon fusion. The next run in early 1988 is expected to reach an integrated luminosity approaching one event/pb, thus opening the window of opportunity to much larger masses.

It is worth considering the prospects for the search for top production at the Fermilab Tevatron collider. FIGURE 1 shows the expected total cross section for heavy quark pair production via gluon fusion. At this cms energy, gluon fusion produces much more top than top from the decay of W's; we will also ignore the possible existence of a heavy Higgs particle, which would enhance the top production at top masses of order one-half of that of the Higgs.

In order to discover top particles, we need sufficient statistics of events with a clean signature. In general, this would require selection of particular decay modes and making cuts that keep only a fraction of events with topology distinguishable from background. For example, one might demand that one of the two tops decay into b + electron + neutrino (probably 2×1 BR), and that the event contain one electron and four jets above some Pt threshold (perhaps 0.3 in efficiency ???) for a total efficiency of 0.06. In order to have 20 signal events, one would need to produce a total of 300 top pairs. We will thus use 300 as the number of top pair events needed to be produced in order to discover it. For top mass above 90 GeV, the top particles will probably decay into b + a real W; this would make the signature easier to recognize, since some pairs

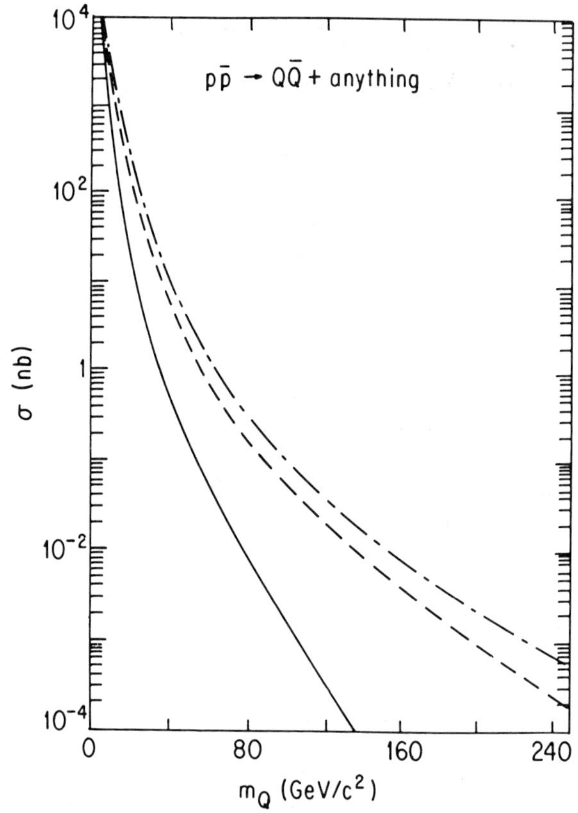

FIGURE 1. The total cross section for heavy-quark pair production via gluon fusion as a function of heavy quark mass, m_Q, for $\bar{p}p$ collisions at $\sqrt{s} = 630$ GeV (solid line), 1.8 TeV (dashed line), and 2.0 TeV (dot-dashed line). The parton distributions of set 2 used. (Based on E. Eichten. Theoretical expectations at collider energies, Fermilab Pub-85/178-T.)

TABLE 1.

Time	Integrated Luminosity	Discovery up to T Mass of
1987 (now)	35/nb	50 GeV
1988	1/pb	80 GeV
1989–1990	10/pb	120 GeV
Early TEV I+ (92 ??)	100/pb	180 GeV
Late TEV I+ (94 ??)	500/pb	250 GeV

of the jet/leptons would have additional mass constraint; thus fewer events might be needed for discovery. Nevertheless, we will conservatively assume 300 produced top pairs are needed for probable (though not guaranteed) discovery.

In TABLE 1, we list the expected integrated luminosity for the next several years at the Tevatron collider, making somewhat optimistic assumptions. In addition, we added two lines for TEV I +, the possible Tevatron collider upgrade that is being studied now. The goal is to achieve a peak luminosity of 5 × 10**31. The discovery limits correspond to the mass at which 300 top pairs would be produced, at the integrated luminosity listed in the table.

Calculation of Lepton Spectra from *W*-Production at the $Sp\bar{p}S$-Collider[a]

C. GREUB AND P. MINKOWSKI

Institute for Theoretical Physics
University of Bern
Bern, Switzerland

INTRODUCTION

Calculations of the process

$$p\bar{p} \rightarrow W^- X \rightarrow e^- \bar{\nu}_e X$$

are presented, including order $\alpha_s(m_W)$ QCD corrections, and are compared with recent results of the UA2 collaboration.[8] Deviations from the theoretical expectations are marginal, but nevertheless tentatively interpreted as a signature of the top quark. The best parameters we extract are $m_W = 82 \pm 1.5$ GeV, $\Gamma_W^{\text{tot}} = 2.43 \pm 0.5$ GeV. (Interpreting marginal deviations as due to the top quark we infer

$$m_{\text{top}} = 60 \left\{ \begin{matrix} +10 \\ -5 \end{matrix} \text{GeV.} \right)$$

THEORETICAL PART

The basis for the amplitudes on the level of quarks and gluons and the subsequent factorized expressions involving the scale-dependent proton (and antiproton) structure functions was laid in the work of Altarelli *et al.*[1] within the general framework of the impact parameter representation as outlined by Collins and Soper.[2]

Parameterization of the Cross Section for W *Production from* $p\bar{p}$ *Collision in the Leptonic Decay Mode*

The multidifferential cross section in question to order α_s (the running QCD coupling constant) has the structure as discussed in approximate form in reference 3:

$$d\sigma = \frac{1}{3s} \operatorname{Res} f\,[q^2, m_W^2, \Gamma_W^{\text{tot}}; G_F] \cdot dX_{\text{IR}} \tag{1}$$

$$s = (p_p + p_{\bar{p}})^2 = \{(630 \text{ GeV})^2 \text{ for numerical analysis})\}$$

[a]This work was supported in part by Schweizerischer Nationalfonds.

$$\text{Res}\, f = 8\, G_F^2 \left| \frac{m_W^2}{q^2 - m_W^2 + i m_W \Gamma_W^{\text{tot}}} \right|^2$$

$$G_F = 1.16637 \cdot 10^{-5}\, (\text{GeV})^{-2}\text{: Fermi constant}$$

$$q = p_{e^-} + p_{\bar{\nu}_e}$$

and corresponds to the reaction kinematics and momenta:

$$p + \bar{p} \rightarrow W^- X \rightarrow e^- \bar{\nu}_e X$$

$$p_p + p_{\bar{p}} = q + p_X = p_{e^-} + p_{\bar{\nu}_e} + p_X$$

$$(p_{e^-} \equiv l). \tag{2}$$

In equation 1, dX_{IR} denotes the infrared regulated differential cross section, where the regularization procedure we use is dimensional-infrared

$$dX_{\text{IR}} = \underset{\epsilon \to 0}{\text{reg}}\, dX_\epsilon$$

$$dX_\epsilon = d\Omega(l) \cdot \{d\Omega_0(q) X_\epsilon^{(0)} + d\Omega_\epsilon(q) X_\epsilon^{(1)}\}. \tag{3}$$

In equation 3, $X^{(0)}$, $X^{(1)}$ denote the cross sections for subprocesses involving the emission of 0 and 1 secondary quanta, respectively:

$$X_\epsilon^{(0)} = \delta(q^2 - 2ql)\delta(q_\perp^2) \frac{1}{\Gamma_\gamma(1+\epsilon)}$$

$$\times (1 - V_{DY}) \sum_{r=1}^{2} \frac{|M_r^{(0)}|^2}{4s} H_r^{(0)}(z_1^0, z_2^0) \frac{1}{z_1^0 z_2^0}. \tag{4}$$

$r = 1$:

$$\begin{array}{ccccc} d \;+ & \bar{u} & \rightarrow W^- \rightarrow & e^- \;+ & \bar{\nu}_e \\ p_d = z_1^0 p_p & p_{\bar{u}} = z_2^0 p_{\bar{p}} & q & l & q - l \end{array}$$

$r = 2$:

$$\begin{array}{ccccc} \bar{u} \;+ & d & \rightarrow W^- \rightarrow & e^- \;+ & \bar{\nu}_e \\ p_{\bar{u}} = z_1^0 p_p & p_d = z_2^0 p_{\bar{p}} & q & l & q - l \end{array}$$

+ quark flavor mixing substitutions.

The subprocesses associated to $X^{(1)}$ are given in TABLE 1. The kinematic variables are defined, including the extension of dimensions to

$$d = 4 + 2\epsilon \qquad (\epsilon \geq 0)$$

where necessary:

$$d\Omega(l) = \frac{1}{16\pi^2} d^2 l_\perp dy_l \qquad (\epsilon = 0 \text{ throughout})$$

$$d\Omega_\epsilon(q) = dq^2 d(q_\perp^2) dy_q d\overline{\omega}_{1+2\epsilon}(\mathbf{n}_\perp) \times \frac{1}{\Gamma_\gamma(1+\epsilon)} \left(\frac{q_\perp^2}{\mu^2}\right)^\epsilon . \tag{5}$$

In equation 5, $\Gamma_\gamma(1+\epsilon)$ denotes the rationalized gamma function

$$\Gamma_\gamma(1+\epsilon) = e^{\gamma_E \cdot \epsilon}\Gamma(1+\epsilon) = \prod_{n=1}^{\infty}\left[\left(1+\frac{\epsilon}{n}\right)e^{-\epsilon/n}\right]^{-1}$$

$$\left[\Gamma_\gamma(1+\epsilon) = 1 + \frac{\pi^2\epsilon^2}{12} + 0(\epsilon^3)\right] \tag{6}$$

γ_E : Euler's constant (= 0.5772. . . .).

TABLE 1. Subprocesses

$r = 1$	$d + \overline{u} \rightarrow W^- + g$	
$r = 2$	$\overline{u} + d \rightarrow W^- + g$	
$r = 3$	$d + g \rightarrow W^- + \overline{u}$	
$r = 4$	$\overline{u} + g \rightarrow W^- + \overline{d}$	
$r = 5$	$g + d \rightarrow W^- + \overline{u}$	
$r = 6$	$g + \overline{u} \rightarrow W^- + \overline{d}$	
$r = 7$	$d + \overline{u} \rightarrow W^-$ + scalar gluon	dim > 4
$r = 8$	$\overline{u} + d \rightarrow W^-$ + scalar gluon	
+ analogous reactions induced by weak mixing angles		

The term $\mathbf{n}_\perp$ denotes the normalized vector on the $(1 + 2\epsilon)$-dimensional unit sphere $S_{1+2\epsilon}$ in the momentum directions transverse to the beams, and $d\overline{\omega}_{1+2\epsilon}$ denotes the normalized surface element on $S_{1+2\epsilon}$ such that

$$\int_{S_{1+2\epsilon}} d\overline{\omega}_{1+2\epsilon} = 1; \qquad \left[d\overline{\omega}_1 = \frac{1}{2\pi}d\varphi\right]. \tag{7}$$

The terms y_l, y_q in equation 5 denote the rapidities of e^- and the $e^-\overline{\nu}_e$ system, respectively:

$$E_e = |E_\perp| \operatorname{ch}(y_l), \qquad l_\parallel = E_\perp \operatorname{sh}(y_l)$$

$$E_\perp = |\boldsymbol{l}_\perp|,$$

$$E_q = m_\perp(q) \operatorname{ch}(y_q), \qquad q_\parallel = m_\perp(q) \operatorname{sh}(y_q)$$

$$[m_\perp^2(q) = q^2 + q_\perp^2], \tag{8}$$

where $l_\parallel$, $q_\parallel$ are taken positive in the proton direction.

Lepton masses and masses of quarks [u, d, s, c, (b)] are neglected throughout.

The terms z_1^0, z_2^0 denote the characteristic momentum fractions of the quarks (antiquarks) associated with the reactions in equation 4:

$$z_1^0 = \frac{q^+}{\sqrt{s}}\bigg|_{q_\perp = 0}, \qquad z_2^0 = \frac{q^-}{\sqrt{s}}\bigg|_{q_\perp = 0}; \qquad q^\pm = E_q \pm q_\parallel. \tag{9}$$

The term V_{DY} in equation 4 denotes the virtual gluon correction to the $(W^- d\bar{u})$ vertex (for $q^2 > 0$):

$$V_{DY} = \left(\frac{8}{3}\kappa\right)\left(\frac{q^2}{\mu^2}\right)^\epsilon \left[\frac{2}{\epsilon^2} - \frac{3}{\epsilon} + 8 - \pi^2\right]. \tag{10}$$

The auxiliary scale μ in equations 5 and 10 is induced by the extension of the QCD coupling constant

$$\kappa = \frac{\alpha_s}{4\pi}$$

to $4 + 2\epsilon$ dimensions

$$\kappa_\epsilon = \kappa_\mu \left(\frac{4\pi}{\mu^2 e^{\gamma_E}}\right)^\epsilon. \tag{11}$$

In equation 4, $M_r^{(0)}$ denotes the amplitude for process r on the quark gluon level: for example,

$$\frac{|M_1^{(0)}|^2}{4s} = \frac{1}{s}(2p_d l_1)(2p_{\bar{u}} l_2)$$

$$l_1 = p_{e^-} = l; \qquad l_2 = p_{\bar{\nu}_e} = q - l. \tag{12}$$

Finally $H_r^{(0)}$ in equation 4 denotes a product of unregulated quark (gluon) structure functions:

$$H_1^{(0)}(z_1^0, z_2^0) = d_p^{(0)}(z_1^0)\bar{u}_{\bar{p}}^{(0)}(z_2^0)$$
$$= d^{(0)}(z_1^0)u^{(0)}(z_2^0)$$
$$[d = d_p, \qquad u = u_p, \qquad \bar{u}_{\bar{p}} = u_p, \ldots]$$
$$H_2^{(0)}(z_1^0, z_2^0) = \bar{u}^{(0)}(z_1^0)\bar{d}^{(0)}(z_2^0). \tag{13}$$

We shall maintain the redundancy of the IR-regularization procedure in its entire generality and discuss our specific choice later.

The structure of the contributions to $X^{(1)}$ in equation 3 is of the form

$$X_\epsilon^{(1)} = \frac{1}{\Gamma_\gamma(1+\epsilon)}\left(\frac{q_\perp^2}{\mu^2}\right)^\epsilon \delta(q^2 - 2ql)$$
$$\times \sum_{r=1}^{2} \kappa\, \bar{d}_r \frac{|M_r^{(1)}|^2}{8s} H_r^{(1)}(z_1, z_2)$$
$$\times \frac{dz_1}{z_1}\frac{dz_2}{z_2}\delta\left[(z_1 - \alpha^+)(z_2 - \alpha^-) - \frac{q_\perp^2}{s}\right], \tag{14}$$

where $\alpha^{\pm} = q^{\pm}/\sqrt{s}$; $\bar{d}_r$: channel-dependent color factor (8/3 or 1). In equation 14 $H_r^{(1)}$ denotes an appropriate product of regulated (anti)quark/gluon structure functions of the proton, and $M_r^{(1)}$ stands for the matrix element involving the emission of an additional (anti)quark/gluon quantum: for example,

$$\frac{|M_1^{(1)}|^2}{8s} = \frac{q^2}{s}\frac{[(2p_d l_2)^2 + (2p_{\bar{u}} l_1)^2]}{(2p_d k)(2p_{\bar{u}} k)}$$

$$1 : d(p_d) + \bar{u}(p_{\bar{u}}) \rightarrow W^-(q) + g(k) \rightarrow e^-(l_1) + \bar{\nu}_e(l_2) + g(k). \quad \mathbf{(15)}$$

Redefining the contributions $X^{(0)}$, $X^{(1)}$ in terms of the reduced expressions

$$\frac{|M_r^{(0)}|^2}{4z_1^0 z_2^0 s} = q^2 F_r^{(0)}$$

$$\frac{|M_r^{(1)}|^2}{8z_1 z_2 s} = \frac{q^2}{q_\perp^2} F_r^{(1)} \quad \mathbf{(16)}$$

we obtain

$$X_\epsilon^{(0,1)} = \delta(q^2 - 2ql)Y_\epsilon^{(0,1)}$$

$$Y_\epsilon^{(0)} = q^2\delta(q_\perp^2)[\Gamma_\gamma(1+\epsilon)]^{-1}(1 - V_{DY})\sum_{r-1}^{2} F_r^{(0)}H_r^{(0)}$$

$$Y_\epsilon^{(1)} = \frac{q^2}{q_\perp^2}[\Gamma_\gamma(1+\epsilon)]^{-1}\left(\frac{q^2}{\mu^2}\right)^{\epsilon}$$

$$\times \int \sum_{r-1}^{2} F_r^{(1)}H_r^{(1)}\,\delta\left[(z_1 - \alpha^+)(z_2 - \alpha^-) - \frac{q_\perp^2}{s}\right]dz_1\,dz_2. \quad \mathbf{(17)}$$

The reduced cross sections $F_r^{(0),(1)}$ in equation 17 are functions of ratios of four-momentum components only.

If no observation is made of the emitted quanta (jets) in the final state, there are five relevant kinematic variables. We choose

$$E_\perp = |P_{e^-\perp}|; \qquad q^2; \qquad q_\perp^2; \qquad y_q; \qquad \Phi_{q_\perp l_\perp} \equiv \Phi. \quad \mathbf{(18)}$$

In terms of the lepton momenta $l_1 = p_{e^-}$, $l_2 = p_{\bar{\nu}_e}$, the set in equation 18 is equivalent to

$$E_\perp = |l_{1\perp}|; \qquad y_{l_1}; \qquad |l_{2\perp}|; \qquad y_{l_2}; \qquad \Phi_{l_\perp \bar{\nu}_\perp}.$$

The transverse mass of the $(e^-\bar{\nu}_e)$ system in terms of the variables in equation 18 is given by

$$m_\perp^2 = 2E_\perp |l_{2\perp}|[1 - \cos(\Phi_{e\bar{\nu}})]$$

$$= 2E_\perp[\sqrt{E_\perp^2 + q_\perp^2 - 2E_\perp q_\perp \cos\Phi} + E_\perp - q_\perp \cos\Phi]. \quad \mathbf{(19)}$$

Factorizing the Lepton Momentum Dependence

The lepton momentum dependence is contained in the distribution functions $F_r^{(0)}$, $F_r^{(1)}$ in equation 17. They are of the form

$$F_r^{(0)} = L_r G_r^{(0)}; \qquad F_r^{(1)} = L_{T(r)} \cdot G_r^{(1)} + \Delta_r \qquad (r = 1, 2)$$

$$L_{T(r)} = \{L_1 \text{ or } L_2 \text{ depending on the subprocess}\}. \tag{20}$$

The leptonic variables are contained in the factors $L_{1,2}$ constructed from the lepton momentum fractions

$$\lambda_1^{\pm} = \frac{(l_1)^{\pm}}{q^{\pm}}; \qquad \lambda_2^{\pm} = \frac{(l_2)^{\pm}}{q^{\pm}}$$

$$L_1 = \frac{1}{2}[(\lambda_1^+)^2 + (\lambda_2^-)^2]; \qquad L_2 = \frac{1}{2}[(\lambda_1^-)^2 + (\lambda_2^+)^2]. \tag{21}$$

We note that virtual as well as near-resonance W-momenta q are treated on the same footing.

The terms Δ_r in equation 20 contain genuine correlations between the W-production and -decay amplitudes. These contributions to the cross section are infrared finite and are integrated numerically over the relevant variables. Their contribution to the $p\bar{p} \rightarrow W^-X$ cross section is, except for extreme kinematical situations, maximally a few percent of the remaining terms. Retaining the terms $L_r G_r^{(0),(1)}$ only and restricting the W-four momentum to the mass shell $q^2 = m_W^2$ in the zero-width approximation is tantamount to the calculation of the W-production cross section multiplied by the final state decay branching fraction.[1]

Integrating Over the Structure Functions of the Beam Particles (p,$\bar{\text{p}}$)

The convolution with the effective subprocess luminosities of the incoming particles are of the form

$$Z_r^{(1)} = \int dz_1\, dz_2\, G_r^{(1)}\left[\frac{\alpha^+}{z_1}, \frac{\alpha^-}{z_2}\right] H_r^{(1)}(z_1, z_2)$$
$$\cdot\, \delta\left[(z_1 - \alpha^+)(z_2 - \alpha^-) - \frac{q_\perp^2}{s}\right]. \tag{22}$$

$$Z_r^{(0)} = G_r^{(0)} H_r^{(0)}(z_1^0, z_2^0).$$

$$G_{1,2}^{(0)} = 1 - \frac{8}{3}\kappa\left(\frac{q^2}{\mu^2}\right)^{\epsilon}\left[\frac{2}{\epsilon^2} - \frac{3}{\epsilon} + 8 - \pi^2\right].$$

$$\alpha^{\pm} = \frac{q^{\pm}}{\sqrt{s}} \qquad z_1^0 = e^{y_q}\sqrt{q^2}; \qquad z_2^0 = e^{-y_q}\sqrt{q^2}. \tag{23}$$

Extraction of the Infrared Relevant Parts

The generic contribution to $Z_r^{(1)}$ in equation 22 is expanded around the critical point at $q_\perp = 0$:

$$
\begin{aligned}
Z &= Z_S + Z_p \\
Z_S &= \int dz_1\, dz_2\, \delta\left[(z_1 - \alpha^+)(z_2 - \alpha^-) - \frac{q_\perp^2}{s}\right] G_s \\
G_S &= G\left(\frac{\alpha^+}{z_1}, \frac{\alpha^-}{z_2}\right) H(z_1, z_2) - G(1,1) H(z_1^0, z_2^0) \\
Z_p &= G(1,1) H(z_1^0, z_2^0) I_p \\
I_p &= \int dz_1\, dz_2\, \delta\left[(z_1 - \alpha^+)(z_2 - \alpha^-) - \frac{q_\perp^2}{s}\right].
\end{aligned} \tag{24}
$$

The integral Z_S in equation 24 is split into the infrared dominant and subdominant parts

$$
\begin{aligned}
Z_S &= Z^{\mathrm{IR}} + \tilde{\delta} \\
Z_S &= Z_S\left(\frac{q^2}{s}, \frac{q_\perp^2}{s}, y_q; H\right) \\
Z^{\mathrm{IR}} &= Z_s\left(\frac{q^2}{s}, q_\perp = 0, y_q; H\right) \\
&= \int dz_1\, dz_2\, \delta[(z_1 - z_1^0)(z_2 - z_2^0)] G_S(\alpha^+ = z_1^0; \alpha^- = z_2^0) \\
\tilde{\delta} &= Z_S - Z^{\mathrm{IR}}.
\end{aligned} \tag{25}
$$

The subtracted contribution from the terms $\tilde{\delta}$ in equation 25 are infrared finite and are integrated numerically, but only in a region $q_\perp^2 < q_1^{2\mathrm{critical}}$, since the following resummation of the infrared relevant parts constitutes an asymptotic expansion to the cross section for $(\Lambda_{\mathrm{QCD}}^2 \ll) q_\perp^2 \ll q^2$ and loses its validity for sizable $q_\perp$, where, however, the straightforward QCD perturbative cross section becomes a good approximation. We found that for $q_\perp \gtrsim 25$ GeV the perturbative contribution starts to dominate over the extrapolated small $q_\perp$ expressions.

Only the main subprocesses $r = 1, 2$

$$d + \bar{u} \rightarrow W^- X \rightarrow e^- \bar{\nu}_e X$$

have a nonvanishing particular contribution Z_p in equation 24

$$
G(1,1) = \begin{cases} \left(\frac{8}{3}\kappa\right) \cdot 2, & \text{for } r = 1, 2 \\ 0, & \text{otherwise} \end{cases}
$$

$$I_p = \log\left[\frac{s}{q_\perp^2}(1-\alpha^+)(1-\alpha^-)\right]. \tag{26}$$

Correspondingly the integrals Z_S for $r = 1, 2$ exhibit the following collinear structure

$$Z_{S1,2} = P_{qq}^{(1)}(\xi) * H_{1,2}\left(\frac{z_1^0}{\xi}; z_2^0\right) + P_{qq}^{(1)}(\eta) * H_{1,2}\left(z_1^0; \frac{z_2^0}{\eta}\right)$$
$$+ \left(\frac{8\kappa}{3}\right) \cdot 2H_{1,2}(z_1^0, z_2^0) \log \frac{z_1^0 z_2^0}{(1-z_1^0)(1-z_2^0)}. \tag{27}$$

The two convolutions in equation 27 involve the Altarelli–Parisi[4] evolution functions and refer to the proton- and antiproton-collinear processes, respectively.

$$f(\xi) * g\left(\frac{z}{\xi}\right) = \int_0^1 f(\xi) g\left(\frac{z}{\xi}\right)\frac{d\xi}{\xi}$$
$$P_{qq}^{(1)} = \frac{8}{3}\kappa\left(\frac{1}{1-\xi}\right)_+ (1+\xi^2) + 0(\kappa^2)$$
$$P_{qq} = P_{qq}^{(1)} + \left(\frac{8}{3}\kappa\right)\frac{3}{2}\delta(1-\xi) + 0(\kappa^2). \tag{28}$$

The remaining local part $(8/3\kappa) \cdot 3/2 \cdot \delta(1-\xi)$ in equation 28 results from combining all contributions $G_r^{(0)}$, Z_{Pr}, Z_{Sr} for $r = 1, 2$, which are all proportional to $H_r(z_1^0, z_2^0)$.

$r = 1, 2$:

$$Z_{Sr} = Z_r^{\mathrm{IR}} + \left(\frac{8}{3}\kappa\right) \cdot 2H_r(z_1^0, z_2^0) \log \frac{z_1^0 z_2^0}{(1-z_1^0)(1-z_2^0)}$$
$$Z_{Pr} = \left(\frac{8}{3}\kappa\right) H_r(z_1^0, z_2^0)\left\{2\log\left[\frac{s}{q_\perp^2}(1-\alpha^+)(1-\alpha^-)\right] - 3\right\}$$
$$Z_r^{\mathrm{IR}} = P_{qq}(\xi) * H_r\left(\frac{z_1^0}{\xi}, z_2^0\right) + P_{qq}(\eta) * H_r\left(z_1^0, \frac{z_2^0}{\eta}\right). \tag{29}$$

The final form of the main subprocesses is given by the following.

$r = 1, 2$:

$$Z_r = Z_r^{\mathrm{IR}}(P_{qq}, H_r) + \left(\frac{8}{3}\kappa\right) H_r(z_1^0, z_2^0)\left[2\log\left(\frac{q^2}{q_\perp^2}\right) - 3\right] + \delta_r^{(0)}$$
$$\delta_r^{(0)} = \left(\frac{8}{3}\kappa\right) H_r(z_1^0, z_2^0) \cdot 2\log\frac{(1-\alpha^+)(1-\alpha^-)}{(1-z_1^0)(1-z_2^0)}. \tag{30}$$

The contributions from $\delta_r^{(0)}$ in equation 30 are infrared finite and are integrated numerically for $q_\perp \leq q_\perp^{\mathrm{critical}}$.

The infrared relevant parts combined to the following expression, which we discuss here ony for $r = 1$: $d + \bar{u} \rightarrow W^- X \rightarrow e^- \bar{\nu}_e X$.

$r = 1$:

$$\frac{1}{q^2} Y_1 = YY_1 + XY_1$$

$$YY_1 = \{[1 + R_1(q^2, s_\perp)]\delta(q_\perp^2) + S_1(q^2, q_\perp^2, s_\perp)\}\, d^{(0)}(z_1^0) u^{(0)}(z_2^0)$$

$$+ \left[\left(\frac{1}{q_\perp^2}\right)_{+,s_\perp} + \frac{1}{\epsilon} + \log \frac{s_\perp}{\mu^2}\right] \delta(q_\perp^2)$$

$$\times \left[\sum_\alpha P_{1\alpha} * H_\alpha(z_1^0, z_2^0)\right] \Theta(q_\perp^{2\,\max}[q^2, y_q, s] - q_\perp^2)$$

$$R_1 = \left(\frac{8}{3}\kappa_\mu\right)\left[\pi^2 - 8 - \log^2\left(\frac{s_\perp}{q^2}\right) - 3\log\left(\frac{s_\perp}{q^2}\right)\right]$$

$$S_1 = \left(\frac{8}{3}\kappa_\mu\right)\left[\left(2\log\left(\frac{q^2}{q_\perp^2}\right) - 3\right)\frac{1}{q_\perp^2}\right]_{+,s_\perp}. \quad (31)$$

In equation 31, $s_\perp$ denotes an auxiliary parameter that fixes the range of the $q_\perp$ subtraction in the distributions $((2 \log (q^2/q_\perp^2) - 3)(1/q_\perp^2)_{+,s_\perp}$ and $(1/q_\perp^2)_{+,s_\perp}$:

$$\int_0^\infty dq_\perp^2 f(q_\perp^2)_{+,s_\perp} g(q_\perp^2) = \int_0^{S_\perp} dq_\perp^2 f(q_\perp^2)[g(q_\perp^2) - g(0)] + \int_{S_\perp}^\infty dq_\perp^2 f(q_\perp^2) g(q_\perp^2). \quad (32)$$

Dimensional regularization is exactly independent of $s_\perp$, which leads to compensations among the different contributions in YY_1 independent of ϵ. The collinearly singular terms proportional to $1/\epsilon$ (to order κ) are process independent and promote the unregulated proton (antiproton) structure functions to be regulated scale-dependent ones. In contrast to these terms, XY_1 contains process-dependent contributions:

$$XY_1 = \sum_\alpha \delta(q_\perp^2) f_{1\alpha}^{DY} * H_\alpha(z_1^0, z_2^0). \quad (33)$$

The universal evolution functions

$$P_{ik} = \begin{pmatrix} P_{qq} & P_{qg} \\ P_{gq} & P_{gg} \end{pmatrix}$$

enter in all expressions on general grounds of the redundancies of the infrared–ultraviolet regularization procedures in the additive form

$$\left[\frac{1}{\epsilon} P_{ik} + f_{ik}^{\text{process dependent}}\right] * H_{\alpha(k)}. \quad (34)$$

The factorization of the different $q_\perp$-distributions in YY and XY is achieved through Fourier transformation to impact parameter space

$$F(q_\perp) \longrightarrow \tilde{F}(b) = \frac{1}{\pi} \int d^2 q_\perp \, e^{i\mathbf{q}_\perp \mathbf{b}} F(q_\perp)$$

$$F(q_\perp) = \frac{1}{4\pi} \int d^2 b \, e^{-i\mathbf{q}_\perp \mathbf{b}} \tilde{F}(b). \tag{35}$$

We do not include the characteristic function $\Theta(q_\perp^{2\max}(q^2, y_q, s) - q_\perp^2)$ in the Fourier–Bessel transform

$$\left(\frac{1}{q_\perp^2}\right)_{+,s_\perp} \longrightarrow L(b) = \int_0^\infty dq_\perp^2 \left(\frac{1}{q_\perp^2}\right)_{+,s_\perp} J_0(q_\perp b) = \log \frac{b_0^2}{b^2 s_\perp}; \qquad [b_0 = 2e^{-\gamma E}]$$

$r = 1, 2$:

$$\tilde{Y}_r(b) = [1 + R_{11}(q^2, s_\perp) + \tilde{S}_1(b^2, q^2, s_\perp)] H_r(z_1^0, z_2^0)$$
$$+ \left\{ \left[\frac{1}{\epsilon} + \log \frac{b_0^2}{b^2 \mu^2} \right] P_{r\alpha} * H_\alpha(z_1^0, z_2^0) + f_{r\alpha}^{DY} * H_\alpha(z_1^0, z_2^0) \right\}. \tag{36}$$

We have reabsorbed the constant term $8/3\, \kappa_\mu(\pi^2 - 8)$ in R_1 into the redefined process-dependent evolution functions f_r^{DY}:

$$R_{11} = -\frac{8}{3}\kappa \left[\log^2 \frac{s_\perp}{q^2} + 3 \log \frac{s_\perp}{q^2} \right]$$

$$\tilde{S}_1 = \frac{8}{3}\kappa \left\{ \int_0^{s_\perp} \frac{dq_\perp^2}{q_\perp^2} [J_0(bq_\perp) - 1] \left(2 \log \frac{q^2}{q_\perp^2} - 3 \right) \right.$$
$$\left. + \int_{s_\perp}^\infty \frac{dq_\perp^2}{q_\perp^2} J_0(bq_\perp) \left(2 \log \frac{q^2}{q_\perp^2} - 3 \right) \right\}. \tag{37}$$

The infrared regularization we used is only identical to the one performed by Altarelli, *et al.*[1] upon substituting $s_\perp \longrightarrow q_\perp^{2\max}$ and dropping the integral from $s_\perp$ to infinity in the expression for $\tilde{S}_1$ in equation 37 and in the integral for $L(b)$ in equations 36 and 37. The ensuing treatment is different from the operations performed in reference 1.

We note that the $s_\perp$ dependence has disappeared from the combination $(1/\epsilon + \log \cdot b_0^2/b^2\mu^2)$ in equation 36. This is not accidental since the dimensional regularization procedure is exactly insensitive to $s_\perp$. It follows that

$$\frac{\partial}{\partial s_\perp} [R_{11}(q^2, s_\perp) + \tilde{S}_1(b^2, q^2, s_\perp)] = 0. \tag{38}$$

For $s_\perp = q^2$, R_{11} vanishes and thus equation 36 becomes as follows.

$r = 1, 2$:

$$\tilde{Y}_r(b) = [1 + \tilde{S}_1(b^2, q^2, s_\perp = q^2)] H_r(z_1^0, z_2^0)$$
$$+ \left[\left(\frac{1}{\epsilon} + \log \frac{b_0^2}{b^2 \mu^2} \right) P_{r\alpha} + f_{r\alpha}^{DY} \right] * H_\alpha(z_1^0, z_2^0). \tag{39}$$

Modulo terms of order κ^2 and guided by the Sudakov form factor exponentiation, we cast equation 39 into the equivalent form as follows.

$r = 1, 2$:

$$\tilde{Y}_r(b) = (\exp \tilde{S}_{11}) \cdot \sum_r$$

$$\sum_r = \left[\delta_{r\alpha} + \left(\frac{1}{\epsilon} + \log \frac{b_0^2}{b^2\mu^2}\right) P_{r\alpha} + f_{r\alpha}^{DY}\right] * H_r(z_1^0, z_2^0)$$

$$\tilde{S}_{11} = \frac{8}{3}\left\{\int_0^{q^2} \frac{dq_\perp^2}{q_\perp^2}\left(2 \log \frac{q^2}{q_\perp^2} - 3\right)[J_0(q_\perp b) - 1]\bar{\kappa}(q_\perp^2)\right.$$

$$\left. + \int_{q^2}^{\infty} \frac{dq_\perp^2}{q_\perp^2}\left(2 \log \frac{q^2}{q_\perp^2} - 3\right) J_0(q_\perp b)\bar{\kappa}(q_\perp^2)\right\}. \quad \textbf{(40)}$$

The term $\kappa(q_\perp^2)$ denotes the running coupling constant of QCD evaluated at the scale $q_\perp$. Finally, modulo terms of order κ^2 and in accordance with the factorization properties of the universal parts of collinear singularities,[5] Σ_r is cast into the factorized form:

$$H_r = \begin{cases} d(z_1^0)u(z_2^0) & (r = 1) \\ \bar{u}(z_1^0)\bar{d}(z_2^0) & (r = 2) \end{cases} + \text{KMC} - \text{mixing terms.}$$

$r = 1$:

$$\sum_1 = \sigma_d^{DY}(z_1^0) \cdot \sigma_u^{DY}(z_2^0) .$$

$$\sigma_d^{DY}(z) = d^{(0)}(z) + \left\{\left(\frac{1}{\epsilon} + \log \frac{b_0^2}{b^2\mu^2}\right)[P_{qq} * d + P_{qg} * g]\right.$$

$$\left. + [f_{qq}^{DY} * d + f_{qg}^{DY} * g]\right\}. \quad \textbf{(41)}$$

Similarly we would find, for example, for the structure function $F_2^{\nu P}$ in deep inelastic neutrino–proton scattering

$$\frac{1}{z} F_2^{\nu P} = \sigma_d^{DI} + \sigma_{\bar{u}}^{DI}$$

$$\sigma_{d,2}^{DI}(z) = d^{(0)}(z) + \left(\frac{1}{\epsilon} + \log \frac{-q^2}{\mu^2}\right)[P_{qq} * d + P_{qg} * g]$$

$$+ f_{2qq}^{DI} * d + f_{2qg}^{DI} * g. \quad \textbf{(42)}$$

We note here the process-specific distributions f_{qq}, f_{qg}:

$$f_{qq}^{DY}(\xi) = \frac{8}{3}\kappa\left[(1 - \xi) + \left(\frac{\pi^2}{4} - 4\right)\delta(1 - \xi)\right]$$

$$f_{qg}^{DY}(\xi) = \kappa \cdot 1$$

$$f_{2qq}^{DI}(\xi) = \frac{8}{3}\kappa\left[\left(\frac{\log(1-\xi)}{1-\xi}\right)_+(1+\xi^2) - \frac{3}{2}\left(\frac{1}{1-\xi}\right)_+ - \frac{1+\xi^2}{1-\xi}\log\xi\right.$$

$$\left. + 3 + 2\xi - \left(\frac{9}{2} + \frac{\pi^2}{3}\right)\delta(1-\xi)\right]$$

$$f_{2qg}^{DI}(\xi) = \kappa\left\{[\xi^2 + (1-\xi)^2]\log\left(\frac{1-\xi}{\xi}\right) + 6\xi(1-\xi)\right\} \tag{43}$$

Using the doublet of quark and gluon structure functions

$$h = (q, g), \qquad \bar{h} = (\bar{q}, g)$$

the expression for Σ_r in equation 41 takes the form

$$\sigma^{DY}(h_\alpha, z) = h_\alpha^{(0)}(z) + \left[\left(\frac{1}{\epsilon} + \log\frac{b_0^2}{b^2\mu^2}\right)P_{\alpha\beta} + f_{\alpha\beta}^{DY}\right] * h_\beta(z). \tag{44}$$

Equation 44 exhibits the characteristic separation into universal and process specific convolutions.

Redundancy of the Regularized Structure Functions

The minimal dimensional regularization corresponds (to order κ) to the substitution

$$h_\alpha^{(0)}(z) = -\frac{1}{\epsilon}P_{\alpha\beta} * h_\beta^{\text{reg}}(z, \mu^2) + h_\alpha^{\text{reg}}(z, \mu^2)$$

$$\sigma^{DY}\left(h_\alpha^{\text{reg}}, z, \frac{b_0^2}{b^2}\right) = h_\alpha^{\text{reg}}(z, \mu^2) + \log\left(\frac{b_0^2}{b^2\mu^2}\right)P_{\alpha\beta} * h_\beta^{\text{reg}}(z, \mu^2)$$

$$+ f_{\alpha\beta}^{DY} * h_\beta^{\text{reg}}(z, \mu^2) + 0(\kappa_\mu^2)$$

$$P, f = P, f(\kappa). \tag{45}$$

The redundancy of the so-defined regularized structure functions is governed by the convolution with an arbitrary set of auxiliary functions

$$h_\alpha'^{\text{reg}} = h_\alpha^{\text{reg}} + \delta f_{\alpha\beta} * h_\beta^{\text{reg}}(z, \mu^2). \tag{46}$$

The change from the structure functions h^{reg} to h'^{reg} amounts to a redefinition of the process specific functions

$$f_{\alpha\beta}^{\text{pr.sp.}} \longrightarrow f_{\alpha\beta}'^{\text{pr.sp.}} = f_{\alpha\beta}^{\text{pr.sp.}} - \delta f_{\alpha\beta} \tag{47}$$

In reference 1 a choice of $\delta f_{\alpha\beta}$ was proposed such that the deep inelastic structure function $(1/z)F_2$ is identified with the structure function itself. This procedure corresponds to

$$\delta f_{\alpha\beta} = f_{2\alpha\beta}^{DI} \longrightarrow f_{2\alpha\beta}'^{DI} = 0, \tag{48}$$

whereby a process-specific cross section is absorbed into the structure function. Of course, only the differences of f-correlation functions

$$f_{\alpha\beta}^{\text{process 1}} - f_{\alpha\beta}^{\text{process 2}} \tag{49}$$

are regularization independent.

Nevertheless we propose here to define the structure function according to the minimal dimensional regularization procedure of equation 45 because of the universality, that is, process independence, of this procedure. The different choices do not affect any prediction expressing one physical cross section by another, for example, giving the Drell–Yan cross section in terms of the deep inelastic structure function F_2 for a complete set of deep inelastic scattering processes.

The different regularization schemes only come into play when simple parametrizations of structure functions are used to approximately guess their form for rescaled momenta at which precise measurements are not available.

Remark on the Adler Sum Rule

Since current algebra is valid in any dimension, the minimally subtracted structure functions respect the Adler sum rule, which, neglecting all weak mixing angles, reads (independently of the choice of structure functions)

$$\int dx \frac{1}{x} [F_2^{\bar{\nu}P} - F_2^{\nu P}](x, q^2) = 1$$

$$\frac{1}{x}(F_2^{\bar{\nu}P} - F_2^{\nu P}) = (u - \bar{u})^{\text{reg}}(x, q^2) - (d - \bar{d})^{\text{reg}}(x, q^2)$$
$$+ f_{2qq}^{DI} * [(u - \bar{u}) - (d - \bar{d})]^{\text{reg}}(x, q^2). \tag{50}$$

We do not consider the structure functions currently used[6] to be evaluated precisely enough in order to take the effect due to f_{2q}^{DI} into account.

The Adler sum rule is independent of f_{2q}^{DI}, since the latter convolution function pertaining to the minimal regularization scheme satisfies

$$\int d\xi\, f_{2qq}^{DI}(\xi, \kappa) = 0 \tag{51}$$

and thus enforces (to all orders in κ) the corresponding relations for the structure functions

$$\int dx(u - \bar{u})^{\text{reg}} = 2; \qquad \int dx(d - \bar{d})^{\text{reg}} = 1. \tag{52}$$

The renormalization group equations for the so-defined structure functions (equation 45), neglecting quark masses, read

$$\left(\mu^2 \frac{\partial}{\partial \mu^2} + \beta(\kappa)\frac{\partial}{\partial \kappa}\right) h_\alpha^{\text{reg}}(z; P^2, \mu^2, \kappa_\mu) + \int \frac{d\xi}{\xi} P_{\alpha\beta}(\xi, \kappa_\mu) * h_\alpha^{\text{reg}}\left(\frac{z}{\xi}; P^2, \mu^2, \kappa_\mu\right) = 0$$
$$\beta(\kappa) = -b_0\kappa^2 + 0(\kappa^3); \qquad b_0 = 11 - \tfrac{2}{3}N_{fl}. \tag{53}$$

In equation 53 the moments of $P_{\alpha\beta}$ are the respective anomalous dimension matrices for the composite operators of leading twist.[4]

The solution to equation 53 has the well-known form

$$h_\alpha(z; P^2, \mu^2, \kappa_\mu) = \left[\text{`exp'} \int_0^{\log(P^2/\mu^2)} dt\, P_{\alpha\beta}(\bar{\kappa}_t) \right] * h_\beta \left(\frac{z}{\xi}; \mu^2, \mu^2, \bar{\kappa}_{P^2} \right)$$

$$t = \log \frac{P^2}{\mu^2}; \qquad \bar{\kappa}_t \text{ is the solution to the equation.} \tag{54}$$

$$\frac{d}{dt} \bar{\kappa}_t = \beta(\bar{\kappa}_t); \qquad \bar{\kappa}_{t=0} = \kappa_\mu . \tag{55}$$

The 'exp' in equation 54 denotes a path-ordered exponentiation.

Having defined the scale-dependent structure functions we return to equation 30, which becomes as follows.

$r = 1, 2$:

$$\tilde{Y}_r(b) = \exp \tilde{S}_{11}(b^2, q^2, \Lambda^2_{\mathrm{QCD}}) H_r \left(z_1^0, z_2^0; \frac{b_0^2}{b^2} \right) + 0(\kappa^2)$$

$$H_r = \begin{cases} \sigma_d^{DY}\left(z_1^0; \frac{b_0^2}{b^2} \right) \sigma_u^{DY}\left(z_2^0; \frac{b_0^2}{b^2} \right) & (r = 1) \\ \sigma_{\bar{u}}^{DY}\left(z_1^0; \frac{b_0^2}{b^2} \right) \sigma_{\bar{d}}^{DY}\left(z_2^0; \frac{b_0^2}{b^2} \right) & (r = 2) \end{cases}$$

+ weak mixing angle extensions

$$\sigma_{q(\bar{q})}^{DY}(z, P^2) = h_{q(\bar{q})}(z, P^2) + f_{q\beta}^{DY} * h_\beta(z, P^2). \tag{56}$$

Substituting the fixed coupling constant $\kappa(q^2)$ as the argument in the Drell–Yan convolution functions $f_{\alpha\beta}^{DY}$ is allowed modulo higher order corrections.

Sudakov Exponentiation[7]

The substitution $1 + \tilde{S}_{11} \to \exp \tilde{S}_{11}$ is only valid in the sense of an asymptotic expansion for

$$\frac{1}{\Lambda^2_{\mathrm{QCD}}} \gg b^2 \gg \frac{1}{q^2}. \tag{57}$$

The expression in equation 40 formally exists also for very small impact parameters where

$$\lim_{b \to 0} (\exp \tilde{S}_{11}) = 0.$$

This implies the sum rule for $Y_r^S(q_\perp^2)$ obtained from $\tilde{Y}_r(b^2)$ by the backward Bessel transform

$$Y_r^S(q_\perp^2, q^2, \Lambda^2_{\mathrm{QCD}}; z_1^0, z_2^0) = 1/4 \int db^2 \tilde{Y}_r(b^2) J_0(bq_\perp); \qquad \int_0^\infty dq_\perp^2 Y_r^S = 0, \tag{58}$$

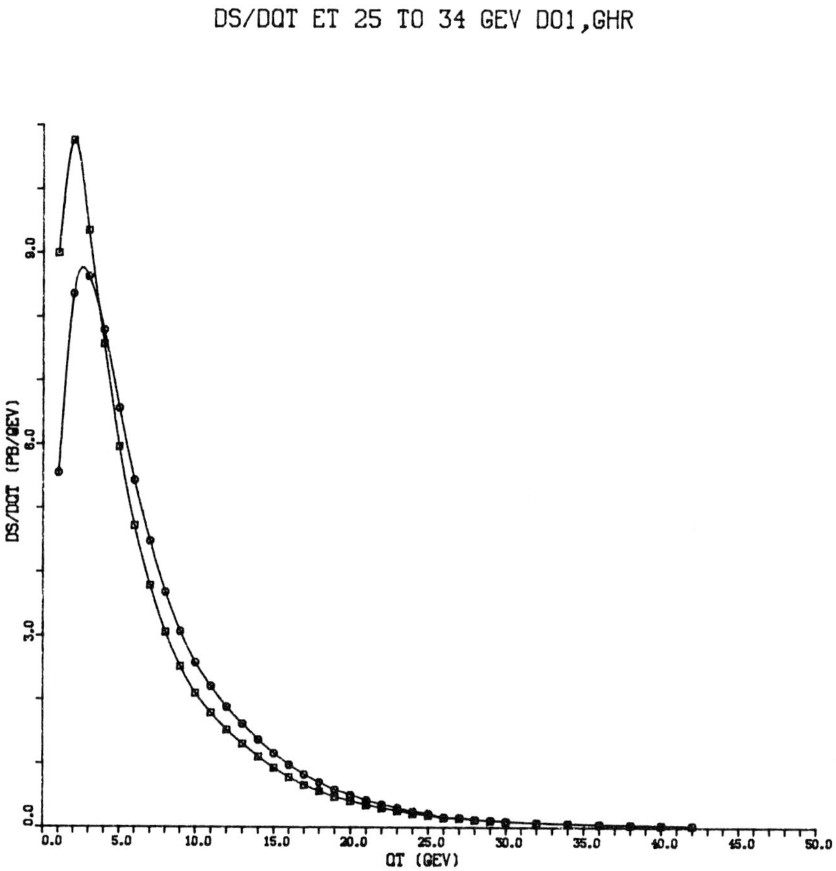

FIGURE 1.

$$\int_{E_{11}}^{E_{12}} dE_\perp \frac{d^2\sigma}{dE_\perp dq_\perp} \qquad \text{vs. } q_\perp$$

$$d\sigma = d\sigma[p\bar{p} \rightarrow W^- X \rightarrow e^- \bar{\nu}_e X].$$

$E_\perp$: transverse momentum of the electron; $q_\perp$: transverse momentum of the (virtual) W^-; (E_{11}, E_{12}): limits of integration of $E_\perp$ in region 1:

$$E_{11} = 25 \text{ GeV}, \qquad E_{12} = 34 \text{ GeV}, \qquad m_W = 82 \text{ GeV}.$$

-□-□-□- DO1 structure functions (Duke and Owens[6]); -⊖-⊖-⊖- GHR structure functions (Glück *et al.*[6]).

where $Y_r^S(q_\perp)$ is independent of the maximally allowed $q_\perp$, and thus for given $q_\perp^{max}$ the sum rule in the equation can well be saturated by a formally negative cross section outside the kinematically allowed range of $q_\perp$. We found that $Y_r^S(q_\perp)$ turns negative, for $\sqrt{s} = 630$ GeV, for $q_\perp \simeq 30$ GeV for values of z_2^0, z_2^0 ($\geq \sqrt{q^2/s}$) for which the structure functions give a sizable contribution to the integrated cross section, well within the physically allowed $q_\perp$ range. This signals the breakdown of the Sudakov

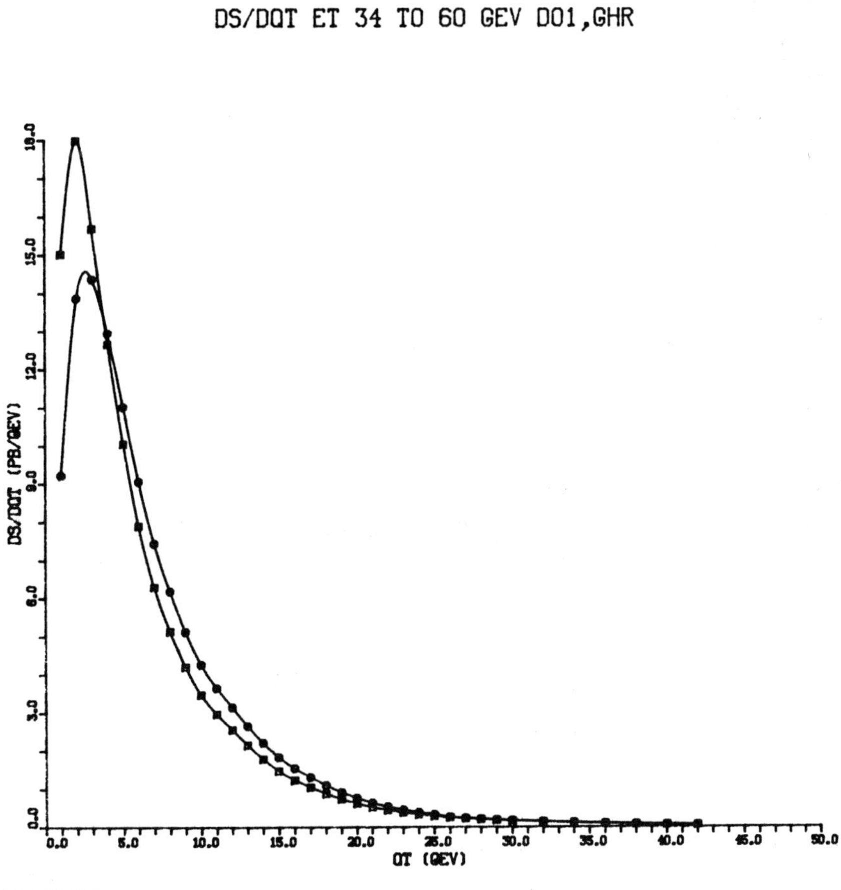

FIGURE 2.

$$\int_{E_{21}}^{E_{22}} dE_\perp \frac{d^2\sigma}{dE_\perp dq_\perp} \qquad \text{vs. } q_\perp.$$

$d\sigma$, $E_\perp$, $q_\perp$ as in FIGURE 1; (E_{21}, E_{22}): limits of integration of $E_\perp$ in region 2:

$$E_{21} = 34 \text{ GeV}, \qquad E_{22} = 60 \text{ GeV}; \qquad m_W = 82 \text{ GeV}.$$

-⊟-⊟-⊟- and -⊖-⊖-⊖- as in FIGURE 1.

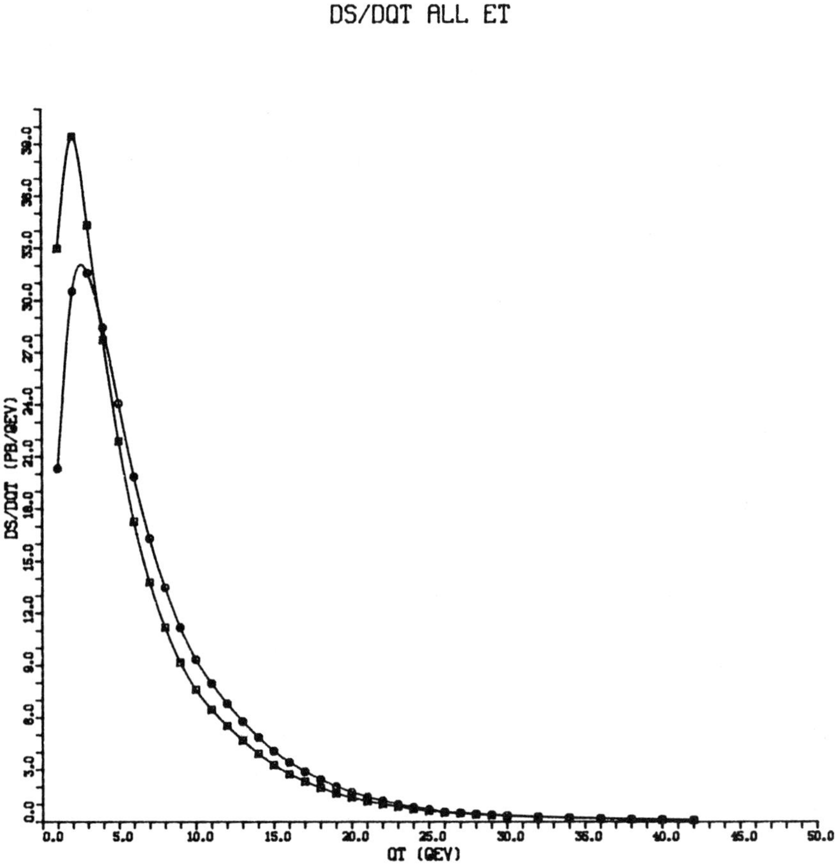

FIGURE 3.

$$\frac{d\sigma}{dq_\perp} = \int_{\text{all } E_\perp} dE_\perp \frac{d^2\sigma}{dE_\perp dq_\perp} \qquad \text{vs. } q_\perp.$$

$d\sigma$, $E_\perp$, $q_\perp$, —□—□—□— , and —○—○—○— as in FIGURE 1.

approximation in the given process for

$$q_\perp > q_\perp^{\text{critical}} \simeq 30 \text{ GeV}.$$

In the preceding region, however, the purely perturbative expressions for the quark/gluon subprocesses [to order $\kappa(q^2)$] are expected to give a good approximation. Thus we adopt the following strategy for the approximation to the physical cross section:

$$Y_r^{\text{Phys}} = Y_r^{\text{Phys}}(q_\perp^2, q^2, \Lambda_{\text{QCD}}^2, z_1^0, z_2^0)$$

$$\hat{Y}_r = Y_r^S + \tilde{\delta}_r + \delta_r^{(0)}$$

$$Y_r^{\text{Phys}} = \begin{cases} \hat{Y}_r, & (q_\perp \ll q_\perp^{\text{crit}}\ (<q_\perp^{\max})) \\ \max\,(\hat{Y}_r, Y_r^{\text{pert}}), & (q_\perp \lesssim q_\perp^{\max}) \end{cases}$$

$$Y^{\text{Phys}} = \sum_r Y_r^{\text{Phys}} \tag{59}$$

This ends the discussion of the theoretical aspects of our calculation of the cross section for W production from $p\bar{p}$ in the leptonic mode.

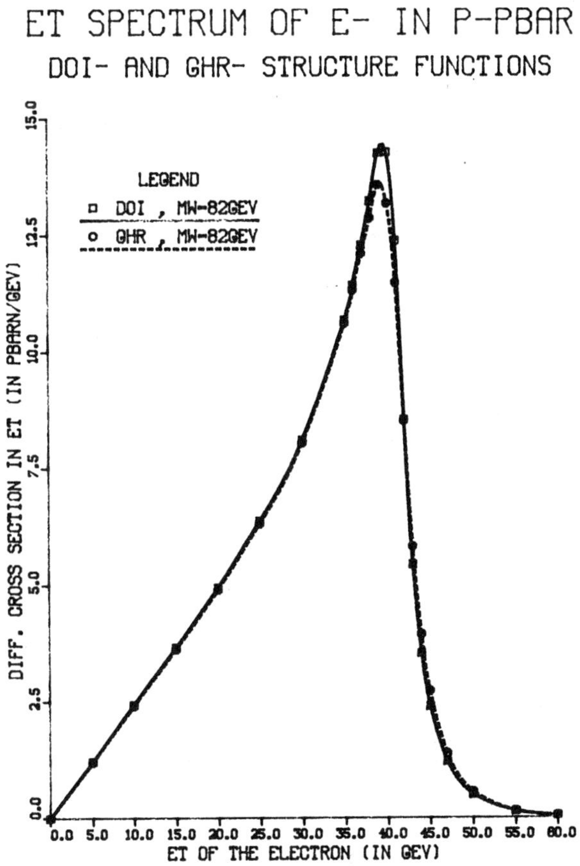

FIGURE 4.

$$\frac{d\sigma}{dE_\perp} = \int_{\text{all } q_\perp} dq_\perp \frac{d^2\sigma}{dE_\perp dq_\perp} \qquad \text{vs. } E_\perp$$

for $m_W = 82$ GeV. $d\sigma$, $E_\perp$, $q_\perp$ as in FIGURE 1. —— DO1 structure functions; - - - GHR structure functions.

NUMERICAL EVALUATIONS

Collecting the results from the preceding section, we present a numerical analysis of the three quantities:

- $$\frac{d\sigma}{dE_\perp} = \frac{d\sigma^{p\bar{p}\to W^- X \to e^- \bar{\nu}_e X}}{dE_\perp^{e^-}}$$

- $$\frac{d\sigma}{dq_\perp} = \frac{d\sigma^{p\bar{p}\to W^- X \to e^- \bar{\nu}_e X}}{dq_\perp^{W^-}}$$

- $$\sigma = \int dE_\perp \frac{d\sigma}{dE_\perp} = \int dq_\perp \frac{d\sigma}{dq_\perp}$$

The relevant formulas pertaining to scalar gluons are presented in the appendix.

Starting from the multidifferential cross section

$$\frac{d^5\sigma}{dE_\perp^e \, dq_\perp^w d\Phi dq^2 dy_q}$$

we performed a numerical integration on the three variables (y_q, q^2, Φ) yielding

$$\frac{d^2\sigma}{dE_\perp dq_\perp}(p\bar{p} \to W^- X \to e^- \bar{\nu}_e X) = \sum (E_\perp, q_\perp; m_w, \Gamma_w^{\text{tot}}).$$

The actual Bessel transformations to and from impact parameter space pose considerable convergence problems that were controlled by comparison of Bessel transforms of test functions with known analytical results.

The aim of the calculation is to provide as precise a determination as possible of the mass of the W boson in fitting the experimental distribution in $E_\perp$ in an appropriate interval of $E_\perp$ values encompassing the region of the Jacobian peak near $E_\perp = m_W/2$.

Thus the differential cross section $\Sigma\ (E_\perp, q_\perp)$ was evaluated for 79 GeV $\leq m_W \leq$ 85 GeV in 25 steps of $\Delta m_W = 250$ MeV. The scope of the integrations proved to be beyond the capacity of the scalar IBM 3083 computers at the computer center in Bern. The calculations were performed on the CRAY-1S machine at the EPFL Computer Center in Lausanne. One set of input parameters ($\sqrt{s} = 630$ GeV, $\Gamma_W = 2.43$ GeV, $m_W = 82$ GeV) demanded approximately six hours CPU time.

We present the results in FIGURES 1 to 8 and in TABLE 2.

CONCLUSIONS

We have presented the methods and definitions used to calculate the multidifferential $W^{-(+)}$ production cross section from $p\bar{p}$ in the leptonic W-decay mode for $\sqrt{s} = 630$ GeV and for 25 masses ranging from $m_W = 79$ GeV in steps of 0.25 GeV to 85 GeV. We have compared the distribution in $E_\perp$, the transverse momentum of the electron (positron) with data of the UA2 collaboration.[8] The comparison in FIGURE 7 shows an

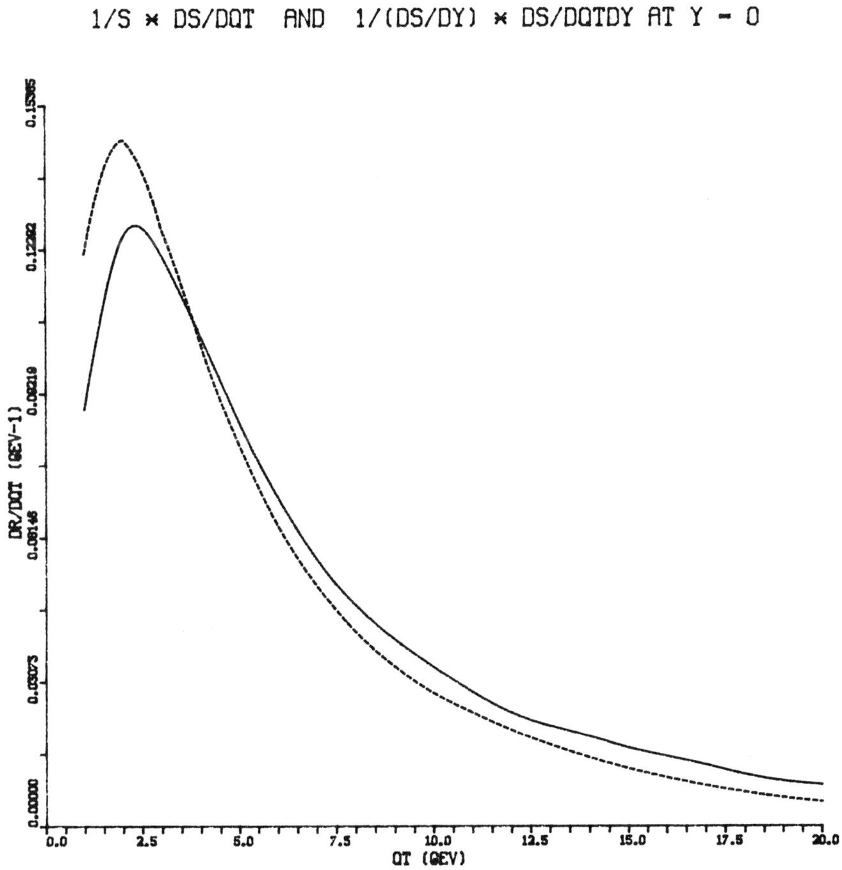

FIGURE 5.

$$\frac{1}{\sigma}\frac{d\sigma}{dq_\perp} = \frac{dR}{dq_\perp} \qquad \text{vs. } q_\perp$$

according to the present calculation compared to

$$\left.\frac{d^2\sigma}{dq_\perp dy_q}\right|_{y_q=0} \cdot \left[\left(\frac{d\sigma}{dy_q}\right)_{y_q=0}\right]^{-1} \qquad \text{vs. } q_\perp$$

according to Altarelli *et al.*[1] —— Altarelli *et al.*; – – – present calculation. Both curves correspond to $m_W = 83$ GeV and involve the DO1 structure functions.

excellent agreement of the calculated distribution with the data in the range 35 GeV $\leq E_\perp \leq$ 45 GeV. The deviation in the bin 45 GeV $\leq E_\perp \leq$ 47 GeV is in our opinion due to six to seven misidentified Z events, where upon the decay of Z in e^+e^-, predominantly one lepton is processed as a jet whereby a fake missing energy results. The excellent sensitivity of the UA2 detector as examplified by the clear separation of the sharp

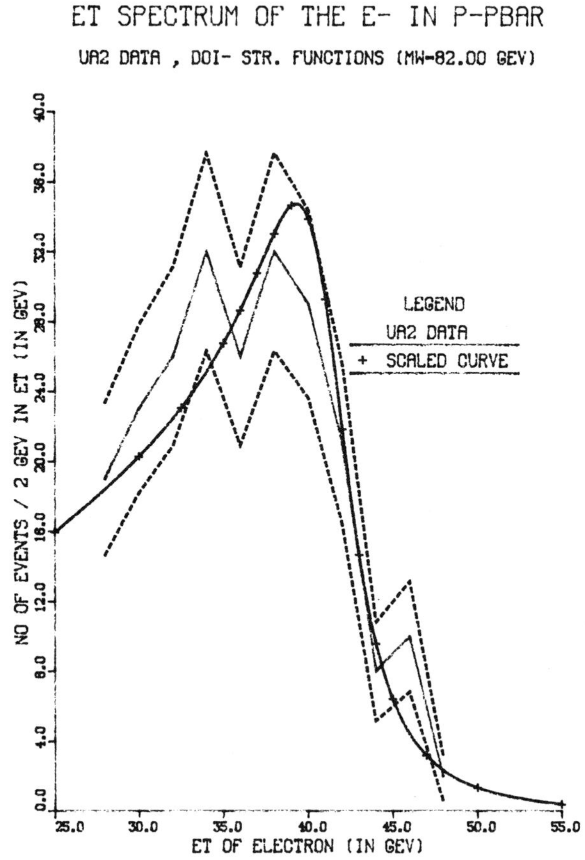

FIGURE 6. $d\sigma/dE_\perp$ scaled to the area of the experimental distribution of the UA2 collaboration[8] vs. $E_\perp$ for $m_W = 82$ GeV, $\Gamma_W^{\text{tot}} = 2.43$ GeV, $\sqrt{s} = 630$ GeV. The data are compared by setting

$$\left(\frac{d\sigma}{dE_\perp}\right)^{\exp}\left(E_\perp = E_{\perp j} + \frac{1}{2}\,\text{Binsize}\right) = \text{const} \cdot N_{e\nu}\,(E_{\perp j} \le E_\perp \le E_{\perp j+1}),$$

where $E_{\perp j}$ (GeV) = (27, 29, 31, 33, 35, 37, 39, 41, 43, 45, 47, 49) and Binsize = 2 GeV. The data are cut by the condition $m_\perp^{e\nu} \geq 50$ GeV

$$(m_\perp^{e\nu})^2 = 2E_\perp P_{\perp\nu}\,(1 - \cos\Phi_{e\nu}).$$

~~XXX~~ Theoretical calculation convoluted with $E_\perp$ resolution $\sigma = 0.14 \cdot E_\perp$ (GeV) scaled to the area of the experimental distribution; —— UA2 data as reported by Roussarie at the 23rd International Conference on High Energy Physics, Berkeley, Calif., 1986; – – – Statistical one standard deviation curves.

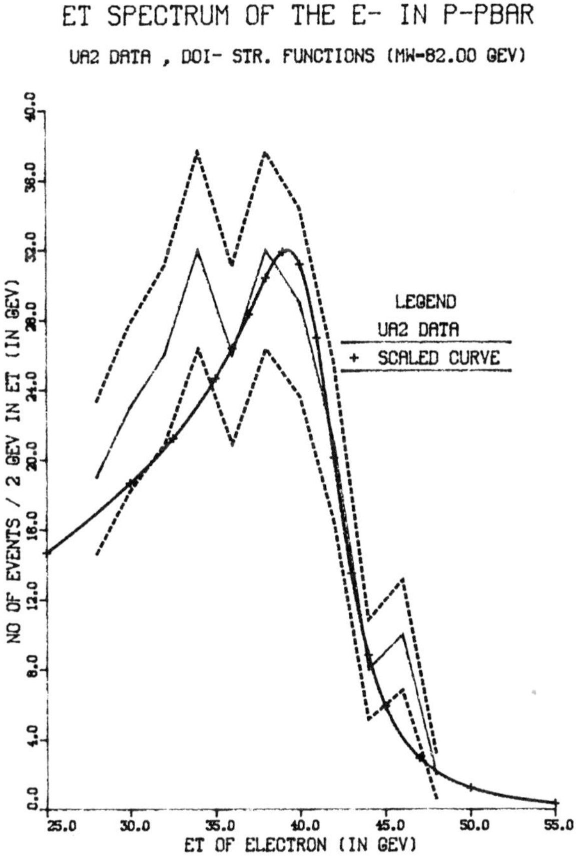

FIGURE 7. $d\sigma/dE_\perp$ scaled approximately taking into account the integrated luminosity, acceptance, and efficiency of the UA2 detector for $m_W = 82$ GeV, $\Gamma_W = 2.43$ GeV. The data are represented as in FIGURE 6. XXX Theoretical calculation convoluted with the $E_\perp$ resolution in FIGURE 6; —— and – – – as in FIGURE 6.

$$\sigma = \int \frac{d^2\sigma}{dE_\perp dq_\perp} dE_\perp dq_\perp |_{W^- \& W^+} = 567 \text{ pb } (+/-20\%)$$

where $\sqrt{s} = 630$ GeV, $\Gamma_W^{\text{tot}} = 2.43$ GeV; $m_{\text{top}} = 60$ GeV; and $m_W = 82$ GeV.

falloff of the genuine W Jacobian peak from the secondary one due to misinterpreted Z events qualifies the deviation of the experimental distribution from the calculated one in the region 27 GeV $\leq E_\perp \leq$ 35 GeV as significant in our interpretation. This significance can be translated into an excess of 20 events in this region out of 100. In the same region we expect 4 to 5 electrons (positrons) from the leptonic decay chain $W^- \rightarrow \tau^- \bar{\nu}_\tau \rightarrow e^- \bar{\nu}_e \bar{\nu}_\tau$. Leaving aside the question of how many background events resulting from misidentified jets fall into this region, it is obvious that the statistical significance of the 15 remaining excess events is considerably less than two standard

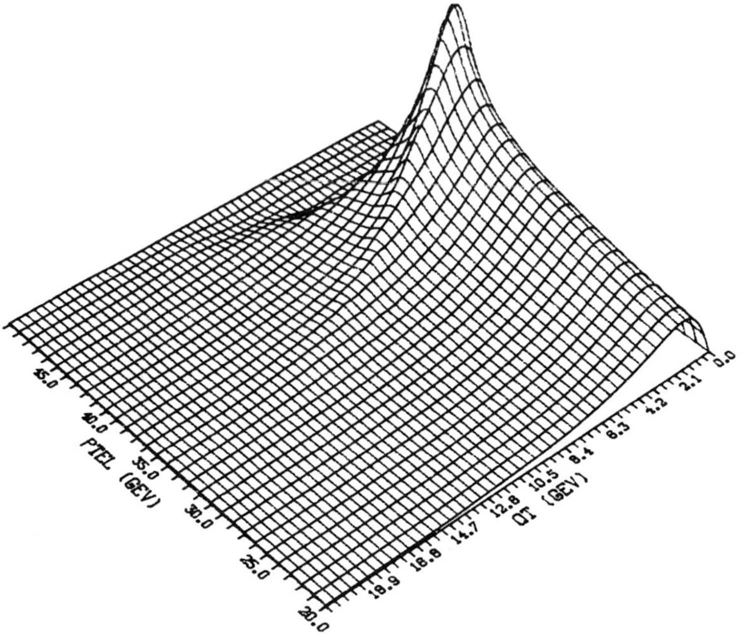

FIGURE 8. $d^2\sigma/dE_\perp dq_\perp$ in the domain

$$20 \text{ GeV} \le E_\perp \le 50 \text{ GeV}$$

$$0 \le q_\perp \le 21 \text{ GeV}$$

drawn in arbitrary units. ($\sqrt{s} = 630$ GeV.)

TABLE 2. Total Cross Section and Average Transverse Momentum of W^+ and W^- Produced from $p\bar{p}$ in the Leptonic Decay Mode ($e\nu_e$) at $\sqrt{s} = 630$ GeV ($m_{\text{top}} = 60$ GeV)

MW	Total	$\langle QT \rangle$
0.79000E + 02	0.59520E + 03	0.61861E + 01
0.79250E + 02	0.59387E + 03	0.61945E + 01
0.79500E + 02	0.59475E + 03	0.62088E + 01
0.79750E + 02	0.59453E + 03	0.62214E + 01
0.80000E + 02	0.59139E + 03	0.62255E + 01
0.80250E + 02	0.58653E + 03	0.62241E + 01
0.80500E + 02	0.58270E + 03	0.62245E + 01
0.80750E + 02	0.58129E + 03	0.62302E + 01
0.81000E + 02	0.58084E + 03	0.62410E + 01
0.81250E + 02	0.57947E + 03	0.62471E + 01
0.81500E + 02	0.57579E + 03	0.62486E + 01
0.81750E + 02	0.57106E + 03	0.62478E + 01
0.82000E + 02	0.56736E + 03	0.62482E + 01
0.82250E + 02	0.56511E + 03	0.62547E + 01
0.82500E + 02	0.56353E + 03	0.62628E + 01
0.82750E + 02	0.56100E + 03	0.62684E + 01
0.83000E + 02	0.55723E + 03	0.62687E + 01
0.83250E + 02	0.55304E + 03	0.62677E + 01
0.83500E + 02	0.54962E + 03	0.62704E + 01
0.83750E + 02	0.54748E + 03	0.62760E + 01
0.84000E + 02	0.54553E + 03	0.62851E + 01
0.84250E + 02	0.54315E + 03	0.62910E + 01
0.84500E + 02	0.54014E + 03	0.62936E + 01
0.84750E + 02	0.53729E + 03	0.62949E + 01
0.85000E + 02	0.53511E + 03	0.62990E + 01

deviations. In spite of this fact we consider this to be an indication of a mechanism producing relatively hard leptons unaccounted for by W production and/or background.

One is tempted to associate such a signal (provided it is real) with the top quark in the decay chain $W^+ \rightarrow t\bar{b} \rightarrow e^+ \nu_e b\bar{b}$; if so, the mass of the top quark is constrained to the interval 55 GeV $\leq m_t \leq$ 70 GeV. In this mass range the propagator effect of the virtual W mediating the leptonic decay of top, starts to favor an effective two-body kinematics for the $e\nu$-system. We wish to acknowledge that this tentative interpretation is categorically refuted by the UA2 collaboration, which, upon a dedicated analysis of their W sample,[8] comes to the conclusion that no indication whatsoever of a signal corresponding to the top quark is present in the data.

APPENDIX

Contribution of Scalar Gluons in Dimensional Infrared Regularization of Collinear Singularities

The virtual gluon corrections to the on-shell $W - q - \bar{q}$ vertex when regulated dimensionally in the infrared contain through the gluon propagator (in Landau gauge)

$$\left[-g^{\mu\nu} + \frac{k^\mu k^\nu}{k^2}\right]\frac{1}{k^2 + i\eta}$$

in $d = 4 + 2\epsilon$ dimensions, contributions (for $\mu, \nu \neq 0, 1, 2, 3$) corresponding to quarks and gluons in 6, 8, . . . , $4 + 2\epsilon$ dimensions. These are only consistently combined with the gluon emission terms if scalar gluons (in the 4-dimensional terminology) are included.

This is achieved in parallel to an $N = 2$ supersymmetric extension of QCD generating $q - \bar{q}$ – scalar gluon Yukawa couplings according to the Lagrangian

$$L_{\text{scalar gluon}} = q_s \bar{q} \frac{\lambda_a}{2} q \cdot G^a_{(\alpha)}$$

$$[a = 1, 2, \ldots, 8; \qquad (\alpha) = 1, 2, \ldots, 2\epsilon] \tag{A.1}$$

where (α) numbers the scalar gluons, one each for every dimension in excess of four.

In the scalar gluon emission cross section a sum over final states includes the sum over the parameter $(\alpha) = 1, \ldots, 2\epsilon$ in equation A.1, which upon continuation of 2ϵ to general real values yields a factor 2ϵ, multiplying all corresponding emission terms. For instance, the scalar gluon emission for the subprocess $r = 7$ in TABLE 2 corresponds to

$$F_7 = \epsilon \frac{1}{q^2}\left(\frac{1}{2p_2 k}\right)(2p_1 l_2)(2p_2 l_1) \tag{A.2}$$

Of course there are also contributions from scalar gluons in the incoming hadrons. These involve a free set of scalar gluon structure functions that only renormalize the physical gluon structure function; hence, explicit inclusion of corresponding contributions is not necessary.

ACKNOWLEDGMENTS

We wish to thank the members of the UA2 collaboration, in particular J. Schacher and L. di Lella, and F. Jegerlehner for clarifying discussions, and also the Centre de Calcul EPFL in Lausanne for providing generous access to their CRAY-1S system.

REFERENCES

1. ALTARELLI, G., R. K. ELLIS & G. MARTINELLI. 1979. Nucl. Phys. **B157:** 461; ALTARELLI, G., R. K. ELLIS, M. GRECO & G. MARTINELLI. 1984. Nucl. Phys. **B246:** 12.
2. COLLINS, J. C. & D. E. SOPER. 1981. Nucl. Phys. **B193:** 381; COLLINS, J. C. & D. E. SOPER. 1982. Nucl. Phys. **B194:** 445; COLLINS, J. C. & D. E. SOPER. 1982. Nucl. Phys. **B197:** 446; CHAO, S. C., D. E. SOPER & J. C. COLLINS. 1983. Nucl. Phys. **B214:** 513.
3. MINKOWSKI, P. 1984. Phys. Lett. **139B:** 431.
4. ALTARELLI, G. & G. PARISI. 1977. Nucl. Phys. **B126:** 298.
5. ELLIS, R. K., H. GEORGI, M. MACHACEK, H. D. POLITZER & G. G. ROSS. 1978. Phys. Lett. **B78:** 281; ELLIS, R. K. *et al.* 1979. Nucl. Phys. **B152:** 285.
6. GLÜCK, M., E. HOFFMANN & E. REYA. 1982. Z. Phys. **C13:** 119; DUKE, D. W. & J. F. OWENS. 1984. Phys. Rev. **D30:** 49.
7. DOKSHITZER, Yu. L., D. I. DYAKONOV & S. I. TROYAN. 1980. Phys. Rep. **C58:** 269.
8. ROUSSARIE, A. 1987. *In* Proc. 23rd Int. Conf. on High Energy Physics (Berkeley, Calif.). S. LOKEN, Ed.: 989. World Scientific. Singapore; ANSARI, R. *et al.* 1987. Phys. Lett. **B186:** 440. (UA2 Collaboration.)

Indirect Searches for Very Heavy Quarks[a]

I. I. BIGI[b]

Stanford Linear Accelerator Center
Stanford University,
Stanford, California 94305

INTRODUCTION

The existence of charm quarks was inferred from rare K^0 decays;[1] CP violation in K_L decays was invoked as evidence for the third family of quarks consisting of bottom and top quarks.[2] One should note that the mass of charm, bottom, and top quarks is much larger than the K mass. History might repeat itself and allow the discovery of yet another family of quarks (or other heavy quanta) in such an indirect way; or at least the scale of the top mass might be obtained this way.

There is hardly a doubt left that a top indeed exists in nature: the cleanest, though still indirect evidence for it so far comes from the observed forward–backward asymmetry of bottom jets produced in e^+e^- annihilation. For the data[3] support the expected assignment of b-quarks into an isodoublet; hence, there is an isopartner—the top.

This argument does not however give any clue as to the value of the top mass. PETRA data yield a lower limit of 22 GeV, whereas a comprehensive analysis of isospin breaking in deep-inelastic lepton nucleon scattering suggests[4] an upper limit:

$$22 \text{ GeV} \leq m_t \lesssim 180 \text{ GeV}. \tag{1}$$

A useful nomenclature is provided by the following distinction:

(1) a "light" top allows $W \to t\bar{b}$ to proceed, that is, $m_t \lesssim 70$ GeV;
(2) for a "heavy" top $t \to Wb$ occurs instead, that is, $m_t > 90$ GeV.

Finding a heavy top hadron as a real on-shell state poses a formidable challenge even for Tevatron experiments. It is our judgment that in the near future there are (at least) two processes that have a very good chance of revealing indirectly the existence of heavy top or even heavier states like quarks from a fourth family:

$$K^+ \to \pi^+\nu\bar{\nu};$$

$$B^0 - \bar{B}^0 \text{ mixing.}$$

These, in particular the second one, are discussed in some detail. There are other reactions like $B \to K^{(*)}\gamma$, $K^{(*)}l^+l^-$ with a similar potential;[5] they will be treated in other papers.[6]

[a]This work was supported by the U.S. Department of Energy under Contract DE-AC03-76SF00515.
[b]Heisenberg Fellow.

Searching for $Z^0 \to b\bar{s} + s\bar{b}$, on the other hand, appears to represent a hopeless task, since it is hard to see how $\mathrm{BR}(Z^0 \to b\bar{s} + s\bar{b})$ could exceed 10^{-7}.

In the end we will make a few short comments on CP violation in B^0 decays.

$K^+ \to \pi^+ \nu\bar{\nu}$

Relating $K^+ \to \pi^+\nu\bar{\nu}$ to $K \to \pi l \nu$ one can make very reliable predictions for $\mathrm{BR}(K^+ \to \pi^+\nu\bar{\nu})$ in terms of m_t and the Kobayashi–Maskawa (KM) parameter $V^*(ts)V(td)$. One finds[7,8]

$$[3.2, 3.7, 4.2] \times 10^{-11} \lesssim \mathrm{BR}(K^+ \to \pi^+\nu\bar{\nu}) \lesssim [1.0, 3.4, 7.4] \times 10^{-10} \tag{2}$$

for

$$m_t = [40, 100, 160]\ \mathrm{GeV}.$$

Thus 10^{-10} provides an important benchmark: if the measured branching ratio (BR) significantly exceeds 10^{-10}, then—in the nomenclature just introduced—top has to be "heavy" or/and a fourth family has to exist. In the latter case, even a branching ratio of $O(10^{-9})$ could be generated.[8]

In passing it should be noted that the Argus findings on $B_d - \bar{B}_d$ mixing that are discussed in the next section strongly suggest that this branching ratio exceeds 2×10^{-10}.

$B^0 - \bar{B}^0$ MIXING

The Argus collaboration has presented highly intriguing preliminary findings on $B_d - \bar{B}_d$ mixing as obtained on the $\Upsilon(4s)$ resonance[9]

$$y_p = \frac{N(l^\pm l^\pm)}{N(l^+l^-)} = \begin{cases} (23.4 \pm 6.7 \pm 3.1)\%, & \text{inclusive } ll \\ (19 \pm 10)\%, & \text{tagged events.} \end{cases} \tag{3}$$

These numbers are rather surprising because of the previous upper bound from CLEO—$y_p \leq 24\%$ (90% C.L.)—and previous theoretical expectations that are given later.

First we want to address some immediate phenomenological issues.

$\mathbf{B_d - \bar{B}_d}$ *vs.* $\mathbf{B_s - \bar{B}_s}$ *Mixing*

In the standard model with three families one obtains in a straightforward manner

$$\frac{\Delta m(B_s)}{\Delta m(B_d)} = \frac{\mathrm{Re}(V(ts))^2}{\mathrm{Re}(V(td))^2} \frac{Bf_B^2(B_s)}{Bf_B^2(B_d)} \tag{4}$$

where Bf_B^2 is a measure of the size of the relevant hadronic matrix element. Different theoretical calculations all agree on[10]

$$Bf_B^2(B_s) \geq Bf_B^2(B_d). \tag{5}$$

The main uncertainty enters via the KM parameters that yield (in the Wolfenstein representation)

$$\frac{\Delta m(B_s)}{\Delta m(B_d)} \geq \frac{\mathrm{Re}(V(ts))^2}{\mathrm{Re}(V(td))^2} = \frac{1}{\lambda^2}\frac{1}{(1-\rho)^2 - \eta^2} \gtrsim 6.5 \tag{6}$$

and therefore

$$r_s = \frac{\Gamma(B_s \rightarrow l^+X)}{\Gamma(B_s \rightarrow l^-X)} \geq 0.80 \tag{7}$$

for $r_d \geq 0.09$.

This point is very important for our later discussion: as long as one limits oneself to the Standard Model with three families, then a 10% (or more) B_d mixing leads quite conservatively to near-maximal B_s mixing.

A scenario with $r_d \geq 0.09$ and $r_s \geq 0.80$ by itself is not inconsistent with other data on $B^0 - \bar{B}^0$ mixing as obtained by UA1, Mark II, and Jade.[11] This statement rests largely on the fact that the relative abundance of B_s is not known independently and a priori could be as small as 14%.

$\tau(\mathbf{B}^{\pm})$ *vs.* $\tau(\mathbf{B}^0)$

While most authors expect the lifetimes and correspondingly also the semileptonic branching ratios of bottom hadrons to agree with each other to within, say, 20%, its should be kept in mind that experimentally a much larger variation is still allowed by CLEO data:

$$\frac{1}{2} \lesssim \frac{b_{SL}(B^{\pm})}{b_{SL}(B^0)} \lesssim 2. \tag{8}$$

Theoretically it is very hard to see how this ratio could be smaller than 1; thus we restrict our analysis to

$$1 \leq R \equiv \frac{b_{SL}(B^{\pm})}{b_{SL}(B^0)} \leq 2. \tag{9}$$

One finds for y_p, the ratio of like-sign to opposite-sign dileptons on the $\Upsilon(4s)$:

$$y_p = \frac{\chi_d}{\dfrac{1-f_0}{f_0}R^2 + (1-\chi_d)} \tag{10}$$

where $\chi = r/(1+r)$ and f_0 denotes the fraction of $B^0\bar{B}^0$ pairs. Therefore

$$\chi_d = \frac{y_p}{1+y_p} \cdot \left(\frac{1-f_0}{f_0}R^2 + 1\right). \tag{11}$$

One reads off from equation 11 that for a given y_p the mixing strength χ_d depends strongly on R. For example, if $y_p \simeq 0.04$, one finds

$$\chi_d \simeq [0.09, 0.15, 0.24] \qquad \text{for} \qquad R = [1, 1.5, 2]. \tag{12}$$

On the other hand, $y_p \simeq 0.09$ leads to

$$\chi_d \simeq [0.19, 0.32, 0.50]. \tag{13}$$

In that case $R \leq 2$ for certain since $\chi \leq 0.5$ must trivially hold.

Later discussion shows that if $R \geq 1.5$ indeed holds, then the case for new physics is significantly strengthened. First we address a more phenomenological issue: when R exceeds unity, one has to increase χ_d correspondingly to reproduce a given ratio of like-sign to opposite-sign dileptons in $\Upsilon(4s) \rightarrow B\bar{B}$. The ratio of like-sign to opposite-sign dileptons in bottom production *well above threshold* receives a relatively small enhancement of roughly 10–20% when R goes from 1 to 2 and $y_p \simeq 0.04$–0.09.

Such a change in R has a considerably larger impact on the forward–backward asymmetry of bottom jets in e^+e^- annihilation where one finds[12]

$$A_{FB}(\text{bottom jets}) = \frac{1}{1 + \bar{r}} A_{FB}(b\bar{b}) \tag{14}$$

with

$$\bar{r} = \frac{2}{R} \frac{\chi_d + f_s \chi_s}{1 + \frac{1}{R}(f_\Lambda + 1 - 2\chi_d + f_s(1 - 2\chi_s))} \tag{15}$$

where $f_s[f_\Lambda]$ denotes the abundance of $B_s[\Lambda_b]$ states relative to that of B^-. Using $f_s = 1/3, f_\Lambda = 0.1$ and $\chi_d = 0.09$ [0.19] one obtains for $R = 1$

$$\bar{r} \simeq 0.24 \; [0.40]. \tag{16}$$

If instead $R = 2$ were to hold, one gets

$$\bar{r} \simeq 0.30 \; [0.64]. \tag{17}$$

Experimentally a 90% C. L. upper bound has been found[12]

$$\bar{r}_{\text{exp}} \leq 0.35. \tag{18}$$

Thus a moderate increase in experimental sensitivity should reveal a nonvanishing $\bar{r}$, in particular if $R \geq 1.5$—unless, of course, there exists new physics that contributes *destructively* to $B_s - \bar{B}_s$ mixing.

THEORETICAL ESTIMATES ON $B_d - \bar{B}_d$ MIXING

The ratio $x = \Delta m/\Gamma$, which is the driving force behind $B^0 - \bar{B}^0$ mixing, can be calculated in terms of three main parameters:

(1) m_t.
(2) The KM parameters $V^*(tb)V(td)$.

(3) The hadronic matrix element $\langle B^0 | J \cdot J | \bar{B}^0 \rangle$, which is conventionally expressed in terms of $B \cdot f_B^2$, $B = 1$ corresponding to "vacuum saturation:"

$$\Delta m(B_d) = f(m_t) |V(td)|^2 B f_B^2(B_d). \tag{19}$$

The parameter $f(m_t)$ is a known function of m_t.[13] Theoretical estimates on B range between 0.5 and 1 and on f_B between 70 and 190 MeV.[10] The different theoretical calculations thus exhibit a much stronger trend to agree than was the case a few years ago—yet even so, one has to reckon with uncertainties of a factor of 2 to 3. A reasonable calibration for theoretical expectations is provided by expressing them in terms of a factor of

$$F = \frac{|V(td)|^2}{(0.01)^2} \frac{Bf_B^2}{(100\ \text{MeV})^2}. \tag{20}$$

As estimate of r_d with the rather conservative range $F = 1$–8 is given in FIGURE 1; from

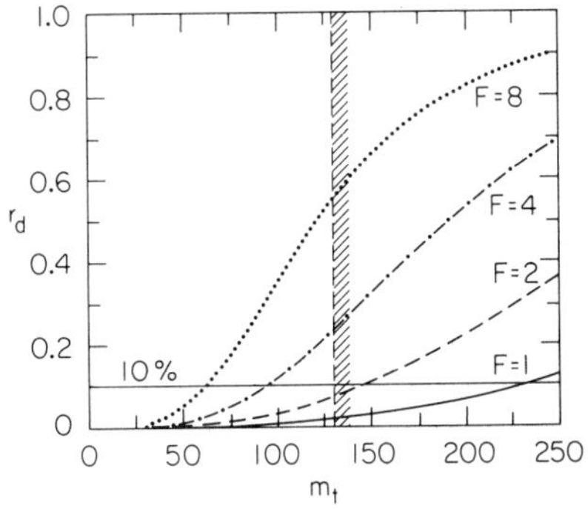

FIGURE 1. r_d as a function of m_t in the standard model with three families; the theoretical uncertainties are expressed in terms of a factor of

$$F = \frac{\text{Re}\,(V(td))^2}{(0.01)^2} \frac{Bf_B^2}{(100\ \text{MeV})^2}.$$

The upper bound on m_t shown here is from reference 4.

it we conclude that if $r_d \geq 0.10$, then

$$m_t \geq 70\ \text{GeV}. \tag{21}$$

It is intriguing to note again the upper bound on m_t, $m_t \lesssim 130$ GeV.[4]

Therefore the Standard Model with three families does not allow $Z^0 \to t\bar{t}$ to proceed if indeed $r_d \geq 0.10$. Observing $Z^0 \to t\bar{t}$, on the other hand, establishes the presence of new physics in $B_d - \bar{B}_d$ mixing. One, but by no means the only, example is given by an ansatz with four families,[8] as shown in FIGURE 2.

CP VIOLATION IN B^0 DECAYS

A CP asymmetry could show up in *semileptonic* B^0 decays:

$$a_{SL} = \frac{\sigma(B^0\bar{B}^0 \to l^+l^+ + X) - \sigma(B^0\bar{B}^0 \to l^-l^- + X)}{\sigma(B^0\bar{B}^0 \to l^+l^+ + X) + \sigma(B^0\bar{B}^0 \to l^-l^- + X)} = \frac{\text{Im}\,\dfrac{\Gamma_{12}}{M_{12}}}{1 + \dfrac{1}{4}\left|\dfrac{\Gamma_{12}}{M_{12}}\right|^2}. \tag{22}$$

In the Standard Model with three families such asymmetries remain unobservably small; adding new physics like a fourth family could produce an a_{SL} on the percent level. However, if at the same time $r_d \geq 0.10$ has to be reproduced, it is an almost inescapable conclusion that $a_{SL} \lesssim 10^{-3}$.

On the other hand, the prospects for observing CP asymmetries in *nonleptonic* B_d decays are greatly enhanced. As explained elsewhere in more detail,[14] the mixing strength optimal for observing a difference between $\Gamma(B^0(t) \rightarrow f)$ and $\Gamma(\bar{B}^0 \rightarrow \bar{f})$—$f$ being a common decay mode of B^0 and $\bar{B}^0$—is $r = 33\%$. Yet also $r_d \geq 10\%$ presents an excellent scenario where CP asymmetries of up to 50% can be realized.

CONCLUSIONS

The history of K decay studies shows that new physics, like parity and CP violation and charm, can be found in an indirect way. There is every reason to believe that

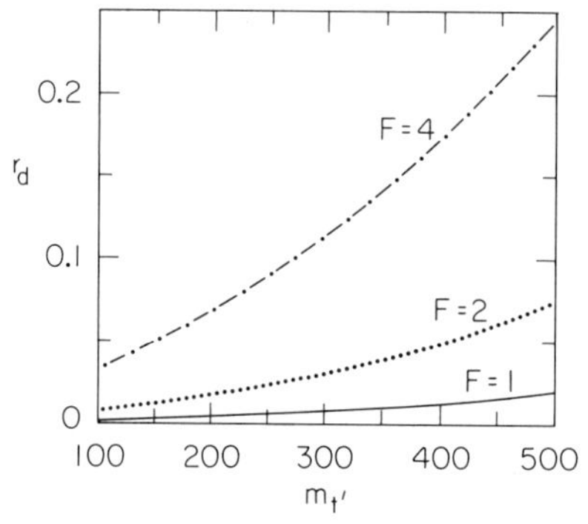

FIGURE 2. r_d as a function of $m_{t'}$, the mass of a fourth family quark, with $m_t = 40$ GeV kept fixed.

detailed studies of weak decays will score more such successes in the future; searches for $K^+ \rightarrow \pi^+\nu\bar{\nu}$ and $B_d - \bar{B}_d$ mixing are just two, though highly promising, examples. In these processes one has sensitivity for mass scales that are beyond the reach of even the Tevatron as far as direct production is concerned.

The recent Argus findings, if they stand the test of further scrutiny, are an eminently intriguing step in such a direction: If $Z^0 \rightarrow t\bar{t}$ is observed or $B_s - \bar{B}_s$ mixing restricted to be less than near-maximal, then one has established the presence of new physics in B^0 decays. The presumed size of the effect—$r_d \geq 0.10$—already "smells" of new physics, yet at the moment we cannot claim for sure that this "new smell" establishes a "new flavor." The preceding discussion given shows (and $B_d - \bar{B}_d$ mixing thus provides a typical case study for the paradigm of indirect searches) that no gain can be achieved without its proper prize: At each step the reliability of the theoretical reasoning has to be gauged in a careful manner. Here it is our understanding of the B meson wave function that has to be cross-examined. This requires more work, both of a theoretical and an experimental nature.

ACKNOWLEDGMENTS

We gratefully acknowledge exchanges with K. Schubert and A. Sanda. We congratulate the organizers of this meeting for creating such a fine atmosphere and providing such spectacular views.

REFERENCES

1. GLASHOW, S., J. ILIOPOULOS & L. MAIANI. 1970. Phys. Rev. **D2:** 1285.
2. KOBAYASHI, M. & T. MASKAWA. 1973. Prog. Theor. Phys. **49:** 652.
3. SAXON, D. H. 1984. Rutherford RAL-84-094. Preprint.
4. MARCIANO, W. 1987. Constraints and speculations on fourth-generation physics. This volume.
5. HOU, G. 1987. Talk given at the APS Meeting, Salt Lake City, Utah, with references to earlier work.
6. HOU, G. W.-S. & A. SONI. 1987. Loop-induced rare B decays and the fourth generation. This volume.
7. GILMAN, F. & J. HAGELIN. 1983. Phys. Lett. **133B:** 433; TÜRKE, U. 1986. Phys. Lett. **168B:** 296.
8. BIGI, I. I. & S. WAKAIZUMI. SLAC-PUB-4180. Submitted for publication in Phys. Lett.
9. ALBRECHT, H. 1987. New results on $B - \bar{B}$ mixing from ARGUS. Presented at the 1st Int. Symp. Fourth Family of Quarks and Leptons (Santa Monica, Calif., Feb. 26–28, 1987).
10. BIGI, I. I. & A. I. SANDA. 1987. Nucl. Phys. **281:** 41. (With references to earlier work.)
11. See for example, ALI, A. 1987. $B - \bar{B}$ mixing: A window on the 4th family of quarks. Presented at the 1st Int. Symp. Fourth Family of Quarks and Leptons (Santa Monica, Calif., Feb. 26–28, 1987). PASCHOS, E. A. 1987. Signatures of the fourth generation at present energies. This volume.
12. BARTEL, W. *et al.* 1984. Phys. Lett. **146B:** 437. BIGI, I. I. 1985. Phys. Lett. **155B:** 125.
13. INAMI, T. & C. S. LIM. 1981. Prog. Theor. Phys. **65:** 297.
14. I. I. BIGI, invited talk given at the UCLA Workshop on $B\bar{B}$ Factories, January 1987, to appear in the Proceedings.

Some New Aspects of Fourth-Generation Physics

V. BARGER

Physics Department
University of Wisconsin
Madison, Wisconsin 53706

INTRODUCTION

The UA1 limit $m_L > 41$ GeV on the mass of a charged heavy lepton L is evaded if ν_L is heavy and decays. The search for fourth-generation leptons and quarks that may be produced in Z decays at SLC and LEP I is briefly addressed. At the superconducting super collider (SSC) the signals for superheavy charged leptons L with massless ν_L and decays $L \rightarrow \nu_L W$ are obscured by backgrounds from single and pair-produced $W^\pm$ and Z bosons. Superheavy quarkonium gives interesting signals for fourth-generation quarks and the Higgs boson at the SSC.

BOUND ON m_L

Fourth-generation leptons can be produced at hadron colliders via $q\bar{q}' \rightarrow W^\pm \rightarrow L\nu_L$. If $m_{\nu_L} \approx 0$, the most promising signature[1] is missing p_T plus jets, coming from $L \rightarrow \nu_L q\bar{q}'$ decays with ν_L undetected. FIGURE 1 shows calculations[2] of this signal and backgrounds at CERN and Tevatron $p\bar{p}$ colliders. The principal background from $W \rightarrow \tau\nu_\tau$ is not a serious problem, since $\tau \rightarrow \nu_\tau$ + hadrons can be identified by the small angular spread and low charged multiplicity of the hadrons. Microvertex detectors will further aid in separating the τ background. The other backgrounds also have different properties from the L-signal, such as jet multiplicity and distributions.

Recently the UA1 collaboration presented a limit $m_L > 41$ GeV, for $m_{\nu_L} \approx 0$, based on an analysis of events with large missing p_T and jets.[3] However, this limit on m_L applies only if ν_L is light and escapes undetected. If ν_L is heavy, there are three different possible scenarios, depending on the mass of ν_L (that we take to be dominantly composed of the fourth mass eigenstate ν_4), each of which allows the UA1 bound to be evaded.[4]

Scenario A: $m_{\nu_L} > M_W - m_L$. Here $L + \nu_L$ production proceeds via virtual W bosons and the event rate is suppressed well below the UA1 sensitivity.

Scenario B: $m_L < m_{\nu_L} < M_W - M_L$. In this case there is real W resonance production. The ν_L decays to L and L decays via mixing to ν_e, ν_μ, or ν_τ by charged currents, as illustrated in FIGURE 2a for $L \rightarrow \nu_\tau$ decay. The final state is $W \rightarrow \nu_\tau\bar{\nu}_\tau W^* W^* W^*$, where the virtual W^* give $l\nu$ or $q\bar{q}'$. In comparison with the $m_{\nu_L} \approx 0$ case, where $W \rightarrow \nu_L\bar{\nu}_L W^*$, the missing p_T in Scenario B is considerably degraded. The

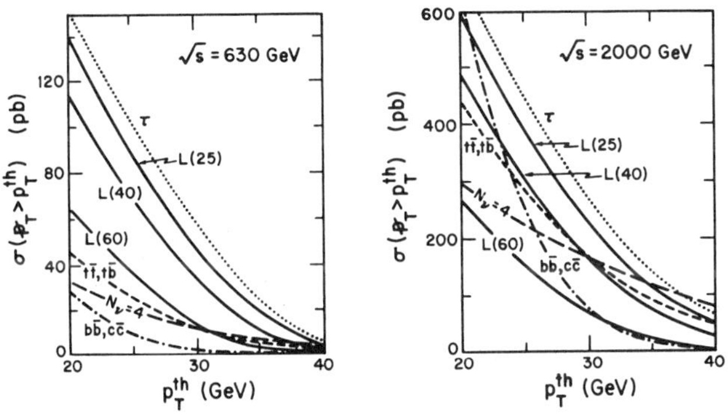

FIGURE 1. The signal and backgrounds for $p\bar{p} \to W \to L\nu_L \to \not{p}_T$ + hadrons for missing p_T (denoted by $\not{p}_T$) greater than a threshold $\not{p}_T^{\text{th}}$ (from reference 2). The t-quark backgrounds are based on $m_t = 40$ GeV.

number of events with missing $p_T > 15$ GeV is degraded by a factor of 3 below the $m_{\nu_L} \approx 0$ result and the UA1 limit is lowered to the e^+e^- limit $m_L > 22.7$ GeV from PETRA.[5]

Scenario C: $m_{\nu_L} < m_L < M_W - m_{\nu_L}$. Here $W \to L\nu_L$ occurs at the real W resonance rate. The L decays to ν_L and ν_L decays via mixing, most likely to τ, as illustrated in FIGURE 2b. If ν_L and $\bar{\nu}_L$ decay close to the interaction point, they give no missing p_T signal in the UA1 experiment and there is again no UA1 m_L limit. The ν_L decay length is governed by the $V_{4\tau}$ mixing matrix element. Following an extrapolation[6] of quark mixings to the fourth generation we suppose $|V_{4\tau}| \sim 0.01$. Then ν_L and $\bar{\nu}_L$ decays occur close to the interaction point for a wide range of L, ν_L masses, as illustrated in FIGURE 3.

We conclude that the UA1 bound on m_L is removed if ν_L is heavy. The present PETRA limit $m_L > 22$ GeV from the $R = \sigma(e^+e^- \to \text{hadrons})/\sigma(e^+e^- \to \mu^+\mu^-)$ measurement is the only firm bound on m_L.

e^+e^- SEARCHES FOR THE FOURTH GENERATION

Any new particles f with mass $m_f \lesssim 46$ GeV that are coupled to the Z boson will be revealed at SLC and LEP I in $e^+e^- \to Z \to f\bar{f}$. FIGURE 4 shows the resonance

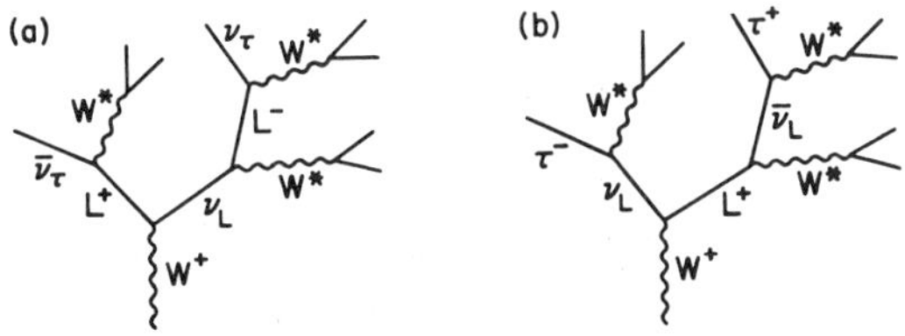

FIGURE 2. Heavy lepton signals from real W production in hadron collisions. (a) Scenario B. (b) Scenario C.

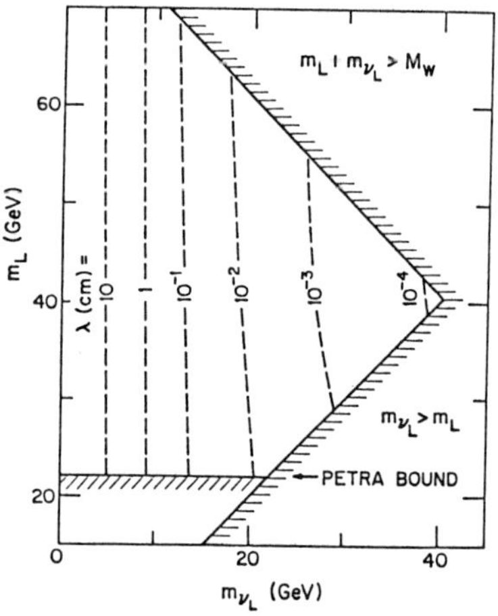

FIGURE 3. Mean decay length λ of the faster member of $\nu_L\bar{\nu}_L$ pair produced in $p\bar{p} \to W \to L\nu_L$ at $\sqrt{s} = 630$ GeV (from reference 4) according to Scenario C assuming $|V_{4\tau}| = 10^{-4}$.

production cross section[7] vs. m_f for a charged heavy lepton L, a heavy neutrino ν_L, a $-1/3$ charge fourth-generation quark v, the t-quark, and an isosinglet fermion h in the 27-plet of E_6 symmetry.

Neutrinos

Massless neutrinos can be "counted" in $e^+e^- \to \nu\bar{\nu}\gamma$ and also in the Z total width, provided that in the latter case other decay contributions are identified. Heavy

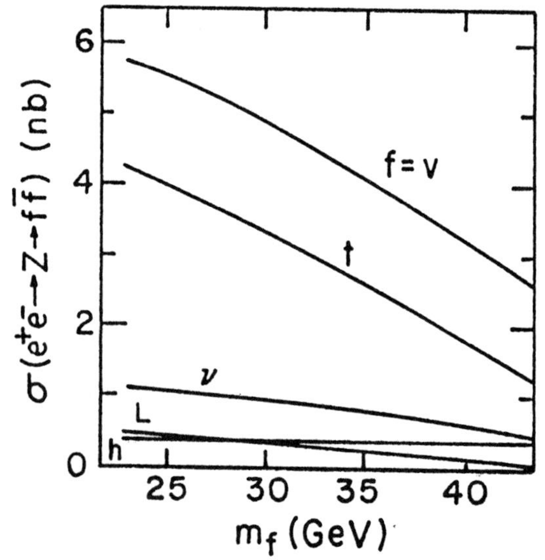

FIGURE 4. Cross sections for $e^+e^- \to Z \to \bar{f}f$ on resonance (from reference 7). For comparison $\sigma(e^+e^- \to Z \to \mu^+\mu^-) = 1.8$ nb.

neutrinos will decay and the possibility of heavy fourth-generation neutrinos can be investigated with vertex detectors.[8] At SLC vertex detectors will probe mixing matrix elements $|V_{4l}|^2 \gtrsim 10^{-10}$ for $m_{\nu_4} < 40$ GeV.[9] Should no evidence for ν_4 be found by neutrino counting or by unstable particle searches, the prospects for finding a fourth generation would be greatly diminished.

Charged Heavy Lepton

An e^+e^- machine operating on the Z resonance could be a copious source of heavy leptons. The two primary classes of decays are $L \to l + \not{p}$ and $L \to q_1\bar{q}_2 + \not{p}$, where $\not{p}$ is missing energy-momentum carried by decays. A clean signal for $L\bar{L}$ is $l_1\bar{l}_2 + \not{p}$, which is about 7% of the $L\bar{L}$ rate. The mass m_L can be determined from the distribution in the dilepton opening angle, illustrated in FIGURE 5. The $lq_1\bar{q}_2 + \not{p}$ signal is about 40% of $\sigma(e^+e^- \to Z \to L\bar{L})$, and in this case m_L can be determined from the $m(q_1\bar{q}_2)$ distribution.

v-*quark*

An interesting possibility[6,7] is that $m_v < m_t$. For instance, the low-thrust events containing muons observed[10] at the topmost energy in the Mark J and Jade experiments could be of $e^+e^- \to \bar{v}v$ origin. To investigate the consequences at SLC, we suppose that the mixings follow the pattern

$$|V_{nm}| = \begin{array}{c} \\ u \\ c \\ t \\ a \end{array} \begin{array}{c} \begin{array}{cccc} d & s & b & v \end{array} \\ \begin{pmatrix} 1- & \theta & \theta^3 & \theta^5 \\ \theta & 1- & \theta^2 & \theta^4 \\ \theta^3 & \theta^2 & 1- & \theta^3 \\ \theta^5 & \theta^4 & \theta^3 & 1- \end{pmatrix} \end{array}$$

based on an extrapolation[6] of the 3 × 3 mixing matrix. The $v \to c$ transitions dominate ($v \to t$ and $v \to a$ are kinematically forbidden) but are strongly suppressed, $|V_{cv}|^2 \sim \theta^8$. Hence, the v-lifetime may be long,[6] possibly of order 10^{-13} sec and measurable.

There would be a high charm multiplicity resulting from the $v \to cs\bar{c}$ decay mode. The multiplicity of charm particles in $v\bar{v}$ events would differ substantially from $b\bar{b}$ events.[7]

$N(c \text{ and } \bar{c}) =$	2 (%)	3 (%)	4 (%)
$v\bar{v}$	~100	44	11
$b\bar{b}$	~100	12	½

The v-jet broadness would determine m_v. FIGURE 6 shows the distribution[7] in Σp_T^i summed over visible particles in a v-jet from $Z \to v\bar{v}$ decays, for several values of the v-mass. The sum of momenta of light decay products normal to the v-jet axis is on

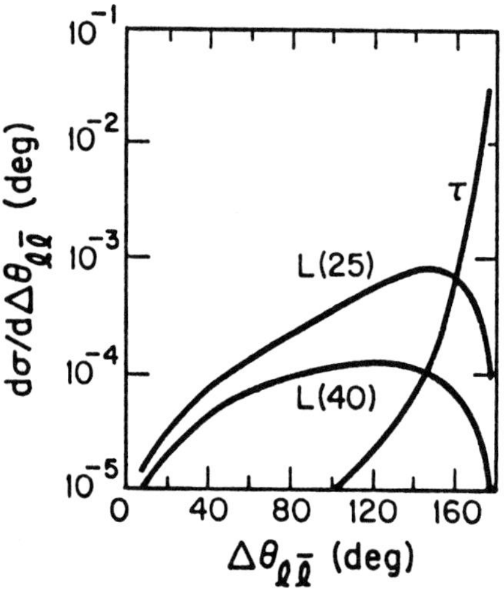

FIGURE 5. Dilepton opening angle distribution from leptonic decays of $L\bar{L}$ produced in e^+e^- collisions at the Z resonance (from second paper of reference 6).

average

$$\left\langle \sum p_T^i \right\rangle \simeq \frac{\pi}{4} m_v.$$

The broad jet results from the large energy release in v-decay. The maximum electron $p_T = (m_v^2 - m_c^2)/(2m_v)$ from $v \rightarrow ce\nu$ decay also reflects the large energy release. The electron from v semileptonic decay is more likely to be "isolated" from hadronic decay products than the electron from $b \rightarrow ce\nu$ decay.

Finally, there may be additional signatures for the v-quark, such as a two-body flavor-changing neutral current decay[7] $v \rightarrow bg$, with branching fraction as high as 5–10%, that has distinctive kinematic features.

HEAVY LEPTONS AT THE SSC

At multi-TeV hadron colliders, such as the SSC, superheavy leptons with masses exceeding 100 GeV, could be produced. Then L decays to real W bosons, $L \rightarrow \nu_L + W$,

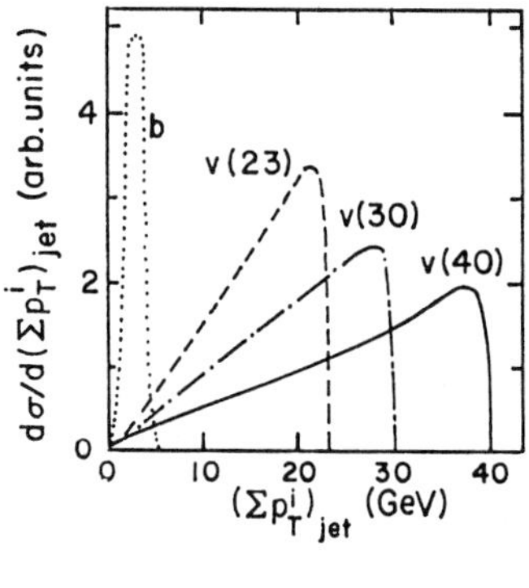

FIGURE 6. Distribution in $\Sigma\, p_T^i$ of momenta transverse to the jet axis for detectable particles in a v jet. The result for a b jet is compared (from reference 7).

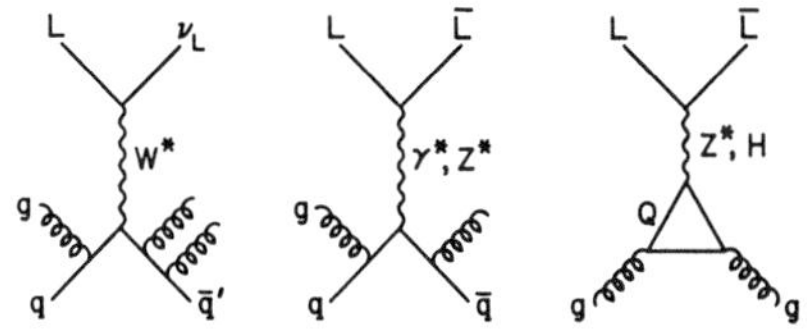

FIGURE 7. Dominant diagrams for production of superheavy charged leptons in pp collisions.

assuming that $m_L > m_{\nu_L} + M_W$. What are the prospects for its detection? In the following we consider the case $m_{\nu_L} \approx 0$.

The dominant subprocesses for L-production in hadron collisions are shown in FIGURE 7. The expected cross sections[11,12] at the SSC are given in FIGURE 8. Substantial event rates at the SSC are expected (note that 1 pb corresponds to 10^4 events per year at the SSC). The signals are $W^* \to L\nu_L \to W\not{p}_T$ and $Z^* \to L\bar{L} \to WW\not{p}_T$, where W^* and Z^* are off-shell. The problem is that there are large backgrounds to these signals from the production of real W and Z bosons, both singly and in pairs, that submerge these L signals. FIGURE 9 illustrates typical background subprocesses. FIGURE 10 compares various possible signals and the associated backgrounds. Attempts to isolate the L signal in theoretical simulations have proved

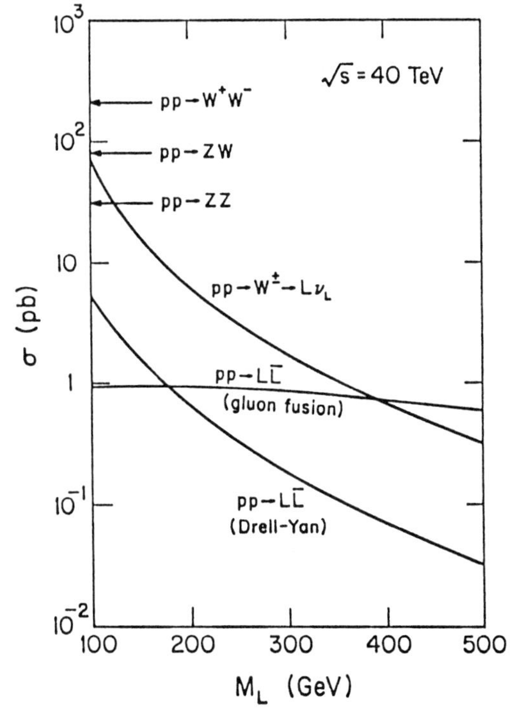

FIGURE 8. Cross sections for superheavy charged lepton production at the SSC (from reference 12). The gluon–gluon fusion cross-section result assumes $m_H = 100$ GeV, $m_v = m_L$, and $m_a - m_v = 250$ GeV.

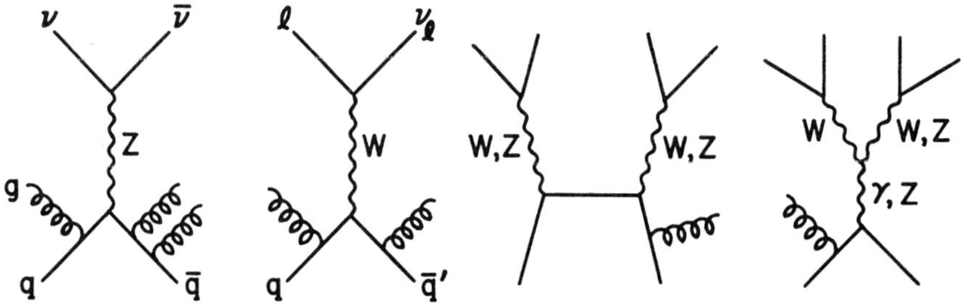

FIGURE 9. Typical subprocesses that give backgrounds to a heavy lepton signal at the SSC.

unsuccessful. We conclude that it is hard to detect a 100–200-GeV charged heavy lepton with $m_{\nu_L} \approx 0$ at a hadron supercollider.

SUPERHEAVY QUARKONIA

The first signals of charm and bottom quarks in hadron collisions came from the leptonic decays of the J/ψ and Υ bound states. Could superheavy quarks also be first discovered at multi-TeV hadron colliders through the decays of their quarkonium bound states?

The partial widths for the decays of $\eta(Q\bar{Q})$ and $\psi(Q\bar{Q})$ to gluons are

$$\Gamma(\eta \rightarrow gg) \simeq 3\text{ MeV}\left(\frac{M}{200\text{ GeV}}\right)$$

$$\Gamma(\psi \rightarrow ggg) \simeq 20\text{ keV}\left(\frac{M}{200\text{ GeV}}\right)$$

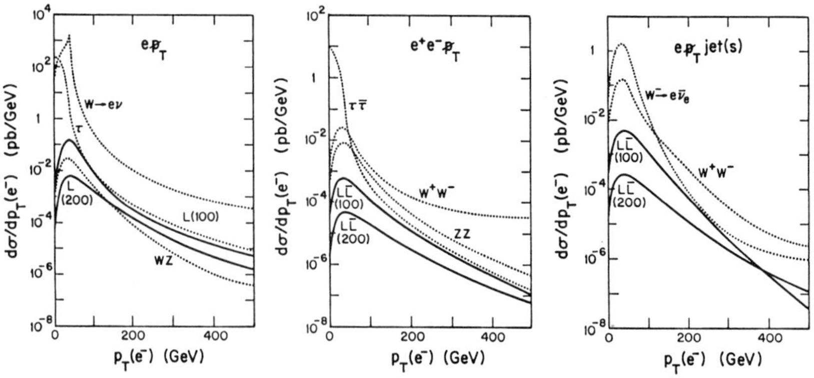

FIGURE 10. Comparison of several heavy lepton signals and corresponding backgrounds at the SSC c.m. energy (from reference 12).

where M is the mass of the quarkonium state. More interesting decays to VV, VH, $f\bar{f}$, where V is a gauge boson, H is a Higgs boson, and f is a fermion have partial width of the generic forms[13]

$$\Gamma(O \rightarrow VV, VH) = K \frac{\alpha^2}{M^2}\left(\frac{M^2}{M_W^2}\right)^x |\psi(0)|^2$$

$$\Gamma(O \rightarrow f\bar{f}) = K' \frac{\alpha^2}{M^2}\left(\frac{m_f^2}{M^2}\right)^y \left(\frac{M^2}{M_Z^2}\right)^x |\psi(0)|^2$$

where K and K' are dimensionless factors that contain threshold and other mass dependence; x represents enhancements from Higgs or longitudinal W, Z axial vector

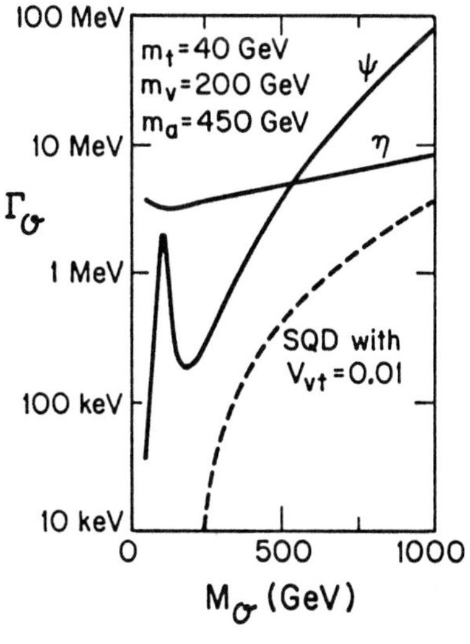

FIGURE 11. Total widths for $\eta(v\bar{v})$ and $\psi(v\bar{v})$ from gluon and gauge boson decays, compared with the widths from single-quark decays $v \rightarrow t + W$. Masses $m_t = 40$ GeV, $m_a = m_v + 250$ GeV, and $m_H = 2000$ GeV are assumed (from reference 13).

couplings to superheavy quarks:

		$f\bar{f}$	$\gamma\gamma$	$Z\gamma, ZZ$	W^+W^-	γH	ZH
η	$x =$	2	0	0	0	—	2
ψ	$x =$	0	—	1	2	1	1

The $x = 2$ modes dominate, but in the case of $f\bar{f}$ only if m_f/M_Z is not small. The parameter y in the $f\bar{f}$ decays is associated with chirality suppression:

$$y = 0 \quad \text{for} \quad \psi \qquad \text{and} \qquad y = 1 \quad \text{for} \quad \eta.$$

Possible competition to these interesting decay modes are the decays $Q \rightarrow q + W$ in which the companion $\bar{Q}$ acts as a spectator. If $Q \rightarrow q + W$ dominates, quarkonium decays are indistinguishable from continuum $Q\bar{Q}$ events.

Fortunately, in the case of the v-quark, the decay $v \rightarrow t + W$ is expected to be suppressed by a factor of $|V_{tv}|^2 \sim 10^{-4}$. Then the single-quark decay rate is well below contributions from gluon and gauge boson decay modes of η and ψ, as shown in FIGURE 11.

The η and ψ states are produced in pp collisions via $gg \rightarrow \eta$ and $gg \rightarrow \psi g$ subprocesses. The cross sections at the SSC are shown in FIGURE 12, for a realistic quarkonium potential. For M_η up to 700 GeV the cross section is $\gtrsim 1$ pb (i.e., $\gtrsim 10^4$ events/year). The ψ cross section is about two orders of magnitude below the η cross section at high quarkonium masses.

The dominant decay of η_v when $M_\eta > 400$ GeV and $m_H < M_\eta - 250$ GeV is $\eta_v \rightarrow ZH$ with a branching fraction that approaches 100%. The large event rate allows the use of a $Z \rightarrow l\bar{l}$ trigger. It may be possible to simultaneously find the Higgs

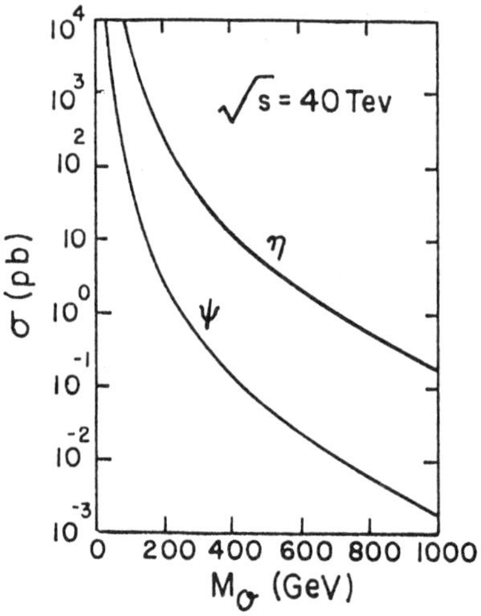

FIGURE 12. Cross sections for η and ψ production in pp collisions at 40 TeV (from reference 13).

boson and η_v quarkonium by reconstruction of the mass from the hadrons of H^0 decay along with the invariant mass of $l^+l^-H^0$. Even for the elusive intermediate mass Higgs boson, with mass $2m_t < M_H < 2M_W$ and decay $H^0 \rightarrow t\bar{t}$, the signal exceeds the background[14] from $q\bar{q} \rightarrow Zt\bar{t}$ for reasonable mass resolution ($\sim \pm 15\%$) on M_{ZH} and M_H.

The best prospect for detection of ψ_v at a hadron collider is through the decay $\psi_v \rightarrow l^+l^-$. The branching fraction is $\mathrm{BF}(\psi_v \rightarrow e^+e^-) \sim 3\%$ for $M_\psi \lesssim 200$ GeV, giving about 100 events per year at the SSC.[13] For an e^+e^- mass resolution of 3% the ψ signal is comparable to the Drell–Yan background. At higher M_ψ, the decay $\psi \rightarrow W^+W^-$ dominates and the Standard Model backgrounds exceed the signal. In fact, all other interesting decays of η_v and ψ_v are unobservable in a hadron collider because of Standard Model backgrounds. These other modes would be observable in a high-energy e^+e^- collider, however.

SUMMARY

In brief summary we have drawn the following conclusions about fourth-generation searches:

(1) The UA1 bound on a charged heavy lepton is relaxed to $m_L > 22$ GeV if ν_L is heavy and decays.

(2) SLC will push fourth-generation mass limits to $m_f > 46$ GeV. If evidence for ν_L is not found, it will be discouraging.

(3) Assuming $m_{\nu_L} \approx 0$, it will be difficult to detect superheavy charged leptons with $L \rightarrow \nu_L W$ decays at the SSC because of overwhelming backgrounds from weak boson production.

(4) Superheavy $\eta(v\bar{v})$ quarkonium could be a new Higgs source.

REFERENCES

1. Barger, V. *et al.* 1983. Phys. Lett. **B133:** 449; Barger, V. *et al.* 1984. Phys. Rev. **D29:** 2020; Cline, D. B. & M. Mohammadi. Univ. Wisconsin Rep. WISC-EX-86-269.
2. Barger, V., J. Ohnemus & R. J. N. Phillips. 1987. Phys. Rev. **D35:** 158.
3. UA1 collaboration: Albajar, C. *et al.* 1986. Phys. Lett. **185B:** 241; Batley, R. CERN-EP/ 86-181.
4. Barger, V. & R. J. N. Phillips. 1987. Phys. Lett. **189B:** 473.
5. See, for example, the survey by Komamiya, S. *In* Proc. 1985 Int. Symp. Lepton and Photon Interactions (Kyoto, Japan).
6. Barger, V., H. Baer, K. Hagiwara & R. J. N. Phillips. 1984. Phys. Rev. **D30:** 947; Baer, H., V. Barger & R. J. N. Phillips. 1985. Phys. Rev. **D32:** 688.
7. Barger, V., R. J. N. Phillips & A. Soni. 1986. Phys. Rev. Lett. **57:** 1518.
8. Barger, V., W.-Y. Keung & R. J. N. Phillips. 1984. Phys. Lett. **141B:** 126.
9. Gilman, F. 1986. Comments Nucl. Part. Phys. **16:** 231.
10. Mark J Collaboration. 1986. MIT Rep. 146; Jade Collaboration. 1986. DESY Rep. 86-052; Branson, J. 1987. Low-thrust hadron events with isolated μ or e from PETRA. This volume.
11. Willenbrock, S. & D. Dicus. 1985. Phys. Lett. **156B:** 429.
12. Barger, V., T. Han & J. Ohnemus. 1987. Univ. Wisconsin Rep. PH/331.
13. Barger, V. *et al.* 1986. Phys. Rev. Lett. **57:** 1672; Barger, V. *et al.* 1987. Phys. Rev. **D35:** 3366.
14. Gunion, J. & Z. Kunzst. Univ. California, Davis. Preprint 86-21.

Cosmological Limit on the Number of Families[a]

DAVID N. SCHRAMM

University of Chicago
Chicago, Illinois 60637

INTRODUCTION

We review the standard Big Bang nucleosynthesis arguments. Using the primordial helium as inferred from the He and C correlations, and using the strengthened Li constraint, the cosmological limit is tightened such that three (or two) neutrino families fit well, but a fourth is beginning to be questionable, although it cannot be totally excluded.

DISCUSSION

Cosmological predictions[1,2] about the number of neutrino families, N_ν, were one of the first examples of the particle–cosmology interface, and are now beginning to be tested with accelerators. This paper reviews those arguments and shows that they continue to be upheld.

The power of Big Bang nucleosynthesis comes from the fact that essentially all of the physics input is well determined in the terrestrial laboratory. The appropriate temperatures, 0.1 to 1 MeV, are well explored in nuclear physics labs. Thus, what nuclei do under such conditions is not a matter of guesswork, but is precisely known. In fact, it is known for these temperatures far better than it is for what nuclei do in the centers of stars like our sun. The center of the sun is only a little over 1 keV. This energy is below the energy where nuclear reaction rates yield significant results in laboratory experiments, and only the long times and higher densities available in stars enable anything to take place. Unfortunately, for stellar astrophysics this means that nuclear reaction rates must be extrapolated to many orders of magnitude below their laboratory-observed values. The Big Bang does not have this problem. It goes through temperatures and densities where the reaction rates were known and well confirmed with laboratory measurements.

To calculate what happens, all one has to do is follow what a gas of baryons with density ρ_b does as the universe expands and cools. As far as nuclear reactions are concerned, the only relevant region is from a little above 1 MeV down to a little below 100 keV. At higher temperatures, no complex nuclei other than single neutrons and protons can exist, and the ratio of neutrons to protons, n/p, is just determined by $n/p =$

[a]This work was supported in part by NSF at the University of Chicago and in part by NASA at Fermilab.

$e^{-Q/T}$, where $Q = 1.3$ MeV. Equilibrium applies because the weak interaction rates are much faster than the expansion of the universe at temperatures much above 10^{10} K. At temperatures much below 10^9 K, the electrostatic repulsion of nuclei prevents nuclear reactions from proceeding as fast as the cosmological expansion separates the particles.

Because of the equilibrium existing for temperatures much above 10^{10} K, for these calculations we don't have to worry about what went on in the universe at higher temperatures. Thus, we can start our calculation at 10 MeV and not worry about speculative physics like the theory of everything (T.O.E.), the grand unifying theories (GUTs), or even the details of the confinement of quarks, as long as a gas of neutrons and protons exists in thermal equilibrium by the time the universe has cooled to 10 MeV. The equilibrium ratio of neutrons to protons is near unity above 10 MeV, and drops to less than one-third by 10 MeV.

After the weak interaction drops out of equilibrium, a little above 10^{10} K, the ratio of neutrons to protons changes more slowly due to free neutrons decaying to protons, and similar transformations of neutrons to protons via interactions with the ambient leptons. By the time the universe reaches 10^9 K (0.1 MeV), the ratio is slightly below 1/7. For temperatures above 10^9 K, no complex nuclei can be produced because the simplest, complex nucleus, ^{2}D, is so weakly bound that the ambient photons fragment any ^{2}D back into its constituent neutrons and protons at temperatures above 10^9 K. Once the temperature drops to about 10^9 K, ^{2}D can survive. Then, the ^{2}D rapidly adds neutrons and protons, making ^{3}T and ^{3}He. These, in turn, add neutrons and protons to produce ^{4}He, or ^{3}T and ^{3}He and also collide to yield ^{4}He. Since ^{4}He is the most tightly bound nucleus in the region, the flow of reactions converts almost all the neutrons that exist at 10^9 K into ^{4}He. The flow essentially stops there because there are no stable nuclei at either mass-5 or mass-8. Since the baryon density at Big Bang nucleosynthesis is relatively low (much less than 1 g/cm^3), only reactions involving two-particle collisions occur. It can be seen that combining the most abundant nuclei neutrons, protons, and ^{4}He via two-body interactions always lead to unstable mass-5. Even when one combines ^{4}He with rarer nuclei like ^{3}T or ^{3}He, we still only get to mass-7, which when hit by a proton, the most abundant nucleus around, yields mass-8. Eventually, ^{3}T radioactively decays to ^{3}He, and any mass-7 made, radioactively decays to ^{7}Li. Thus, Big Bang nucleosynthesis makes ^{4}He with traces of ^{2}D, ^{3}He, and ^{7}Li. (Also, all the protons left over that did not capture neutrons remain as hydrogen.) All other chemical elements are made later in stars and in related processes. (Stars jump the mass-5 and -8 instability by having gravity compress the matter to sufficient densities that three-body collisions can occur and jump the mass-5 and -8 gaps.) A neutron/proton ratio of $\sim 1/7$ yields a ^{4}He primordial mass fraction,

$$Y_p = \frac{2n/p}{n/p + 1} \approx \frac{1}{4}.$$

The only parameter we can easily vary in such calculations is in the density of the gas that corresponds to a given temperature. From the thermodynamics of an expanding universe we know that $\rho_b \propto T^3$; thus, we can relate the baryon density at 10^{11} K to the baryon density today, when the temperature is about 3 K. The problem is, we don't know today's ρ_b, so the calculation is carried out for a range in ρ_b. Another aspect of the density is that the cosmological expansion rate depends on the total mass-energy density associated with a given temperature. For cosmological temperatures much above 10^4 K, the energy density of radiation exceeds the mass-energy density of the baryon gas. Thus, during Big Bang nucleosynthesis, we need the radiation density as well as the baryon density. The baryon density determines the density of the nuclei and thus their interaction rates, and the radiation density controls the expansion rate of the universe at those times. The density of radiation is just proportional to the number of types of radiation. Thus, the density of radiation is not a free parameter if we know how many types of relativistic particles exist when Big Bang nucleosynthesis occurred.

Assuming that the allowed relativistic particles at 1 MeV are photons, e, μ, and τ neutrinos (and their antiparticles) and electrons (and positrons), we have calculated the Big Bang nucleosynthetic yields for a range in present ρ_b, going from less than that observed in galaxies to greater than that allowed by the observed large-scale dynamics of the universe. The ^{4}He yield is almost independent of the baryon density, with a very slight rise in the density due to deuterium's ability to hold together at slightly higher temperatures and at higher densities, thus enabling nucleosynthesis to start slightly earlier, when the neutron/proton ratio is higher. No matter what assumptions one makes about the baryon density, it is clear that ^{4}He is predicted by Big Bang nucleosynthesis to have to be around 25% of the mass of the universe. This was first noted by Hoyle and Taylor[2], and later found by Peebles[3] and by Wagoner *et al.*[4] The current results do not differ in any qualitative way from Wagoner *et al.*

The fact that the observed helium abundance in all objects is about 20–30% was certainly a nice confirmation of these ideas. Since stars produce only a yield of 2% in all the heavy elements combined, stars cannot easily duplicate such a large yield. While the predicted Big Bang yields of the other light elements were also calculated in the 1960s, they were not considered important at that time, since it was assumed in the 1960s that these nuclei were made in more significant amounts in stars.[5] However, work by our group at Chicago,[6] and others, thoroughly established Big Bang nucleosynthesis and enabled it to be a tool for probing the universe, as opposed to a consistency check. This was done by showing that other light-element abundances had major contributions from the Big Bang, and that the effects of stellar contributions, where relevant, could be removed by appropriate techniques to obtain constraints on the Big Bang yields for those isotopes. Thus, Big Bang predictions for all the four light isotopes are now very relevant.

In particular, it was demonstrated in the early 1970s that contrary to the ideas of the 1960s, deuterium could not be made in any significant amount by *any* astrophysical process other than the Big Bang itself.[7] The Big Bang deuterium yield decreases rapidly with an increase in ρ_b. Since at high densities the deuterium gets more completely converted to heavier nuclei, this quantitatively means that the present density of baryons must be below $\sim 5 \times 10^{-31}$ g/cm^3 in order for the Big Bang to have produced enough deuterium to explain the observed abundances. Similar, though more complex arguments were also developed for ^{3}He, and most recently[8] for ^{7}Li, so that it can be said that only if the baryon density is between 2×10^{-31} and 5×10^{-31} g/cm^3 can all the observed light element abundances be consistent with the Big Bang yields. If the baryon density were outside of this range, a significant disagreement between the Big Bang and the abundance observations would result. To put this in perspective, it should be noted that for this range in densities, the predicted abundances for the four separate species fall within a range from 25% to one part in $\sim 10^{10}$. (In fact, for lithium to get agreement requires an abundance just at 10^{-10}, and that is just what the latest observations show.[8]) The Big Bang yields all agree with only one freely adjustable parameter, ρ_b. Recent attempts to circumvent this argument,[9] by having variable n/p ratios, fail to fit the Li even when numerous additional parameters are added and fine-tuned.

This narrow range in baryon density for which agreement occurs is very interesting. Let us convert it into units of the critical cosmological density for the allowed range of Hubble expansion rates. From the Big Bang nucleosynthesis constraints[6,10] the dimensionless baryon density, Ω_B, that fraction of the critical density that is in baryons, is less than 0.12 and greater than 0.03; that is, the universe *cannot be finished with baryonic matter*. If the universe is truly at its critical density, then nonbaryonic matter is required. This agrument has led to one of the major areas of research at the particle–cosmology interface, namely, the search for nonbaryonic dark matter.

Another important conclusion regarding the allowed range in baryon density is that it is in very good agreement with the density implied from the dynamics of galaxies, *including their dark halos*. An early version of this argument, using only deuterium, was described over ten years ago.[11] As time has gone on, the argument has strengthened, and the fact remains that galaxy dynamics and nucleosynthesis agree at about 10% of the critical density. Thus, if the universe is indeed at its critical density, as many of us believe, it requires most matter to not be associated with galaxies and their halos, as well as to be nonbaryonic.

With the growing success of Big Bang nucleosynthesis, the finer details of the results were put into focus. In particular, the ^{4}He yield was looked at in detail, since it is the most abundant of the nuclei, and thus in principle it is the one that observers should be able to measure to higher accuracy. In addition, it is very sensitive to the n/p ratio. The more types of relativistic particles, the greater the energy density at a given temperature, and thus a faster cosmological expansion. A faster expansion yields the weak-interaction rates being exceeded by the cosmological expansion rate at an earlier, higher temperature; thus, the weak interaction drops out of equilibrium sooner, yielding a higher n/p ratio. It also yields less time between dropping out of equilibrium and nucleosynthesis at 10^9 K, which gives less time for neutrons to change into protons, thus raising the n/p ratio. A higher n/p ratio yields more ^{4}He.

In the standard calculation we allowed for photons, electrons, and the three known neutrino species (and their antiparticles). However, by doing the calculation for additional species of neutrinos we can see when ^{4}He yields exceed observational limits while still yielding a density consistent with the ρ_b bounds from ^{2}D, ^{3}He, and now ^{7}Li. (The new ^{7}Li value gives the same constraint on ρ_b as the others, thus strengthening the conclusion.) The bound on ^{4}He comes from observations of helium in many different objects in the universe. However, since ^{4}He is not only produced in the Big Bang but in stars as well, it is important to estimate what part of the helium in some astronomical object is primordial—from the Big Bang—and what part is due to stellar production after the Big Bang. To do this we[12] have found that the carbon content of the object can be used to track the additional helium. Carbon is made in the same mass as stars that also produce ^{4}He; thus, as the carbon abundance increases, so must the helium. (Other heavy elements, such as oxygen, have been tried for this extrapolation, but these tend to not focus their production as well on the same type of stars as those that also produce helium.) The extrapolation of helium to zero carbon content in an object should be a good estimate of the primordial helium.

We obtain ~ 0.235 as our best estimate for the mass fraction of helium produced in the Big Bang. The upper bound is what is important here. We formally estimate a three standard deviation bound as 0.247. In particular, it seems clear that the primordial ^{4}He was at least a little less than 25%. Since objects (like our sun) have heavy elements and possibly some associated extrastellar-produced helium, and still have helium abundances of 25%, this certainly seems like a *very* safe upper bound. In fact, if anything our estimates are on the high side due to possible systematic errors (e.g., collisional

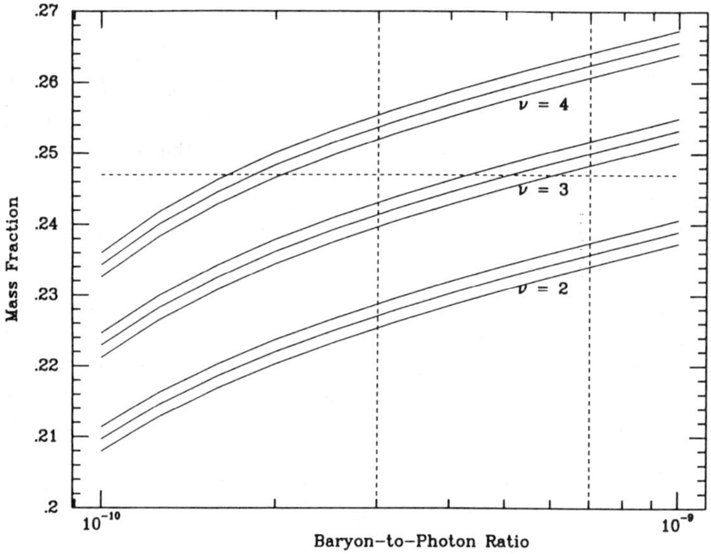

FIGURE 1. Helium mass fraction vs. the baryon-to-photon ratio. The lower bound of 2×10^{-10} is consistent with the ^{3}He + D and ^{7}Li constraints, and the upper bound of 7×10^{-10} with the D and ^{7}Li constraints. The three lines for each neutrino family correspond to neutron half-lives of 10.4, 10.5, and 10.6 minutes.

excitation of helium lines) yielding slight overestimates. (Pagel[13] finds collisional excitation reduces the 0.235 to 0.233.)

We find (see FIGURE 1) that three (or two) types of neutrinos fit the data well, and a fourth only marginally sneaks in if helium exceeds the 3-σ upper bound. Any more neutrinos are strictly prohibited. Since each family contains a neutrino, we are saying that the total number of families is three or at most four. Thus, all the fundamental families of elementary particles may have been discovered already.

It is nice to hear that particle accelerators are beginning to probe this level of sensitivity, and that soon we will know whether or not cosmological theory is able to make reliable predictions about fundamental physics. (It is also intriguing that the recent supernova, 1987a, can set a limit[14] of $\lesssim$ 7 families, or otherwise the $\bar{\nu}_e$ flux would be too diluted to be detectable. Thus we have an independent indication that the number of neutrino families is small.)

ACKNOWLEDGMENT

We would like to acknowledge the hospitality of the Max Planck Institut für Physik und Astrophysik, where this manuscript was written, and the Alexander von Humboldt Foundation for support while in West Germany. We would also like to thank Larry Kawano for his help in the recent calculations, and Gary Steigman for many discussions on the subject.

REFERENCES

1. Steigman, G., D. Schramm & J. Gunn. 1977. Phys. Lett. **B66:** 202.
2. Hoyle, F. & R. Taylor. 1964. Nature **203:** 1108.
3. Peebles, P. J. E. 1966. Phys. Rev. Lett. **16:** 410.
4. Wagoner, R., W. Fowler & F. Hoyle. 1967. Astrophys. J. **148:** 3.
5. Fowler, W., J. Greenstein & F. Hoyle. 1962. Geophys. J.r.A.S. **6:** 148.
6. Yang, J., M. Turner, G. Steigman, D. Schramm & K. Olive. 1984. Astrophys. J. **281:** 493. (And references therein.)
7. Epstein, R., J. Lattimer & D. Schramm. 1976. Nature **263:** 198.
8. Kawano, L., D. Schramm & G. Steigman. Astrophys. J. In press.
9. Scherrer, R., J. Applegate & C. Hogan. 1987. *In* Proc. 13th Texas Symp. Relativistic Astrophysics. M. Ulmer, Ed. World Scientific, Singapore.
10. Freese, K. & D. Schramm. 1984. Nucl. Phys. **B233:** 167.
11. Gott, J., J. Gunn, D. Schramm & B. Tinsley. 1974. Astrophys. J. **194:** 543.
12. Gallagher, J., G. Steigman & D. Schramm. Submitted for publication in Astrophys. J.
13. Pagel, B. 1987 *In* Proc. 1987 Recontre de Moriond, Astrophysics. T. Thran, Ed.
14. Schramm, D. 1987. *In* Proc. 1987 Recontre de Moriond, Astrophysics, T. Thran, Ed.

Neutrino Counting from $\bar{p}p$ and e^+e^- Machines

N. G. DESHPANDE

Institute of Theoretical Science
University of Oregon
Eugene, Oregon 97403

INTRODUCTION

Three methods have been helpful in setting useful limits on N_ν, the number of neutrino species. These methods are complementary to each other and do not measure the same quantity except in the Standard Model. The methods are as follows.

Nucleosynthesis and Cosmology

This method relates primordial helium abundance to the number of neutrino species. It is sensitive to light neutrinos of mass less than 1 MeV, and the sensitivity to right-handed neutrinos depends on their coupling strength. The traditional bound from this method is $N_\nu < 4$.

Electron–Positron Colliders

The data come from $e^+ + e^- \rightarrow \gamma + \bar{\nu} + \nu$ reaction at $\sqrt{s} = 29$ GeV and 42 GeV. It is sensitive in principle to $m_\nu < \sqrt{s/2}$, although there is a cut on photon energy that reduces the mass limit somewhat. It is sensitive to all species of neutrinos that occur in $e^+ + e^- \rightarrow \bar{\nu} + \nu$ reaction.

Proton–Antiproton Collider

This is an indirect method of determining Γ_Z/Γ_W. It is sensitive to $m_\nu < m_Z/2$ and to only those neutrinos that occur in $Z \rightarrow \bar{\nu}\nu$ decay.

In this paper we focus only on the latter two methods.

N_ν FROM ELECTRON–POSITRON COLLIDERS

The method depends on the process $e^+ + e^- \rightarrow \gamma + \bar{\nu} + \nu$ first suggested by Ma and Okada.[1] The background for the process comes from $e^+ + e^- \rightarrow e^+ + e^- + \gamma$, with $e^+ + e^-$ along the beam direction. This can be eliminated by cuts on $\theta_\gamma^{\min}$ and $p_{T_\gamma}^{\min}$. The cross section can be written in the form

$$\frac{d\sigma}{dx} = f(E_\gamma, p_{T_\gamma}^{\min}, \theta_\gamma^{\min})\sigma(e^+e^- \rightarrow \bar{\nu}\nu)(\tilde{s}) \qquad (1)$$

where $x = 2E_\gamma/\sqrt{s}$ and $\tilde{s} = s(1 - x)$. If there are extra gauge bosons Z_i coupled to the usual neutrinos or to extra neutrinos that occur, for example, in models based on the E_6 group, then these can be incorporated in the $e^+ + e^- \rightarrow \bar{\nu} + \nu$ cross section following Barger *et al.*[2] When the limits on extra Z boson mass from neutral current data are incorporated the extra neutrino count ΔN_ν is always less than 0.6 for the most favorable case with three families of light right-handed neutrinos. This is too small to obtain any useful bounds at present.

The present experimental limits from different experiments as well as combined limits are presented from Levine[3] (see TABLE 1).

We thus have a combined limit of

$$N_\nu < 4.8 \text{ (90\% CL)}. \tag{2}$$

This limit is already useful to put bounds on Majorons, however. The model by Gelmini and Roncadelli[4] has a Higgs triplet with surviving particles χ^{++}, χ^+, χ^0 and M^0, where M^0 is a pseudoscalar Goldstone particle associated with lepton number violation and χ^0 is a scalar particle whose mass is less than 100 keV from astrophysical considerations. $Z_0 \rightarrow M^0\chi^0$ decay yields.[5]

$$\Gamma(Z \rightarrow M^0\chi^0) = 2\Gamma(Z \rightarrow \bar{\nu}\nu). \tag{3}$$

Since M^0 and χ^0 will escape from the detector like neutrinos, the existence of Majorons increases N_ν by two units. We then see that $N_\nu = 5$ is not compatible with the data at 90% CL.

LIMITS FROM $S\bar{p}pS$

Neutrino counting at $S\bar{p}pS$ depends on estimate of Γ_Z/Γ_W using theoretical information on production of $W^\pm$ vs. Z^0. The original method suggested by Halzen and Mursula,[6] and Cline and Rholf[6] has been used by Deshpande *et al.*[7] to limit new physics. The method follows from the equation:

$$\begin{aligned} R_{\text{expt}} &= \frac{\text{number of } e\nu \text{ from } W^\pm}{\text{number of } e^+ + e^- \text{ from } Z^0} \\ &= [\sigma(W^+ + W^-)/\sigma(Z)][\Gamma(W \rightarrow e\nu)/\Gamma(Z \rightarrow e^+e^-)][\Gamma_Z/\Gamma_W]. \end{aligned} \tag{4}$$

The first two ratios in the right-hand side of the equation are determined by using proton structure functions and the Standard Model with $x_W = 0.23$. These are, respectively, 3.3 ± 0.2 and 2.685. For R_{expt} we use the most recent limits[8] obtained by UA2 with the combined data of 142 nb^{-1} at 546 GeV and 768 nb^{-1} at 630 GeV. The limits are

$$\begin{aligned} R_{\text{expt}} &= 7.2 \begin{cases} +1.7 \\ -1.2. \end{cases} \\ &< 9.52 \text{ (90\% CL)} \\ &< 10.42 \text{ (95\% CL)}. \end{aligned} \tag{5}$$

TABLE 1. Experimental Limits[a]

Search	$\sqrt{s}$ (GeV)	Acceptance Cuts	$L\epsilon$ (pb^{-1})	Expected Yield	Observed Yield	N_ν (90% C.L.)
MAC	29.0	$E_{\perp\gamma} > 4.5, 2.0$ 2.6 GeV $\theta_\gamma > 40$	27, 50, 42	1.1	1	17
ASP	29.0	$E > 1$ GeV $\theta_\gamma > 20$	66	2.2	1	7.5
CELLO	42.6	$E_\gamma > 2.1$ GeV $\theta_\gamma > 34$	13	0.7	0	15
Combined				4	2	4.8

[a]From T. L. Lavine.[3]

When combined with theoretical uncertainties we have

$$\frac{\Gamma_Z}{\Gamma_W} < 1.16 \text{ at } 90\% \text{ CL}$$

$$< 1.27 \text{ at } 95\% \text{ CL.} \tag{6}$$

We discuss various consequences of these limits in the next three sections. The discussion is largely based on a recent paper by Barger *et al.*[9]

N_ν and the Mass of the t-Quark

Because Γ_W is a function of t mass, the ratio Γ_Z/Γ_W is a function of m_t and N_ν. We calculate Γ_Z using Standard Model fermion couplings with $x_W = \sin^2\theta_W = 0.23$, and $M_Z = 91.9$ GeV. The contribution of each neutrino, charged lepton, and quark are

$$\Gamma(Z \rightarrow \nu\nu) = \Gamma_Z^0 = (G_F M_Z^3)/(12\pi\sqrt{2}) = 0.17 \text{ GeV} \tag{7a}$$

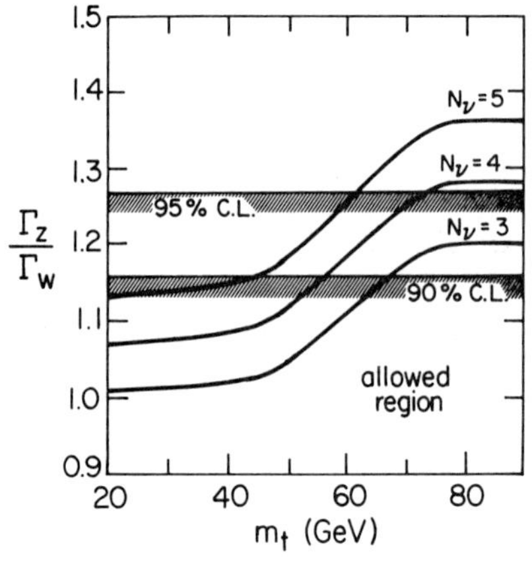

FIGURE 1. Γ_Z/Γ_W predictions assuming no fourth-generation charged fermions compared with the 90% and 95% experimental upper limits; results are shown vs. the t-quark mass, for $N_\nu = 3$, 4, and 5 light-neutrino species.

$$\Gamma(Z \rightarrow \bar{L}L) = \Gamma_z^0 F(m_L^2/M_z^2) \quad \text{(7b)}$$

$$\Gamma(Z \rightarrow \bar{Q}Q) = 3\Gamma_z^0 F(m_Q^2/M_z^2)(1 + \alpha_s/\pi) \quad \text{(7c)}$$

where

$$F(r) = 8\beta[g_V^2(1 + 2r) + g_A^2(1 - 4r)]. \quad \text{(8)}$$

Here

$$\beta = (1 - 4r)^{1/2}, \qquad g_V = T_3/2 - x_W Q, \qquad g_A = -T_3/2.$$

We then find

$$\Gamma_Z = 2.04 + 0.17\, N_\nu + 0.51\, F_t \quad \text{(9)}$$

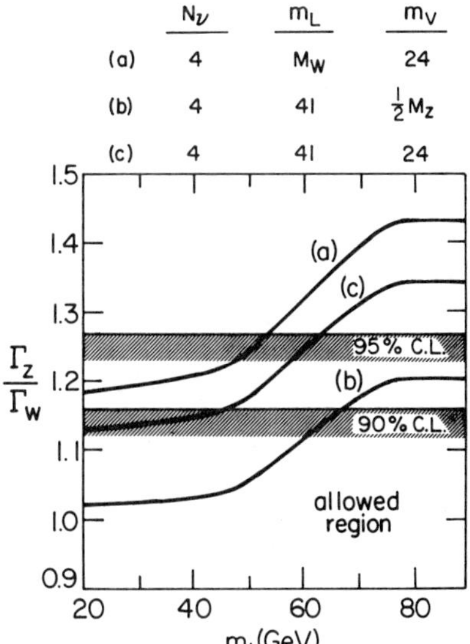

FIGURE 2. Γ_Z/Γ_W predictions for three illustrative cases of fourth-generation fermions. (a) $m_\upsilon = 24$ GeV, $m_L > M_W$. (b) $m_L = 41$ GeV, $m_\upsilon > M_Z/2$. (c) $m_\upsilon = 24$ GeV, $m_L = 41$ GeV. 90% and 95% limits are shown for comparison.

where $F_t = [0.59 - 1.93r_t][1 - 4r_t]^{1/2}$ and $r_t = (m_t/m_Z)^2$.

We calculate Γ_W using $m_W = 80.6$ GeV and the standard formulas

$$\Gamma(W \rightarrow \bar{e}\nu) = G_F M_W^3/6\pi\sqrt{2} = \Gamma_W^0 = 0.23 \text{ GeV} \quad \text{(10a)}$$

$$\Gamma(W \rightarrow \bar{L}\nu) = \Gamma_w^0 H(m_L^2/m_w^2) \quad \text{(10b)}$$

$$\Gamma(W \rightarrow \bar{t}b) = 3\Gamma_W^0(1 + \alpha_s/\pi)H(m_t^2/m_W^2). \quad \text{(10c)}$$

where $H(r) = 1 - 3r/2 + r^3/2$ for $r \leq 1$ and $H(r) = 0$ for $r > 1$. The formula can be summarized as

$$\Gamma_W = 2.12 + 0.72\, H_t. \quad \text{(11)}$$

In FIGURE 1 we plot Γ_Z/Γ_W as a function of m_t for different values of N_ν. The bounds on Γ_Z/Γ_W from the data are also plotted. We note that there is no limit on m_t at 95% CL, while at 90% CL there is a limit $m_t < 68$ GeV. Our conclusions are more conservative than those reached by Halzen,[6] who used bounds that are more stringent. The limit on N_ν, assuming that there are no additional quarks or leptons and assuming $m_t \geq 45$ GeV, is

$$N_\nu < 5 \quad 90\%\ \text{CL}$$

$$N_\nu < 7 \quad 95\%\ \text{CL}. \tag{12}$$

Mass Limits on Fourth-Generation Quarks and Leptons

We now assume there is an additional family with a lepton of mass m_L accompanied by a light neutrino and an additional quark v of charge $-\frac{1}{3}$ and mass m_v. We

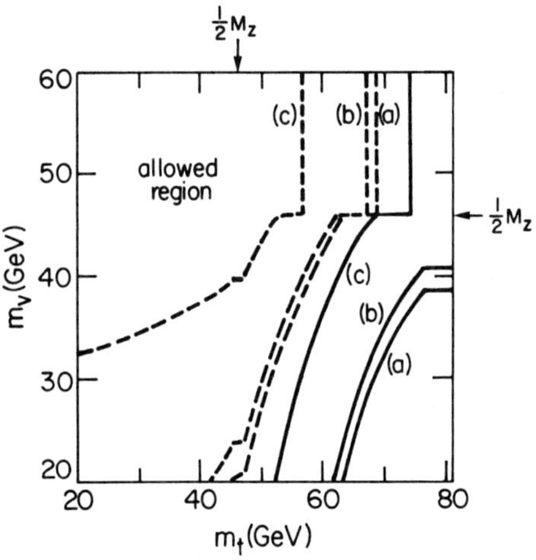

FIGURE 3. Ninety percent limits (dashed curves) and 95% limits (solid curves) in the (m_t, m_v) plane for three illustrative cases. (a) $m_L = 24$ GeV, $m(\nu_L) = 5$ GeV. (b) $m_L = 41$ GeV, $m(\nu_L) = 0$ GeV. (c) $m_L > M_W$. The allowed regions are to the left of the curves.

assume that the additional $\frac{2}{3}$ charge quark is heavier than m_w. The widths for Z and W are now given by

$$\Gamma_Z = 2.04 + 0.17\,N_\nu + 0.51\,F_t + 0.51\,F_v + 0.17\,F_L\ (\text{GeV}) \tag{13}$$

$$\Gamma_W = 2.12 + 0.72\,H_t + 0.23\,H_L\ (\text{GeV}) \tag{14}$$

where $F(r)$ and $H(r)$ are defined as before and $N_\nu = 4$. In FIGURE 2 we plot Γ_Z/Γ_W as a function of m_t for three cases: (1) when $m_L \geq m_W$, $m_v = 24$ GeV, (2) $m_L = 41$ GeV and $m_v = M_Z/2$, and (3) $m_L = 41$ GeV and $m_v = 24$ GeV. We arrive at the interesting conclusion that if $m_v = 24$ GeV, then $m_t < 45$ GeV (60 GeV) regardless of m_L at 90% CL (95% CL).

In FIGURE 3 we plot the allowed regions in the m_v vs. m_t plane for different values of m_L at 90% CL and 95% CL.

CONCLUSIONS

We have reviewed the recent limits on the number of neutrino species N_ν obtained from e^+e^- colliders and from $\bar{p}p$ colliders. In addition, we have studied the consequences of a fourth family of quarks and leptons on neutrino counting at $\bar{p}p$ colliders. As a result, we have drawn useful conclusions at 90% and 95% CL.

1. From $e^+ + e^-$ data $N_\nu < 4.8$ at 90% CL. Majoron excluded at 90% CL.
2. From $\bar{p}p$ data, if $m_t = 45$ GeV and no other light particles, $N_\nu < 5$ (7) at 90% CL (95% CL).
3. If $N_\nu = 3$, $m_t < 68$ GeV at 90% CL and no limit 95% CL.
4. If $N_\nu = 4$ and no other light particles, $m_t < 55$ GeV (75 GeV) at 90% CL (95% CL).
5. If $N_\nu = 4$ and $m_\upsilon = 24$ GeV and m_L arbitrary, $m_t < 45$ GeV (60 GeV) at 90% CL (95% CL).

REFERENCES

1. MA, E. & J. OKADA. 1978. Phys. Rev. Lett. **41:** 287; GAEMERS, K. J. F., R. GASTMANS & F. M. RENERD. 1979. Phys. Rev. **D19:** 1605.
2. BARGER, V., N. G. DESHPANDE & K. WHISNANT. 1986. Phys. Rev. Lett. **57:** 2109.
3. LAVINE, T. L. 1986. Univ. Wisconsin Ph.D. Dissertation **WISC-EX-86/275.**
4. GELMINI, G. B. & M. RONCADELLI. 1981. Phys. Lett. **99B:** 411; GEORGI, H. M., S. L. GLASHOW & S. NUSSINOV. 1981. Nucl. Phys. **B193:** 297.
5. BARGER, V., H. BAER, W.-Y. KEUNG & R. J. N. PHILLIPS. 1982. Phys. Rev. **D26:** 218.
6. HALZEN, F. & K. MURSULA. 1983. Phys. Rev. Lett. **51:** 857; CLINE, D. & J. RHOLF, unpublished.
7. DESHPANDE, N. G., G. EILAM, V. BARGER & F. HALZEN. 1985. Phys. Rev. Lett. **54:** 1757.
8. UA2 COLLABORATION. CERN-EP/87-05.
9. BARGER, V., T. HAN, N. G. DESHPANDE & R. J. N. PHILLIPS. 1987. Univ. Wisconsin Preprint MAD/PH/335. Submitted for publication in Phys. Lett.

Limits on Fourth-Generation Neutrino Masses and Majoron Emission from a Double Beta Decay Search[a,b]

DAVID O. CALDWELL,[c] R. M. EISBERG,[c]
F. W. GOULDING,[d] D. GRUMM,[c] A. R. SMITH,[d]
AND M. S. WITHERELL[c]

[c]*Physics Department*
University of California,
Santa Barbara, California 93106

[d]*Lawrence Berkeley Laboratory*
Berkeley, California 94023

INTRODUCTION

For a fourth-generation neutrino to be a Majorana particle, its mass as a function of its coupling to the first-generation neutrino can be limited by the search for neutrinoless double beta decay ($\beta\beta_{0\nu}$). If it were a Dirac particle, it is a possible candidate to be the constituent of dark matter, and then masses of 20 GeV to 5 TeV can be excluded from the nonobservation of its scattering from nuclei in the Ge detectors that are used to search for $\beta\beta_{0\nu}$. A particular form of $\beta\beta_{0\nu}$ induced by the emission of a Majoron has recently been reported, but the results given here disagree with that observation.

DISCUSSION

A low-energy process like neutrinoless double beta decay[1,2] ($\beta\beta_{0\nu}$) seems a very unlikely place to look for signs of a fourth generation, but in fact this second-order weak process can set important limits on the mass of a fourth-generation Majorana neutrino as a function of its coupling to a first-generation neutrino. Even the double beta decay apparatus can set interesting limits on the mass of a fourth-generation Dirac neutrino, if that particle also constitutes the dark matter in our galaxy. Raby and West[3] have proposed a variation on that idea in which the same particle also solves the solar neutrino problem, making the limit even more interesting. At the request of the organizers of the Symposium for the Fourth Family of Quarks and Leptons, we have added a topic only indirectly related to the fourth generation to our discussion: Some remarks on whether the UCSB/LBL experiment confirmed the reported[4] possible observation of $\beta\beta_{0\nu}$ induced by majoron emission.

[a]This work was supported in part by the U.S. Department of Energy.
[b]The authors of this paper are listed in alphabetical order.

The nucleus ^{76}Ge is a candidate for $\beta\beta$ decay and constitutes 7.8% of normal Ge, which can be made into an excellent detector of electron energy. The $\beta\beta$ decay would then be observed as the sum of the energies of the two electrons emitted. The ~0.1% energy resolution of a Ge detector is particularly useful in the search for the $\beta\beta_{0\nu}$ decay, $^{76}\text{Ge} \rightarrow {}^{76}\text{Se} + 2e^-$, the evidence for which would be a spike at the end-point energy of 2.041 MeV. The energy resolution of the Ge is of little use in searching for the $\beta\beta_{0\nu,B}$ decay, $^{76}\text{Ge} \rightarrow {}^{76}\text{Se} + 2e^- + B$, which gives a three-body spectrum that peaks at about 1.55 MeV. Here B is the massless Goldstone boson, which we shall hereafter refer to as a Majoron. The main reason any meaningful result can be presented for the $\beta\beta_{0\nu,B}$ decay is that remarkably low backgrounds have been achieved with the Ge detectors.

Among the Ge experiments the main distinction in method of background suppression is whether or not an active NaI shield is used. NaI and associated phototubes are not as free of radioactivity as other materials near the Ge detectors, and despite the self-vetoing, systems with NaI tend to display more and larger full-energy peaks than do those without NaI. However, the peaks themselves do not interfere with observing $\beta\beta$ decay. Rather it is the low-energy tails from those peaks that raise backgrounds, and the NaI provides typically an order of magnitude suppression of the Compton tail. This comes about not only because the Compton scattered photon in most cases exits from the Ge and enters the NaI to veto the event, but also because many of the initial photons are part of a cascade decay, and any of the other time-coincident photons can also veto the event. This suppression is important because the number of counts in a Compton tail can be many times the number of counts in the peak, and this peak/Compton ratio is difficult to model, since it is very dependent on the location of the source relative to the Ge and the presence of any intervening material. The UCSB/LBL group has made extensive measurements with sources in different locations with respect to the Ge to determine the shapes of the low-energy tails of peaks resulting from both Ge detector charge-collection inefficiencies and Compton scattering.

The UCSB/LBL experiment[5] 200-m underground in the powerhouse of the Oroville, California, dam has used between 4 and 8 Ge detectors, averaging about 160 cm^3 (0.9 kg) of fiducial volume each, inside a 15-cm-thick complete Pb shield. The data presented here represent a total mass × live-time of 7.75 kg · y. This is by far the largest data sample available for studying the $\beta\beta_{0\nu}$ process. On the basis of the fluctuation allowed in the background of 1.4 counts/keV · kg · y in an energy region around the expected peak at 2.041 MeV we get a 68% C.L. lifetime limit $T_{1/2}(\beta\beta_{0\nu}) > 6 \times 10^{23}$ y. However, there is a dip in the energy region where the peak is expected, so a maximum likelihood analysis results in a larger lifetime limit. Using the analysis procedure recommended by the Particle Data Group,[6] we obtain a limit of 8×10^{23} y.

The result is most often thought of as giving a *lower* limit on the effective mass of the neutrino in the $\beta\beta_{0\nu}$ process. However, if a positive result were observed at this lifetime value (and we choose 6×10^{23} y to be conservative), then a neutrino would have to exist with this or a *larger* neutrino mass.[7] To show the effect of using different nuclear matrix elements (calculated in the references given) to interpret the result, we give the neutrino mass limits to more significant figures than the uncertainty in the lifetime warrants: $\langle m_\nu \rangle < 1.7$ (reference 1), 1.2 (reference 8), 0.66 (reference 9) eV. These values apply to the case in which this is the only contributor to $\beta\beta_{0\nu}$.

Of more interest here is the application of this result to putting constraints on a fourth-generation neutrino. Because lepton number is not conserved in the $\beta\beta_{0\nu}$ process, the neutrino involved in the intermediate state must be self-conjugate; that is limits are set on Majorana neutrinos. Because the two electrons emitted would both be left-handed, the $\beta\beta_{0\nu}$ process also requires a helicity reversal in the intermediate state, and this can be provided if the neutrino has mass. The fourth-generation neutrino can supply that mass, M_ν, if it mixes with the first-generation neutrino with a probability $(U_e^L)^2$. Thus the result given previously also sets a limit on these parameters, as shown in FIGURE 1. Values above and to the left of the diagonal line (which uses the calculation of reference 1) are excluded, and these are generally much better limits than those from other experiments, also shown in the figure.

If the fourth-generation neutrino were a Dirac particle, the $\beta\beta_{0\nu}$ process would provide no constraints, but the apparatus used to search for $\beta\beta_{0\nu}$ still can provide limits if this particle were also the main constituent of the dark matter in our galaxy. The background rates at the lowest energies measured in the Ge detectors sets limits on the scattering of the dark matter candidates from the Ge nuclei.[10] Such a massive particle gives a low value of energy deposition, E_d, only a portion of which results in ionization energy, E_i, which is registered by the Ge detector. Fortunately, measurements[11] of E_i/E_d have been made by neutron scattering in Ge down to $E_i = 2$ keV ($E_d \approx 10$ keV), and the results agree with theory.[12]

The Dirac neutrino would interact via Z^0 exchange, giving coherence in the nucleus, and we assume it is in the class of weakly interacting massive particles (WIMPs) having initial particle–antiparticle asymmetry. Counting rates vs. E_i are plotted in FIGURE 2 for neutrino masses of 15 and 20 GeV using a halo density of 7×10^{-25} g/cm^3, a Maxwellian velocity distribution with a root mean square average of 300 km/s, and the known[10] interaction cross section. These can be compared with the data shown, which were taken in a two-week test run with one detector near which a half gram of In had been removed, since it had been found that the rapidly rising background at low energies was due to the In 486 keV β^- of half-life 4×10^{14} y! The result was a spectrum, part of which is shown in FIGURE 2, which was flat from 400 keV to 14 keV at a level of ½ a count/keV · kg · day, except for peaks. Below the Ga and Zn X-ray peaks, the noise spectrum rises, but the four lowest energy points (and

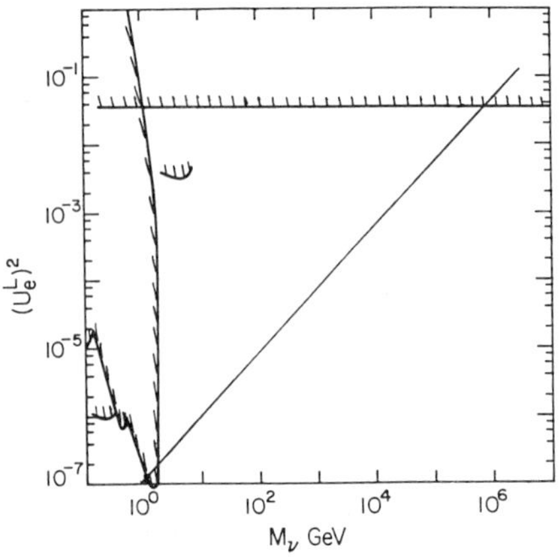

FIGURE 1. Limits on the mass of a heavy Majorana neutrino coupled to a left-handed W boson as a function of its probability for mixing with an electron neutrino.

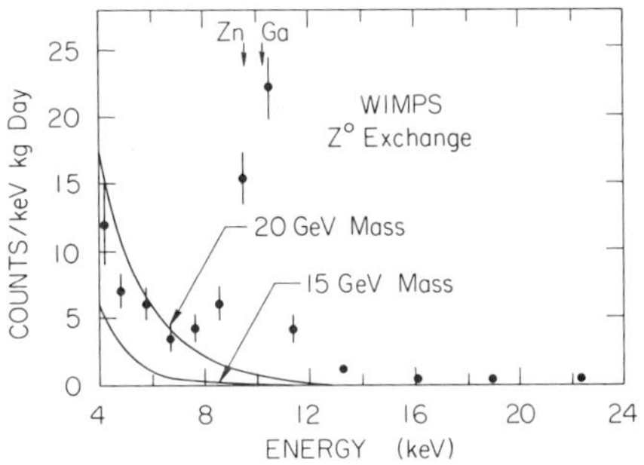

FIGURE 2. Ionization energy detected in one Ge counter compared to that expected for WIMPs of 15 and 20 GeV.

especially the second) lie below the 20-GeV curve, giving that mass as a limit. The mass value is actually double valued, so that the mass region excluded is up to about 5 TeV.

A variant on the standard fourth-generation Dirac neutrino was proposed by Raby and West[3] to have the same particle be the main component of dark matter and also explain[13] the solar neutrino problem. Such a WIMP trapped in the sun would transport energy from the core to larger solar radii, thereby slightly cooling the central region from which most of the neutrinos originate. This mechanism requires cross sections with protons between 10^{-38} and 10^{-35} cm^2, a low enough annihilation rate, and masses roughly in the range of 4 to 10 to 20 GeV, the lower limit to prevent too rapid evaporation and the upper to allow them to travel far enough from the core to cool it efficiently. Such particles could also explain[13] the discrepancy between observations of solar oscillations and the standard solar model, but the required properties are not possessed by conventional or even previously proposed particles. The Raby–West[3] solution is to postulate a neutrino with a large magnetic moment arising from the existence nearby in mass of its charged partner and associated Higgs particle. The resulting electromagnetic interaction with protons has an appropriate cross section $\sim 10^{-36}$ cm^2, but for a nucleus like Ge there is little sensitivity to the electromagnetic cross section for masses as large as 20 GeV. The corresponding slightly higher curve excludes masses larger than ~17 GeV. The difference increases, though, for lower masses, which we hope to reach soon by lowering the threshold below 4 keV.

Neutrinos would be given mass by the majoron,[14] and it can be searched for via $\beta\beta_{0\nu,B}$ decay. The possible observation of that extremely important lepton number violating decay has been reported.[4] One of our eight detectors has a remarkably similar spectrum to that of the detector of reference 4, as shown in FIGURE 3. In both cases there is prominent evidence for an α at 5.3 MeV, which gives a large continuum extending down to the vicinity of 2 MeV. The degraded α's add to the normal background that falls with energy in such a way as to give a nearly flat background from 1.8 to above 3 MeV, making the hump below 1.8 MeV (ascribed to $\beta\beta_{0\nu,B}$) appear more significant than it would if only the falling background were seen, as is the case in FIGURE 4, where data from our other detectors with much lower α background are shown.

In all of these plots full-energy peaks are removed, but their low-energy tails in our data contribute about half the background. About one-fifth of that background comes from ^{68}Ga. The ^{68}Ga activity probably originates from ^{70}Ge(n, $3n$)^{68}Ge, with the ^{68}Ge

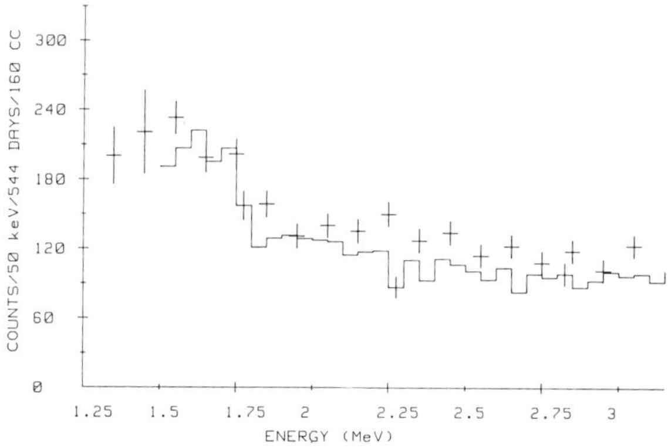

FIGURE 3. Data from the UCSB/LBL detector with the largest α background compared to the PNL/USC data normalized to the same counting time and mass of Ge. The PNL/USC data are shown with error bars, whereas for clarity only one typical error bar is shown for the UCSB/LBL data, which are plotted as a histogram.

decaying by electron capture (280-day half-life) and giving a characteristic Ga X-ray. The daughter, ^{68}Ga, decays with a 68-min half-life, emitting a 1.89 MeV β^+. We first detected this background source by its domination of the region from 0.8 to 1.7 MeV in the coincidence spectrum that results when the annihilation γ's escape and are registered in the NaI. Confirmation of this important source of background is obtained from (1) the size of the Ga X-ray peak, and (2) the rate at which the background level has decreased since the detectors were placed underground, this being consistent with a 280-day half-life. *All* Ge detectors must have this background activity, since it is produced both by cosmic rays when the Ge is aboveground, and probably also because old neutron-damaged Ge detectors may go back into the pool of starting material and contaminate newly produced crystals.

A fit to the data is given in FIGURE 4 using the known ^{68}Ga and tail contribution plus a constant term, although replacing the constant with a term linear or even exponential in energy does not change the results significantly. This one-fourth to one-third of the data represents our ignorance. With more statistics we would very likely identify other peaks or beta spectra. The fit is surprisingly good considering the

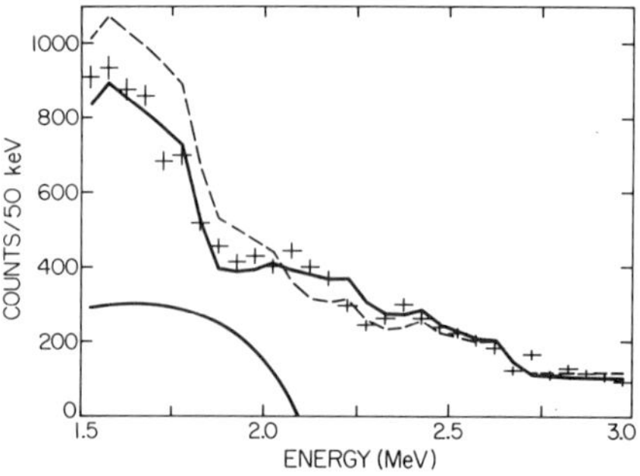

FIGURE 4. Data from this experiment averaged over 50-keV bins with a fit (solid line) to the known background plus a constant term. The dashed curve is a fit requiring the addition of a Majoron-induced decay of $T_{1/2} = 6 \times 10^{20}$ y. The other curve shows the expected shape of the latter decay alone.

complexity of the quite irregular background, the difficulty in modeling the low-energy tails of peaks, and the likelihood that small peaks are still being missed. *Thus, no Majoron-induced decay is needed to explain the data.* To take into account the fact that the systematic errors in describing the background are somewhat larger than the statistical errors, we scale up the statistical errors by a factor of 2.0, which would make the χ^2 per degree of freedom equal to 1. Using these conservatively inflated errors, we get a 90% confidence level upper limit for $T_{1/2}(\beta\beta_{0\nu,B}) > 1.4\ 10^{21}$ y. This would correspond[2] to a limit on the coupling of the Majoron to the electron neutrino of $<7 \times 10^{-4}$ using the matrix elements of reference 1 and $<3 \times 10^{-4}$ using those of reference 9.

If we try to fit the data by including a contribution of Majoron-induced decay at the level of $T_{1/2}(\beta\beta_{0\nu,B}) = 6 \times 10^{20}$ y suggested by the data of reference 4, we get the dashed curve of FIGURE 4. However, this requires that the tail contribution be as small as the errors on the peak heights allow and that the ^{68}Ge contribution go to zero,

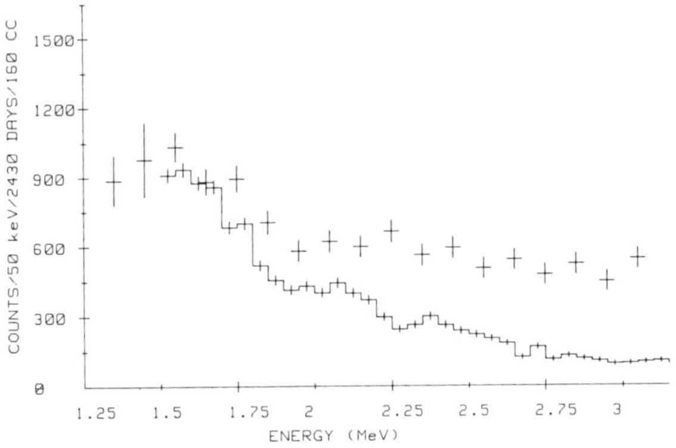

FIGURE 5. Data from the UCSB/LBL detectors (with the one eliminated that was used for FIGURE 3) shown as a histogram compared with the PNL/USC data normalized to the same counting time and mass of Ge. The latter is plotted in 100-keV bins and the former in 50-keV bins.

contrary to fact. Even with these unrealistic conditions, the probability of this curve fitting the data is 10^4 times smaller than the fit with no Majoron-induced decay component.

It will also be noticed that the data do not show the steep rise from the endpoint toward lower energies expected from Majoron-induced decay, as shown by the curve in FIGURE 4. This is the case for both the data of reference 4 (although the statistics obscure the point) or our own, which has an order of magnitude more kilogram · year in this energy range.

We believe that the data of Avignone *et al.*[4] contain more backgrounds than the authors realize, as can be seen by a comparison with our data in FIGURE 5. While they exclude ^{68}Ga as being the *sole* source of the rise in the spectrum, they certainly have some of this activity, as seen by the Ga X-ray in their low-energy data. They also may not identify other γ-ray peaks because of insufficient statistics, and in their case with no Compton suppression the tails of those peaks would contribute many more counts than the peaks themselves.

These various results emphasize the power of double beta decay and the techniques developed to search for it in probing for physics beyond the Standard Model. Unfortunately, so far only minimum lifetime limits have been obtained, and our results do not support a discovery as exciting as double deta decay induced by Majoron emission.

ACKNOWLEDGMENT

The experimental results could not have been obtained without the important help of D. Hale, D. Landis, N. Madden, and R. Pehl. B. Sadoulet was an essential collaborator on the dark matter work.

REFERENCES

1. HAXTON, W. C. & G. J. STEPHENSON, JR. 1984. Prog. Part. Nucl. Phys. **12:** 409.
2. DOI, M., T. KOTANI & E. TAKASUGI. 1985. Prog. Theo. Phys. Suppl. **No. 83**
3. RABY, A. & G. B. WEST. 1987. Nucl. Phys. **B292:** 793; RABY, A. & G. B. WEST. 1987. Phys. Lett. **B194:** 557.
4. AVIGNONE, F. T., III *et al.* 1987. Univ. South Carolina Preprint. (Unpublished.)
5. CALDWELL, D. O. *et al.* 1986. Phys. Rev. Lett. **54:** 281; CALDWELL, D. O. *et al.* 1986. Phys. Rev. D **33:** 2737; CALDWELL, D. O. *et al.* 1987. Phys. Rev. Lett. **59:** 419.
6. See p. 55 of Rev. Part. Prop., Part. Data Group. 1986. (Phys. Lett. **170B:** 1).
7. KAYSER, B. 1986. *In* Proc. 23rd Int. Conf. High Energy Physics. Berkeley, Calif.
8. TOMADA, T. *et al.* 1986. Nucl. Phys. **A452:** 591.
9. GROTZ, K. & H. V. KLAPDOR. 1985. Phys. Lett. **153B:** 1.
10. GOODMAN, M. W. & E. WITTEN. 1985. Phys. Rev. D **31:** 3059.
11. CHASMAN, C. 1970. Penetration of charged particles in matter—A symposium. Nat. Acad. of Sciences, p. 16, and references therein.
12. LINDHARD, J. *et al.* 1963. Kgl. Dan. Vidensk. Selsk., Mat. Fys. Medd. **33:** 10 and 14.
13. GILLILAND, R. L., J. FAULKNER, W. H. PRESS & D. N. SPERGEL. 1986. Astrophys. J. **306:** 703, and references therein.
14. CHIKASHIGE, Y., R. N. MOHAPATRA & R. D. PECCEI. 1981. Phys. Lett. **98B:** 265; GELMINI, G. B. & M. RONCADELLI. 1981. Phys. Lett. **99B:** 411.

$Z^0 \rightarrow N_4\bar{\nu} \rightarrow$ Monojet[a]

SUN HONG RHIE

Stanford Linear Accelerator Center
Stanford University
Stanford, California 94305

INTRODUCTION

Historically, charged leptons have been the first particles to be discovered in each new generation. The charged leptons are lighter than the corresponding quarks, and they give clear signals in detectors. But if the fourth-generation neutrino is massive and decays, it may well be the first to be discovered among the fourth-generation members.

With this motivation, we would like to discuss fourth-generation neutrinos that can be produced from on-shell Z^0 decays at the SLC or LEP. We refer to neutrinos as heavy or light depending on whether the mass is larger or smaller than half of the mass of the Z^0. Let N denote the heavy neutrino.

LIGHT NEUTRINOS

The production rate of light neutrinos from Z^0 decay is

$$\frac{\Gamma(Z^0 \rightarrow \nu_4\bar{\nu}_4)}{\Gamma(Z^0 \rightarrow \nu\bar{\nu})_{|m_\nu=0}} = \begin{cases} \dfrac{1}{4}\beta(3+\beta^2), & \text{Dirac} \\ \beta^3, & \text{Majorana,} \end{cases} \tag{1}$$

where $\beta^2 = 1 - 4m_4^2/M_Z^2$. These neutrinos can be searched for by the increased Z^0 width these decays cause, and more directly by tagging Z^0 decays through detection of the bremsstrahlung photon in the process of $e^+e^- \rightarrow \gamma Z^0 \rightarrow \gamma\bar{\nu}\nu$. If the neutrino is unstable and decays in the detector, we will be able to see a double vertex.

HEAVY NEUTRINOS

Since in this case N is assumed to be heavier than half of the Z^0 mass, $N(\bar{N})$ can be produced, not in pairs, but only in association with a light neutrino $\bar{\nu}(\nu)$. Because in the Standard Model the neutral current is diagonal in flavor at tree level (GIM mechanism), this flavor-changing process can occur only through diagrams involving loops. In these loop calculations, the Higgs content (which determines the type of the

[a]This work was supported by the Department of Energy under Contract DE-AC03-76SF00515.

mass of the neutrino) needs to be specified. Here we assume the following. The gauge group is the standard $SU(2)_L \times U(1)$. The standard Higgs doublet implements the spontaneous symmetry breaking mechanism and also generates Dirac masses for all the fermions, including the neutrinos. The field content of the lepton sector is given as follows:

$$\begin{pmatrix}\nu_e\\ e\end{pmatrix}_L \begin{pmatrix}\nu_\mu\\ \mu\end{pmatrix}_L \begin{pmatrix}\nu_\tau\\ \tau\end{pmatrix}_L \begin{pmatrix}\nu_L\\ L\end{pmatrix}_L$$

$$\nu_{1R} \quad \nu_{2R} \quad \nu_{3R} \quad \nu_{4R} \tag{2}$$

$$e_R \quad \mu_R \quad \tau_R \quad L_R$$

where

$$\nu_\alpha = \sum_i U_{\alpha i}\nu_i \qquad \begin{cases}\alpha = e, \mu, \tau, L, & \text{gauge eigenstate index}\\ i = 1, 2, 3, 4, & \text{mass eigenstate index.}\end{cases} \tag{3}$$

$N \equiv \nu_4$ (heavy), and U is a unitary mixing matrix. We look at the process at one-loop level, where the relevant diagrams in the unitary gauge are shown in FIGURE 1.

THE CALCULATION OF $Z^0 \rightarrow \nu\bar{N}$ OR $\bar{\nu}N$

We can quickly estimate the partial decay width as

$$\frac{\Gamma_i(Z^0 \rightarrow N\bar{\nu}_i)}{\Gamma(Z^0 \rightarrow \nu\bar{\nu})_{|m_\nu=0}} \simeq \left(\frac{g^2}{16\pi^2}\right)^2 \left(\sum_{\alpha=e}^{L} U^*_{\alpha N} U_{\alpha i}\frac{m_\alpha^2}{M_W^2}\right)^2. \tag{4}$$

The factor g^2 is from two extra vertices, $1/16\pi^2$ is a factor one gets from the integral over loop-momentum, and the charged lepton mass factor is due to the coupling of the longitudinal mode of the W to fermion masses. Because the first factor is about 8×10^{-6}, an appreciable production rate can come only when the mass of the heavy charged lepton is much bigger than Mw. The contribution from the light charged leptons can be ignored.

The present experimental upper bounds on the light neutrino masses are known to be[1-3] $m_{\nu_e} \leq 18$ eV, $m_{\nu_\mu} \leq 250$ keV, and $m_{\nu_\tau} \leq 70$ MeV, which are all very light compared to m_N. So neglecting the masses of the light neutrinos, the effective $\nu_i\bar{N}$ neutral current is

$$\epsilon \cdot J = \bar{\nu}_{iL}\left(\epsilon \cdot \gamma F_E + \frac{q \cdot \gamma}{M_Z}\epsilon \cdot \gamma F_M\right)N, \tag{5}$$

where ϵ and q are the polarization and momentum vectors of Z^0. The decay rate that follows from equation 5 is

$$\Gamma_i(Z^0 \rightarrow N\bar{\nu}_i) = \frac{M_Z}{48\pi}(1 - z^2)^2(|F_E|^2(2 + z^2) + 6\,\mathrm{Re}\,(F_E F_M^*)z + |F_M|^2(1 + 2z^2)), \tag{6}$$

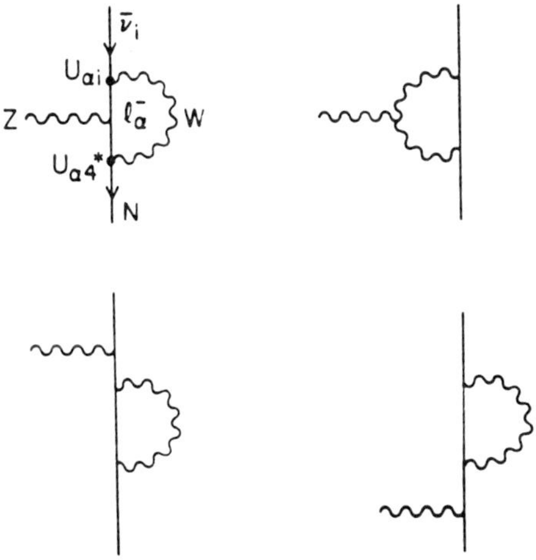

FIGURE 1. Feynman diagrams for $Z^0 \to N\bar{\nu}$ in the unitary gauge.

where $z \equiv m_N/M_Z$. The form factors F_E and F_M are helicity-preserving and helicity-flipping, respectively. We need to compute[4] the diagrams in FIGURE 1 to obtain F_E and F_M. It's well known that the dimensional regularization method is the simplest to use for weak interaction loop calculations, and the apparent incompatibility of the chirality of the theory and the generalized dimension of the regularization scheme can be overcome operationally by using the convention $\{\gamma_5, \gamma_\mu\} = 0$. The loop integrals can be expressed in terms of elementary functions, that is, polynomials, logarithmic functions, and Spence functions (bilogarithmic functions).[5] One can show that the decay amplitude is divergence free in the general linear gauge. Unfortunately, the expression of the amplitude in the general case in terms of elementary functions is so long that a computer is needed to evaluate the expression. FIGURE 2 shows a result of the full calculation at $M_N = 51$ GeV. The cusp is due to the effect of the threshold. The

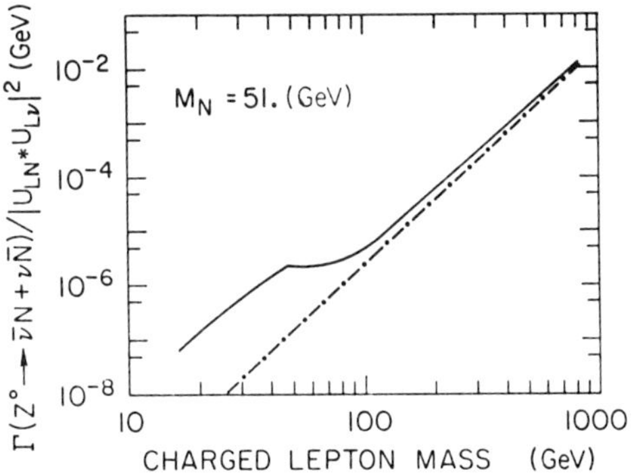

FIGURE 2. The production rate of heavy neutrinos ($m_N > M_Z/2$) from Z^0 decay.

TABLE 1. The Number of Heavy Neutrinos from 10^6 Z^0's

	m_L	
θ	500 GeV	200 GeV
0.22	13	0.4
0.05	0.7	0.02

term F_M is proportional to M_N, and small compared to F_E. In the asymptotic limit, $m_L^2 \to \infty$, $F_M = 0$, and F_E becomes independent of the electric charges of fermions. Therefore all fermions have the same asymptotic behavior, and we have checked that our result agreed with the calculation in the quark sector.[6]

$$\frac{\Gamma_i(Z^0 \to N\bar{\nu}_i)_{m_L^2 \to \infty}}{\Gamma(Z^0 \to \nu\bar{\nu})_{|m_\nu = 0}} = \left(\frac{g^2}{32\pi^2}\right)^2 \left(U_{LN}^* U_{Li} \frac{m_L^2}{M_W^2}\right)^2. \quad (7)$$

The asymptotic value (as $m_L^2 \to \infty$) of equation 7 at $M_N = 51$ GeV is shown by the dash-dot line in FIGURE 2.

There have been similar calculations for quarks[7] and charged leptons.[8] Because of the "symmetry" between the quark sector and lepton sector, the calculations for $\Gamma(Z^0 \to \nu\bar{N})$ and $\Gamma(Z^0 \to q\bar{Q})$ are almost identical. A little difference comes from the different electric charges, and the color degrees of freedom enhance the quark production rate by a factor of 3.

THE NUMBER OF THE HEAVY NEUTRINOS FROM Z^0

Since $U_{LN} \sim 1$, then $|U_{Li}U_{LN}^*|^2 \sim |U_{Li}|^2$. Then the total production rate of heavy neutrinos is

$$\Gamma = \sum_i 2\Gamma_i, \quad (8)$$

which is proportional to $\Sigma_i |U_{Li}|^2$ ($\equiv \theta^2$). TABLE 1 shows the number of heavy neutrinos that can be produced from 10^6 Z^0's for different charged lepton masses m_L and mixing angles θ.

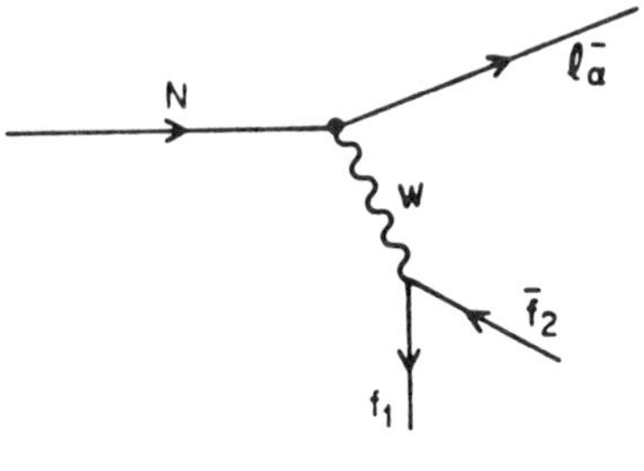

FIGURE 3. The Feynman diagram for heavy neutrino decay.

MONOJET SIGNATURE

Since the light neutrino will be unobserved, the decay of a heavy neutrino will have the signature of a monojet. Depending on the lifetime of the heavy neutrino, we may be able to see a striking signature. So let us look at the lifetime of the heavy neutrino. FIGURE 3 shows the decay diagram. Because of the GIM mechanism, the Z^0 exchange diagram does not contribute to the decay.

The heavy neutrino decay process is essentially the same as μ decay,

$$\frac{\Gamma(N \longrightarrow l_\alpha^- e^+ \nu)}{\Gamma(\mu \longrightarrow e\bar{\nu}\nu)} = \frac{m_N^5}{m_\mu^5} |U_{\alpha N}|^2. \tag{9}$$

The total decay width of the heavy neutrino can be obtained by including all the degrees of the freedom in $N \longrightarrow l_\alpha f_1 \bar{f}_2$ (top is not included).

$$\Gamma_N \simeq 9 \frac{m_N^5}{m_\mu^5} \Gamma_\mu \sum_\alpha |U_{\alpha N}|^2. \tag{10}$$

Therefore the lifetime of the heavy neutrino is

$$\tau_N \simeq \frac{\tau_\mu m_\mu^5}{9 m_N^5 (\Sigma_\alpha |U_{\alpha N}|^2)} \simeq 10^{-22} \frac{M_Z^5}{m_N^5} \left(\sum_\alpha |U_{\alpha N}|^2 \right)^{-1} \text{sec}. \tag{11}$$

For any mixing angles for which there will be a detectable number of $Z^0 \longrightarrow N\bar{\nu}$ decays, the lifetime of the heavy neutrino is extremely small, and the decay not separable from the production vertex. Therefore if a heavy neutrino is to be observed, the signal is a monojet—a jet of decay particles into one hemisphere with the other hemisphere empty.

SUMMARY

We have discussed flavor-changing Z decays involving a heavy fourth-generation neutrino with mass bigger than $M_Z/2$.

REFERENCES

1. FRITSCHI, M. *et al.* 1986. Phys. Lett. **173B:** 485.
2. ALBELA, R. *et al.* 1984. Phys. Lett. **146B:** 431.
3. ALBRECHT, H. *et al.* 1985. Phys. Lett. **163B:** 404.
4. RHIE, S. H. Ph.D. Dissertation, Stanford Univ., Stanford, Calif. To be published.
5. 'T HOOFT, G. & M. VELTMAN. 1979. Nucl. Phys. **B153:** 365.
6. MA, E. & A. PRAMUDITA. 1980. Phys. Rev. **D22:** 214.
7. AXELROD, A. Nucl. Phys. **B209:** 349; GANAPATHI, V. *et al.* 1983. Phys. Rev. **D27:** 579.
8. MANN, G. & T. RIEMANN. 1983. Ann. Phys. (Leipzig) **40:** 334.

Fourth-Family Quark Masses and the Kobayashi–Maskawa Phenomenology from the Regularities in the Structure of Quark Mass Matrices[a]

MICHAEL SHIN[b]

Physics Department
Brown University
Providence, Rhode Island 02912

INTRODUCTION

In 1984, UA1 group at CERN reported[1] the discovery of the long-awaited t-quark in the mass range $m_t = 30$–50 GeV. This experimental fact has not been further confirmed yet, but if it is indeed true, then the final quark of the Standard Model with three families has been found and the search has been completed. However, the Standard Model (and its unified versions, including "the theory of everything") does not predict the number of families, and there exists an open possibility of the existence of the fourth family of quarks (and leptons).

In this paper, we address the following two questions, assuming the existence of these hypothetical fourth-family quarks (which we call t' and b' with charges $+2/3$ and $-1/3$, respectively).

(1) Where should we find t' and b' ($m_{t'} = ?$, $m_{b'} = ?$)?
(2) Can we see any indirect evidence for these quarks in Kobayashi–Maskawa (KM) phenomenology even before the direct observation of these quarks?

One way of determining the allowed mass ranges of t' and b' quarks is to exploit the observed[2] strengths of the KM[3] mixing angles in the first three-generation sector. This could give us some information about the fourth-generation (=family) quarks, since their off-diagonal couplings with the lighter generations in the quark mass matrices affect the KM mixing angles in the lighter three-generation sector. Once a plausible form of the quark mass matrices is given, this task is possible.

TWO ATTRACTIVE FORMS OF QUARK MASS MATRICES

At this time, there exist two attractive forms of quark mass matrices in the literature, which seem to predict the entire 3×3 KM matrix correctly in terms of the quark mass ratios alone and reflect some symmetry principles present beyond the Standard Model. We discuss them briefly in the following subsections.

[a]This work was supported in part by the U.S. Department of Energy under Contract DE-AC02-76ER03130.A020—Task A.

[b]Present address: Department of Physics, University of California, Irvine, Irvine, California 92717.

TABLE 1. Features and Predictions of Two Theoretical Anzatz on Quark Mass Matrices Given in Equations (1) and (2) for the Three-Family Case

	Fritzsch–Shin		Gronau *et al.*
Features	Nearest neighbor interaction ⊗ S.S.B. (spontaneous symmetry breaking) as the origin of CP-violation		Symmetric or antisymmetric structure in flavor space ⊗ nearest neighbor interaction
Predictions on KM angles	$\lvert V_{us}\rvert = \sqrt{\frac{m_d}{m_s} + \frac{m_u}{m_c}}$		$\lvert V_{us}\rvert \simeq \sqrt{\frac{m_d}{m_s}}$
	$\lvert V_{cb}\rvert = \sqrt{\frac{m_s}{m_b}} - \sqrt{\frac{m_c}{m_t}}$		$\lvert V_{cb}\rvert = \sqrt{\frac{m_s}{m_b} - \frac{m_c}{m_t}}$
	$\lvert V_{ub}\rvert = \sqrt{\lvert V_{cb}\rvert^2\left(\frac{m_u}{m_c}\right) + \left(\frac{m_d}{m_s}\right)\left(\frac{m_s}{m_b}\right)^3}$		$\lvert V_{ub}\rvert \simeq \sqrt{\frac{m_u}{m_c}}\lvert V_{cb}\rvert$
	$\delta_{\text{Maiani}} \simeq \frac{\pi}{2}$		$\delta_{\text{Maiani}} \simeq \frac{\pi}{2}$
Prediction on t-quark mass as a function of $\lvert V_{cb}\rvert$	$\lvert V_{cb}\rvert$	m_t^p (≡Constituent Mass)	
	0.04	$42.6^{-8.3}_{+17.6}$ GeV	$28.1^{-4.5}_{+9.0}$ GeV
	0.05	$48.6^{-10.1}_{+22.2}$ GeV	$29.1^{-4.9}_{+9.6}$ GeV
	0.06	$56.3^{-12.6}_{+23.7}$ GeV	$30.1^{-5.2}_{+10.6}$ GeV

Fritzsch–Shin Structure

This structure is due to Fritzsch[4] and Shin,[5] and takes the following form:

$$M_u = \begin{bmatrix} 0 & a & 0 \\ a & 0 & b \\ 0 & b & c \end{bmatrix}, \qquad a = \sqrt{m_u m_c},\quad b = \sqrt{m_c m_t},\quad c = m_t$$

$$M_d = \begin{bmatrix} 0 & ia' & 0 \\ ia' & 0 & b' \\ 0 & b' & c' \end{bmatrix}, \qquad a' = \sqrt{m_d m_s},\quad b' = \sqrt{m_s m_b},\quad c' = m_b \tag{1}$$

where a, b, c, a', b', c' are all real, and the factor of i's in M_d generates the observed CP-violation in $K^0 - \overline{K^0}$ as well as the observed value of the Cabibbo angle

$$\left(\sin\theta_c \simeq \sqrt{\frac{m_d}{m_s} + \frac{m_u}{m_c}}\right).$$

Gronau–Johnson–Schechter Structure

This structure is due to Gronau *et al.*[6] (and is also often called the Stech[7]–Fritzsch[4] hybrid form), and takes the following form:

$$M_u = \begin{bmatrix} 0 & a & 0 \\ a & 0 & b \\ 0 & b & c \end{bmatrix}, \quad M_d = (\text{real constant}) \times M_u + i \begin{bmatrix} 0 & A & 0 \\ -A & 0 & B \\ 0 & -B & 0 \end{bmatrix} \tag{2}$$

where a, b, c, A, B are all real, and the antisymmetric piece in M_d is solely responsible for all of the KM mixing angles and the CP-violating phase. The predictions on the entire KM angles and the t-quark mass, which follow from the structures given in equations 1 and 2, are summarized in TABLE 1. For the computation of the constituent t-quark mass m_t^p($\equiv$ physical t-quark mass observed in colliders), we have used $\Lambda_{\overline{MS}} =$ 100 MeV, and the running quark masses normalized at 1 GeV, m_c(1 GeV) = 1.35 GeV, m_b(1 GeV) = 5.3 GeV, and m_s(1 GeV) = 175 ± 55 MeV. These running quark masses are taken from reference 8. For the central value of m_s(1 GeV) = 175 MeV, m_t^p is predicted to be 42–56 GeV for the first structure, while $m_t^p \simeq$ 28–30 GeV for the second. Higher values of m_s(1 GeV) give lower values of m_t^p for both cases.

GENERALIZATION TO THE FOUR-FAMILY CASE

The structures of the quark mass matrices given in equations 1 and 2 are generalized[9,10,11] to the four-family case as follows:

$$M_u = \begin{bmatrix} 0 & a & 0 & 0 \\ a & 0 & b & 0 \\ 0 & b & 0 & c \\ 0 & 0 & c & d \end{bmatrix}, \qquad \begin{aligned} a &= \sqrt{m_u m_c} \\ b &= \sqrt{m_c m_t} \\ c &= \sqrt{m_t m_{t'}} \\ d &= m_{t'} \end{aligned}$$

$$M_d = \begin{bmatrix} 0 & ia' & 0 & 0 \\ ia' & 0 & b' & 0 \\ 0 & b' & 0 & c' \\ 0 & 0 & c' & d' \end{bmatrix}, \qquad \begin{aligned} a' &= \sqrt{m_d m_s} \\ b' &= \sqrt{m_s m_b} \\ c' &= \sqrt{m_b m_{b'}} \\ d' &= m_{b'} \end{aligned} \tag{3}$$

for the first case, while

$$M_u = \text{same as equation 3}$$

$$M_d = \text{real constant} \times M_u + i \begin{bmatrix} 0 & A & 0 & 0 \\ -A & 0 & B & 0 \\ 0 & -B & 0 & C \\ 0 & 0 & -C & 0 \end{bmatrix} \tag{4}$$

for the second case.

Given these structures, the entire 4 × 4 KM matrix V_{KM} is completely determined by the quark mass ratios alone, implying that V_{KM} is a function of just two unknowns ($m_{b'}$, $m_{t'}$) if we accept the UA1's t-quark mass[1] $m_t^p = 30$–50 GeV. Since the first 3×3 sector of V_{KM} is affected by ($m_{b'}$, $m_{t'}$) and we have rather stringent experimental constraints[2] on the mixing angles in this sector at this time, this leads to some constraints on ($m_{b'}$, $m_{t'}$). Moreover, the KM phenomenologies ($B^0 - \overline{B^0}$ mixing and CP-violations, in particular) can be much different from those of the three-family case due to the contributions from these extra quarks. They thus serve as a probe for the indirect evidence for the existence of these quarks.

The analyses were given in reference 9 for the structure shown in equation 3, and in references 10 and 11 for equation 4. The results are (see reference 9, figs. 1–10),

(i) large possibilities in ($m_{b'}$, $m_{t'}$) plane;
(ii) $m_{b'} \geq 32$ GeV, $m_{t'} \geq 80$ GeV for $|B_K| \geq 0.1$;
(iii) plenty of room for $m_{b'} \leq 50$ Ge V, and $m_{t'} \leq 190$ GeV;
(iv) $B_d^0 - \overline{B_d^0}$ mixing is 5–6 times larger than in the 3-family case for $m_{b'}, m_{t'} \leq 200$ GeV, and may be observable in the near future;
(v) $B_s^0 - \overline{B_s^0}$ mixing will be definitely observable ($R_{odd} \equiv (N^{++} + N^{--})/N^{+-} + N^{-+}) \geq 50\%$) regardless of the fourth-family quark masses;
(vi) CP-violating parameter $l^{\pm}$ ($\equiv (N^{+} - N^{-})/N^{+} + N^{-}$)) in B_d^0 and B_s^0 are both too small to be observed ($l^{\pm} \simeq 0(10^{-4}$–10^{-6})),

for the structure shown in equation 3, while

(i) $m_{b'} \leq 60$ GeV;
(ii) $m_t \leq 40$ Ge V (more probably $m_t \leq 30$ GeV);
(iii) $125 \text{ GeV} \leq m_{t'} \leq 375 \text{ GeV}$, $(m_t/m_b)m_{b'} \leq m_{t'} \simeq (m_t/m_b)\ \sqrt{(m_b/m_s)(m_c/m_t)}m_{b'}$, implying a very narrow region in the ($m_{b'}/m_{t'}$) plane. (See reference 10, figs. 1 and 2).
(iv) for B_s^0, $l^{\pm} \simeq 0(10^{-3}$–10^{-2}), implying a large CP-violation (see TABLE 1 or reference 10);
(v) $B_s^0 - \overline{B_s^0}$ mixing will be definitely observable regardless of the fourth-family quark masses,

for the structure shown in equation 4.

CONCLUSION

In this paper, we have discussed the theoretical expectations on the hypothetical fourth-family of quarks implied by the regularities present in the quark mass matrices. The implications are rich and will be tested by future experiments. In the meantime, we strongly encourage our colleagues with theoretical predilections to test the KM mixing angles predicted by the structures presented here by applying them to all aspects of KM phenomenologies, and narrow down the allowed mass ranges ($m_{b'}$, $m_{t'}$).

SUMMARY

We have given a brief review on two recent theoretical ideas that successfully relate the observed KM angles to the quark mass ratios. We have also presented generalizations to the case with four families of quarks, which allow one to determine the entire 4×4 KM matrix in terms of two unknowns (the fourth-family quark masses $m_{b'}$ and $m_{t'}$). The implications of such hypotheses are rich and will be tested by future experiments.

APPENDIX

In this appendix, we present the formulas for the KM angles predicted by the structure of quark matrices given in equation 3, which we hope will be useful in the future. Defining the running quark mass ratios,

$$x_1 \equiv \frac{m_u}{m_c}, \qquad x_2 \equiv \frac{m_c}{m_t}, \qquad x_3 \equiv \frac{m_t}{m_{t'}},$$

$$y_1 \equiv \frac{m_d}{m_s}, \qquad y_2 \equiv \frac{m_s}{m_b}, \qquad y_3 \equiv \frac{m_b}{m_{b'}}, \qquad \textbf{(A.1)}$$

the entire nine parameters of the 4×4 KM matrix, in the notation of reference 12, are predicted to be

$$S_w \equiv |V_{14}| \simeq \left[\left\{\sqrt{\frac{x_1 x_2}{1+y_3}}\left(\sqrt{\frac{x_3}{1-x_3}} - \sqrt{y_3(1-x_3)}\right) - \sqrt{\frac{y_2 y_3^3}{1-y_3^2}}\,\sqrt{x_1}\right\}^2 + \left(\frac{y_1 y_2^3 y_3^5}{1-y_3^2}\right)\right]^{1/2}$$

$$S_v \equiv \frac{|V_{24}|}{c_w} \simeq \left|\sqrt{\frac{x_2}{1-y_3}}\left\{\sqrt{\frac{x_3}{1-x_3}} - \sqrt{y_3(1-x_3)}\right\} - \sqrt{\frac{y_2 y_3^3}{1-y_3^2}}\right|$$

$$S_u \equiv \frac{|V_{34}|}{c_v c_w} \simeq \frac{1}{c_v}\left|\frac{1}{\sqrt{1+x_3}}\,\frac{1}{\sqrt{1+y_3}}\left\{\sqrt{x_3} - \sqrt{y_3} + \sqrt{\frac{x_2 y_2 y_3^3}{(1-x_3)(1-y_3)}}\right\}\right|$$

$$S_z \equiv \frac{|V_{13}|}{c_w} \simeq \sqrt{\left\{\sqrt{x_1}\left(\sqrt{\frac{y_2}{1-y_3^2}} - \sqrt{\frac{x_2(1-x_3)}{1+y_3}} - \sqrt{\frac{x_2 x_3}{1-x_3}\,\frac{y_3}{1+y_3}}\right)\right\}^2 + \left(\frac{y_1 y_2^3}{1-y_3^2}\right)}$$

$$S_z \equiv \frac{|V_{12}|}{c_w c_z} \simeq \sqrt{y_1 + x_1}$$

$$S_y \equiv \frac{|V_{23}|}{c_v} \simeq \frac{1}{c_v}\left|\sqrt{\frac{y_2}{1-y_3^2}} - \sqrt{\frac{x_2(1-x_3)}{1+y_3}} - \sqrt{\frac{x_2 x_3}{1-x_3}\,\frac{y_3}{1+y_3}}\right|$$

$$\phi_1 \simeq -\frac{\pi}{2} + \tan^{-1}\left[\sqrt{\frac{x_1}{y_1}}\right] - \tan^{-1}[\sqrt{x_1 y_1}] - \tan^{-1}$$

$$\cdot \left[\frac{\sqrt{(y_1 y_2^3/1 - y_3^2)}}{\sqrt{x_1}\,(\sqrt{[y_2/(1 - y_3^2)]} - \sqrt{[x_2(1 - x_3)/(1 + y_3)]} - (\sqrt{[x_2 x_3/(1 - x_3)][y_3/(1 + y_3)]})}\right.$$

$$\phi_2 \simeq \frac{\pi}{2} + \tan^{-1}\left[\sqrt{\frac{x_1}{y_1}}\right] - \tan^{-1}[\sqrt{x_1 y_1}] - \tan^{-1}$$

$$\cdot \left[\frac{\sqrt{(y_1 y_2^3 y_2^5/(1 - y_3^2)}}{\sqrt{[x_1 x_2/(1 + y_3)]}\,(\sqrt{[x_3/(1 - x^3)]} - \sqrt{y_3(1 - x_3)}) - \sqrt{[y_2 y_3^3/(1 - y_3^2)]}\,\sqrt{x_1}}\right]$$

$$\phi_3 \simeq \pi. \qquad \textbf{(A.2)}$$

ACKNOWLEDGMENTS

It is our great pleasure to thank R. Sekhar Chivukula, Jonathan Flynn, and Kyungsik Kang for their collaboration on this subject, and Howard Georgi and Shelly Glashow for valuable discussions when the original work on this subject was done. We also thank David Cline, James Kolonko, and Amarjit Soni for organizing this wonderful conference and letting me chair a session. Finally, we thank Harald Fritzsch for inspiring us through his beautiful summary talk.

REFERENCES

1. Arnison, G. 1984. Phys. Lett. **147B:** 493–508.
2. Gilman, F. J. & K. Kleinknecht. 1986. Phys. Lett. **170B:** 74–75.
3. Kobayashi, M. & T. Maskawa. 1973. Progr. Theor. Phys. **49:** 652–657.
4. Fritzsch, H. 1978. Phys. Lett. **73B:** 317–322.
5. Shin, M. 1984. Phys. Lett. **145B:** 285–291.
6. Gronau, M., R. Johnson & J. Schechter. 1985. Phys. Rev. Lett. **54:** 2176–2179.
7. Stech, B. 1983. Phys. Lett. **130B:** 189–193.
8. Gasser, J. & H. Leutwyler. 1982. Phys. Rep. **87:** 77–169.
9. Shin, M., R. S. Chivukula & J. M. Flynn. 1986. Nucl. Phys. **B271:** 509–530.
10. Kang, K. & M. Shin. 1985. Phys. Lett. **165B:** 383–388.
11. Johnson, R., J. Schechter & M. Gronau. 1986. Phys. Rev. **D33:** 2641–2652.
12. Bottella, F. J. & L. L. Chau. 1986. Phys. Lett. **168B:** 97–104.

Fritzsch Mass Matrix with the Fourth-Generation Quarks on the Renormalization Group Equations

MORIMITSU TANIMOTO

Science Education Laboratory
Ehime University
790 Matsuyama, Japan

INTRODUCTION

We have studied the effect of the fourth-generation quarks on the lighter generation quarks. We take the Fritzsch form for the Yukawa couplings at the grand unified scale, and investigate three cases: the one Higgs doublet model; the two Higgs doublet model; and the supersymmetric model, for the Higgs sector. We calculate the evolutions of mixings among the second-, third-, and the fourth-generation quarks by using the renormalization group equations. We then obtain rather large mixings on the electroweak scale.

DISCUSSION

In the standard grand unified theories (GUTs), the origin of the fermion masses lies in the Yukawa couplings between fermions and the Higgs bosons at the GUT scale, M_X. The mass matrices at low energies (e.g., the electroweak scale, M_W) are connected with the Yukawa couplings at M_X by the renormalization group equations (RGEs) if the perturbative unification and the desert are assumed. As is well known, the Yukawa couplings converge to the infrared fixed points controlled by RGEs.[1,2]

If the Yukawa couplings have large values in the framework of the perturbative unification, they reach near the infrared fixed point at the physical low energy limit, M_W. Since the fourth-generation quarks are expected to have large Yukawa couplings at M_X, their masses are near the value that is given by the fixed point. On the other hand, their mixings with the lighter generation quarks depend on the structure of the Yukawa couplings in the generation space at M_X. Once the matrices of the Yukawa couplings in the generation space are given at M_X, the generation mixings are obtained through the RGEs at the low energy, M_W. Of course, one has not yet found the final answer for the Yukawa coupling matrices, but we know very attractive matrices, that is, the Fritzsch form[3] of the Yukawa couplings, which is given theoretically.

In this paper, we give a priori the Fritzsch form for the Yukawa couplings in the framework of four generations and then study the effect of the heavy fourth-generation quarks on the lighter generation ones. One of the authors has already done the numerical studies of the mixing between the fourth- and third-generation quarks using the Fritzsch matrix.[4] Here we calculate mixing among second, third, and fourth generations. Then we have found that the top quark should exist in the restricted mass regions. Because the fourth-generation quark mass is controlled by the fixed point, our results are different from other analyses[5] using arbitrary fourth-generation quark masses, even if the Fritzsch mass matrices are used.

We deal with the case of the two Higgs doublet model[1] and the supersymmetric model,[2] as in the Higgs sector, in addition to the one Higgs doublet model. In the first two models, we have found that the evolution of the mixing between the fourth-generation quarks and the second (third) ones is clearly contrasted to that of the one Higgs doublet model.

We consider the standard theory based on the group $SU(3)_c \times SU(2) \times U(1)$ with the fourth generation. In the following calculations, we work with the three heavy generations: the second, the third, and the fourth generations. Since the first-generation fermions have extremely small Yukawa couplings, they have practically no effect on our results.

The 3×3 matrices of the Yukawa couplings in the generation space for the up-quarks, the down-quarks, and the charged leptons are denoted by U, D, and E, respectively. Then the RGEs are given in the one-loop approximation as[6]

$$\frac{1}{U}\frac{dU}{dt} = G_U - \mathrm{Tr}\,(3U^\dagger U + 3aD^\dagger D + aE^\dagger E) - \frac{3}{2}(bU^\dagger U - cD^\dagger D),$$

$$\frac{1}{D}\frac{dD}{dt} = G_D - \mathrm{Tr}\,(3aU^\dagger U + 3D^\dagger D + E^\dagger E) + \frac{3}{2}(cU^\dagger U - bD^\dagger D),$$

$$\frac{1}{E}\frac{dE}{dt} = G_E - \mathrm{Tr}\,(3aU^\dagger U + 3D^\dagger D + E^\dagger E) - \frac{3}{2}bE^\dagger E, \tag{1}$$

with

$$t = \frac{1}{16\pi^2}\ln\left(\frac{M_X}{M}\right), \tag{2}$$

where M denotes the running energy scale. The coefficients (a, b, c) are (1, 1, 1), (0, 1, $-1/3$), and (0, 2, $-2/3$) for the one Higgs doublet model, the two Higgs doublet model, and the supersymmetric model, respectively. The values of G_U, G_D, and G_E are given by the linear combinations of the squares of the gauge coupling constants g_1^2, g_2^2, and g_3^2.

We take the quark masses at M_W, which is denoted by m_q^W, as

$$m_b^W = 3.5 \sim 3.7\ \text{GeV}, \qquad m_c^W = 0.76 \sim 0.83\ \text{GeV}, \qquad m_s^W = 70 \sim 130\ \text{MeV}. \tag{3}$$

The τ-lepton and muon masses are, respectively, $m_\tau^W = 1.78$ GeV and $m_\mu^W = 106$ MeV by neglecting the QED correction. The top-quark mass is taken as a free parameter in this paper. Furthermore, the observed Kobayashi–Maskawa (KM) mixing matrix element V_{cb} is taken as $|V_{cb}| = 0.035 \sim 0.049$[7] on the M_W scale.

The Yukawa coupling matrices in the generation space have not been credibly determined at M_X by any theory, but a very attractive matrix is known, that is, the Fritzsch matrix.[4] After a suitable redefinition of the phases of the quark fields, the

Yukawa couplings between up-quarks and the Higgs scalar ϕ are written as

$$(\bar{u}_0, \bar{c}_0, \bar{t}_0\, \bar{a}_0)_L \begin{pmatrix} 0 & \delta & 0 & 0 \\ \delta & 0 & \gamma & 0 \\ 0 & \gamma & 0 & \beta \\ 0 & 0 & \beta & \alpha \end{pmatrix} \phi \begin{pmatrix} u_0 \\ c_0 \\ t_0 \\ a_0 \end{pmatrix}_R + \text{h.c.} \qquad (4)$$

This symmetric matrix U can be diagonalized by a suitable rotation in the quark space described by an orthogonal matrix R_U. Then the KM matrix V[8] at the M_X scale is given by

$$V = [R_U]^T \begin{pmatrix} 1 & 0 & 0 & 0 \\ 0 & e^{i\sigma} & 0 & 0 \\ 0 & 0 & e^{i\tau} & 0 \\ 0 & 0 & 0 & e^{i\eta} \end{pmatrix} R_D, \qquad (5)$$

where σ, τ, and η are phases that are not absorbed into quark fields.

In the following calculations, we take $\sigma = \tau = \eta = \pi/2$ $(-\pi/2)$ for the four-generation scheme as used in reference 5. Furthermore, we neglect the first-generation fermions because their effect on the RGEs is very small.

First, the smallest model with one Higgs doublet is studied. Let us show the numerical results. The evolutions of $|V_{ab}|$ and $|V_{as}|$ are shown in FIGURE 1, where U_{33} $(M_X) = D_{33}$ $(M_X) = 3.0$, m_t (physical) $= m_t^p = 50$ GeV, and the relevant m_s^W are taken. As the energy scale M decreases from M_X to M_W, both magnitudes of the mixings V_{ab} and V_{as} are reduced to about 60% at M_W. The behavior of the mixing V_{ab} was already shown in our previous paper.[4]

Since the calculated value of V_{cb} depends fairly on the top-quark mass, the allowed region of the top-quark mass is obtained by use of the experimental data of $|V_{cb}|$ $(0.035 \sim 0.049)$.[7] Two allowed regions are obtained as follows:

$$23\text{ GeV} \lesssim m_t^p \lesssim 30\text{ GeV}, \qquad 38\text{ GeV} \lesssim m_t^p \lesssim 95\text{ GeV}, \qquad (6)$$

because the s-quark mass is allowed in a wide region due to the large experimental error.

Next, we investigate the two Higgs doublet model, in which a scalar doublet couples to leptons and down-quarks, and another doublet couples to up-quarks. The evolution of the mixings are shown in FIGURE 1. These behaviors are similar to the supersymmetric model. Two regions of m_t^p are obtained, as in the case of the one Higgs doublet model, as follows:

$$23\text{ GeV} \lesssim m_t^p \lesssim 28\text{ GeV}, \qquad 36\text{ GeV} \lesssim m_t^p \lesssim 73\text{ GeV}. \qquad (7)$$

These regions are somewhat dependent of the value v_D/v_U. If we take a smaller value, $v_D/v_U = 0.2$, the allowed ranges are rather wide, as

$$23\text{ GeV} \lesssim m_t^p \lesssim 31\text{ GeV}, \qquad 39\text{ GeV} \lesssim m_t^p \lesssim 89\text{ GeV}. \qquad (8)$$

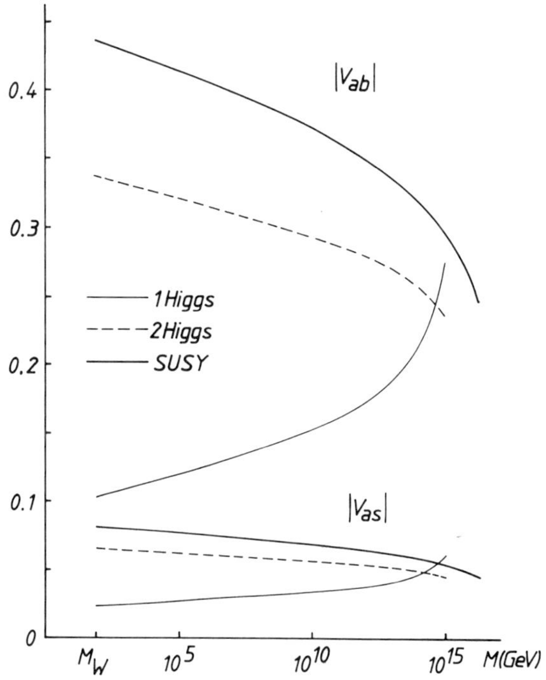

FIGURE 1. Evolutions of V_{as} and V_{ab} for the three models. Here $m_t^p = 50$ GeV and U_{33} $(M_X) = D_{33}$ $(M_X) = 3.0$ are taken.

In the last step, we discussed the case of the supersymmetric model with minimum two Higgs doublets. The evolutions of mixings are shown in FIGURE 1, where we have assumed that $U_{33}\,(M_X) = D_{33}\,(M_X) = 3.0$, $m_t^p = 50$ GeV, and $v_D/v_U = 1.0$. As the energy scale M decreases from M_X to M_W, both magnitudes of the mixings V_{ab} and V_{as} increase to about 180%, which is similar to the two Higgs doublet model.

We show the allowed regions of the physical top-quark mass in the case of $v_D/v_U =$

TABLE 1. The Allowed Ranges of V_{as} and V_{ab} for the Three Models

	$m_t^p = 40$ GeV	$m_t^p = 50$ GeV	$m_t^p = 70$ GeV
One Higgs	$(0.5 \leqq D_{33}(M_X) \leqq 3.0)$	$(0.5 \leqq D_{33}(M_X) \leqq 3.0)$	$(0.5 \leqq D_{33}(M_X) \leqq 3.0)$
	$0.006 \leqq \|V_{as}\| \leqq 0.021$	$0.009 \leqq \|V_{as}\| \leqq 0.023$	$0.013 \leqq \|V_{as}\| \leqq 0.027$
	$0.027 \leqq \|V_{ab}\| \leqq 0.088$	$0.045 \leqq \|V_{ab}\| \leqq 0.103$	$0.072 \leqq \|V_{ab}\| \leqq 0.126$
Two Higgs	$(0.2 \leqq v_D/v_U \leqq 1.0)$	$(0.2 \leqq v_D/v_U \leqq 1.0)$	$(0.2 \leqq v_D/v_U \leqq 1.0)$
	$0.018 \leqq \|V_{as}\| \leqq 0.061$	$0.026 \leqq \|V_{as}\| \leqq 0.068$	$0.038 \leqq \|V_{as}\| \leqq 0.077$
	$0.084 \leqq \|V_{ab}\| \leqq 0.282$	$0.137 \leqq \|V_{ab}\| \leqq 0.339$	$0.220 \leqq \|V_{ab}\| \leqq 0.440$
Supersymmetric	$(0.2 \leqq v_D/v_U \leqq 1.0)$	$(0.2 \leqq v_D/v_U \leqq 1.0)$	$(0.2 \leqq v_D/v_U \leqq 1.0)$
	$0.020 \leqq \|V_{as}\| \leqq 0.076$	$0.032 \leqq \|V_{as}\| \leqq 0.083$	$0.048 \leqq \|V_{as}\| \leqq 0.097$
	$0.098 \leqq \|V_{ab}\| \leqq 0.358$	$0.168 \leqq \|V_{ab}\| \leqq 0.437$	$0.286 \leqq \|V_{ab}\| \leqq 0.636$

1, which are

$$23 \text{ GeV} \lesssim m_t^p \lesssim 28 \text{ GeV}, \qquad 34 \text{ GeV} \lesssim m_t^p \lesssim 61 \text{ GeV}. \tag{9}$$

These regions are also dependent of the value of v_D/v_U. If we take $v_D/v_U = 0.2$ instead of $v_D/v_U = 1$, we get the allowed region, as

$$23 \text{ GeV} \lesssim m_t^p \lesssim 30 \text{ GeV}, \qquad 38 \text{ GeV} \lesssim m_t^p \lesssim 81 \text{ GeV}. \tag{10}$$

In TABLE 1, we summarize the allowed range of V_{ab} and V_{as} for the three models.

SUMMARY

We find that the renormalization effect is very important if the heavy fourth-generation quarks exist. It is noticed that the magnitude of the mixings V_{ab} and V_{as} depends on the models under consideration as to the Higgs sector. In the case of large Yukawa couplings, they reach the fixed point of RGEs, and the evolutions of the mixings V_{ab} and V_{as} are remarkable. The order of the mixing V_{ab} is $O(\lambda) \sim O(\lambda^2)$ in the one Higgs doublet model, and $O(\lambda)$ in both the two Higgs doublet model and the supersymmetric one, where $\lambda(= 0.23)$ is the Cabbibo angle. On the other hand, the mixing V_{as} is $O(\lambda^2) \sim O(\lambda^3)$ in the one Higgs doublet model and $O(\lambda^2)$ in both the two Higgs doblet model and the supersymmetric one. Our result is, of course, not general in the four-generation scheme because it depends on the specific form of the mass matrix, the Fritzsch form. However, we believe that the successful model in the three-generation scheme could be enlarged to the four-generation scheme. So, assuming the fourth generation, we have examined the model to the extent it gives the mixings V_{ab} and V_{as}, and then have indeed obtained the large mixings according to expectation. It is worth noticing that these mixings may affect the $B_s^0 - \overline{B}_s^0$ mixing significantly. Also, since the mass value of the top quark is highly restricted in the Fritzsch matrix, the observation of the top quark is very important for our scheme. We expect to observe the top quark and the fourth-generation fermions at TRISTAN, SLC, and LEP in the near future.

ACKNOWLEDGMENT

We are deeply indebted to Mr. Y. Suetake and Prof. K. Senba for the numerical works.

REFERENCES

1. PENDLETON, B. & G. G. ROSS. 1981. Phys. Lett. **B98:** 291; KOMATSU, H. 1981. Progr. Theor. Phys. **65:** 779; HILL, C. T. 1981. Phys. Rev. **D24:** 691; BAGGER, J., S. DIMOPOULOS & E. MASSÓ. 1985. Phys. Lett. **B156:** 357; BAGGER, J., S. DIMOPOULOS & E. MASSÓ. 1985. Phys. Rev. Lett. **55:** 1450; BAGGER, J., S. D. DIMOPOULOS & E. MASSÓ. 1985. Nucl. Phys. **B253:** 397; PASCHOS, E. A. 1984. Z. Phys. **C26:** 235; HALLEY, J. W., E. A. PASCHOS & H. USLER. 1985. Phys. Lett. **B155:** 107; TANIMOTO, M. 1986. Phys. Lett. **181B:** 311; SASAKI, K. 1986. Z. Phys. **32C:** 149.
2. BAGGER, J., S. DIMOPOULOS & E. MASSÓ. 1985. Phys. Rev. Lett. **55:** 397; ENQUVIST, K.,

D. V. Nanopoulos & F. Zwirner. 1985. Phys. Lett. **B164:** 321; Enquvist, K., D. V. Nanopoulos & F. Zwimer. 1986. Phys. Lett. **B167:** 73; Cvetic, M. & C. R. Preitschopf. 1985. SLAC-PUB-3685.
3. Fritzsch, H. 1975. Nucl. Phys. **155B:** 189; Fritzsch, H. 1978. Phys. Lett. **73B:** 317.
4. Tanimoto, M., T. Hayashi, R. Najima & S. Wakaizumi. 1986. Progr. Theor. Phys. **76:** 1098.
5. Shin, M., R. S. Chivukula & J. M. Flynn. 1986. Nucl. Phys. **271B:** 509; Chen, T. P. & L. F. Li. 1985. Phys. Rev. Lett. **55:** 2249.
6. Chen, T. P., E. Eichten & L. F. Li. 1974. Phys. Rev. **D9:** 2259. (See also references 1 and 2.)
7. Review of Particle Properties. 1986. Phys. Lett. **170B:** 74.
8. Kobayashi, M. & T. Maskawa. 1973. Progr. Theor. Phys. **49:** 652.

Charmed Baryon Decay[a]

A. GARCIA, R. HUERTA, AND P. KIELANOWSKI[b]

Departamento de Física
Centro de Investigación y de Estudios Avanzados
Instituto Politécnico Nacional
México City, México

INTRODUCTION

Detailed comparison of the theoretical prediction for $\Lambda_c^+ \rightarrow \Lambda^0 e^+ \nu$ decay and the currently available measurement for $\Lambda_c^+ \rightarrow e^+ \Lambda^0 X$ strongly suggest the existence of a fourth quark generation.

DISCUSSION

The observation of semileptonic decays of baryons containing heavy quarks has barely begun. Only the semi-inclusive branching ratios of Λ_c^+ decays into e^+X, e^+pX, and $e^+\Lambda^0X$ have been measured. In this paper we are interested in comparing the theoretical predictions for $\Lambda_c^+ \rightarrow \Lambda^0 e^+ \nu$ with the measurement[1] of the transition rate of $\Lambda_c^+ \rightarrow e^+ \Lambda^0 X$,

$$R_{\text{exp}} = \left(0.48 \begin{array}{l} +0.36 \\ -0.39 \end{array}\right) \times 10^{11}\ \text{sec}^{-1}. \quad (1)$$

As we show, a sharp discrepancy is seen. Apparently, a reasonable way to understand it is to relax the limitation to three generations of the Standard Model. It then seems that charmed baryon semileptonic decays give a strong signal for the existence of a fourth generation.

Integrating standard formulas,[2] the transition rate for $\Lambda_c^+ \rightarrow \Lambda^0 e^+ \nu$ in terms of form factors is

$$R_0 = |V_{cs}|^2[1.0341 f_1^2 + 0.7648 f_2^2 + 0.6583 f_1 f_2 + 2.7148 g_1^2 + 2.0617 g_2^2 - 4.1439 g_1 g_2] \times 10^{11}\ \text{sec}^{-1}. \quad (2)$$

We have neglected the contributions of the induced scalar and pseudoscalar form factors, f_3 and g_3, and we have used $M_{\Lambda_c} = 2.2812$ GeV, $M_\Lambda = 1.1156$ GeV. In equation 2, f_2 and g_2 are in units of GeV^{-1}.

To obtain the values of the relevant form factors we use the nonrelativistic quark model. For this purpose we follow references 3 and 4. The mass difference of the c and s

[a]This work was supported in part by COSNET (Mexico).

[b]The work of one of the authors (P. K.) was supported in part by Project MR-17 of the Polish Ministry of Science and Technology while the author was at the University of Warsaw, Warsaw, Poland.

quarks is important in determining f_2. According to reference 4, and using $m_c = 1.5$ GeV and $m_s = 0.5$ GeV, we get, $f_2 = 0.33\ \text{GeV}^{-1}$. The weak-electric form factor g_2, which should be zero if flavor $SU(4)$ were exact, must be taken into account now, because $SU(4)$ breaking cannot be ignored. A first estimate for it can be obtained also from reference 4 (but see below). We get $g_2 = 0.22\ \text{GeV}^{-1}$. Then

$$R_0 = 3.07 \times 10^{11}\ \text{sec}^{-1}, \tag{3}$$

where we have used $V_{cs} = 0.974$. This value comes from the current range for it,[5]

$$0.973 < |V_{cs}| < 0.975. \tag{4}$$

To this, so to speak, bare transition rate we must incorporate corrections from quantum chromodynamics (QCD). These have been discussed by different authors.[6,7] For definiteness, we follow reference 7. The corrected transition rate becomes

$$R_{\text{QCD}} = R_0 \times R(r)^{\text{QCD}}, \tag{5}$$

where $r = m_s/m_c$, and $R(r)^{\text{QCD}} = 1 - 2f(r)\alpha_s(m_c^2)/3\pi$. We again take $m_s = 0.5$ GeV, $m_c = 1.5$ GeV, and the renormalization point at 0.5 GeV, as recommended in reference 7. Then the radiative correction factor $f(r)$ is 2.4 and the running coupling constant $\alpha_s(m_c^2)$ is 0.686. All this makes $R(r)^{\text{QCD}} = 0.65$, which together with equation 3, gives[c]

$$R_{\text{QCD}} = 2.00 \times 10^{11}\ \text{sec}^{-1}. \tag{6}$$

Comparing equations 1 and 6, we find a gross discrepancy between theory and experiment, amounting to 4.2 standard deviations. That is, the theoretical prediction for $\Lambda_c^+ \to \Lambda^0 e^+ \nu$ not only saturates but overwhelms the experimental value of $\Lambda_c^+ \to e^+\Lambda^0 X$.

The prediction of equation 6 is most likely to be a minimum. Looking at R_0 in equation 2, we notice that its dependence on g_2 is very strong. In determining the value of g_2 we ignored a rather subtle point.[8] The term g_3 cannot directly contribute to R_0 significantly, but as discussed in detail in reference 8, it reappears indirectly through g_2. Actually, it makes g_2 more negative, because it subtracts from g_2 and g_3 from *PCAC*, which is expected to have the opposite sign to g_1, that is, a negative sign. Thus taking g_3 into account will make R_0 grow appreciably. Also the q^2-dependence of the form factors, other than the one already taken into account through $f(r)$, is very important. This is usually parametrized by phenomenologically introducing a monopole or a dipole q^2-dependence with masses close to those of the relevant vector and axial vector mesons—in this case, around 2 GeV. Such contributions make R_0 grow again, as much as by 30% or 40%.[c] One correction that should make R_0 decrease is the

[c] By way of comparison, we have calculated the prediction for R_0 in the Cabibbo theory. The term g_2 is taken to be zero. Without q^2-dependence of f_1, f_2, and g_1 we obtain $R_0 = 3.76 \times 10^{11}$ sec^{-1}. Assuming a dipole for each of them with $M_V = 2$ GeV, common to f_1 and f_2, and $M_A = 2.2$ GeV for g_1, R_0 becomes $5.05 \times 10^{11}\ \text{sec}^{-1}$. If $M_V = 1.5$ GeV and $M_A = 1.7$ GeV, $R_0 = 6.88 \times 10^{11}$ sec^{-1}. Thus the Cabibbo theory without $SU(4)$ breaking predicts a value for R_0 that is close to the nonrelativistic quark model and also very far from equation 1. The q^2-dependence of the contributing form factors makes the values of R_0 grow appreciably.

wave function mismatch of the initial and final hadrons.[9] This effect has been estimated to be around 10%, but it could be as large as 50% in $\Lambda_c^+ \longrightarrow \Lambda^0 e^+ \nu$. Other nonperturbative effects, such as the Fermi motion, have been shown[7] to be negligibly small at the level of R_0. Therefore, one can reasonably expect that the theoretical prediction of equation 6 is not likely to decrease.

One can then ask: What is the meaning of the preceding discrepancy? The key to the answer to this question may be in equation 2, not in the form factors f_1, f_2, etc., but in the overall coefficient V_{cs}. The value of 0.974 we have used for it is only meaningful if, indeed, only three generations exist; otherwise, its value is loosely determined.[5] In our case, it should come from the equation

$$R_{\text{exp}} = |V_{cs}|^2 R'. \tag{7}$$

Taking $R' = 4.12 \times 10^{11}\ \text{sec}^{-1}$, which comes from equation 6[d] without the factor $(0.974)^2$, we obtain

$$|V_{cs}| = (0.34\ {}^{+0.11}_{-0.19}). \tag{8}$$

This is a surprisingly low value, strongly suggesting the existence of a fourth generation.

This last number may be too stringent, and it may be wiser to get an upper bound. On the one hand, the experimental result may be subject to changes when more refined measurements are available, so it is better to limit oneself to agreement only within three standard deviations. On the theoretical side, however, it is better to assume that the wave function mismatch along with further QCD corrections can make R_{corr} decrease by as much as 50% more, bringing it down to $1.77 \times 10^{11}\ \text{sec}^{-1}$. Accepting this number as a much more conservative lower bound for the theoretical prediction and allowing up to three standard deviations, we obtain

$$|V_{cs}| < 0.91. \tag{9}$$

If a value of V_{cs} as low as the one indicated in equation 8 obtains in the end, then a situation analogous to the detection of the Cabibbo angle in strange hyperon decays would repeat, where its existence was conjectured form low statistics experiments in $\Lambda \longrightarrow pe\nu$ well in advance of a reliable theory to calculate form factors was introduced, that is, low-energy measurements of charmed baryon decays should bring out a systematic suppression factor due to a small value of V_{cs}, around 0.5. On the contrary, if V_{cs} is to be around 0.9, then very high statistics experiments, as well as very refined theoretical calculations will be required to decide from charmed baryon decays whether a fourth generation exists or not. This only reflects the idea that in order to determine precisely the Kobayashi–Maskawa (KM) matrix elements, a good knowledge of form factors is required, and in order to test the theory through form factor prediction, those matrix elements must be known in advance. Of course, all these remarks apply equally well to charmed meson decays, and will be present with three or four generations. We hope that the preceding analysis stimulates more activity in this field, which tends to be neglected theoretically and experimentally.

[d]Considering other corrections to equation 6, see A. García, R. Huerta & P. Kielanowski. CINVESTAV. (Preprint 1987.)

REFERENCES

1. VELLA, E. *et al.* 1982. Phys. Rev. Lett. **48:** 1515. Of course, the agreement between theory and experiment quoted here should be understood in the order-of-magnitude sense, which invites a more detailed comparison between theory and experiment.
2. PIETSCHMANN, H. 1974. Formulae and Results in Weak Interactions. Springer-Verlag. New York. His G_2 is our g_3 and his G_3 is our g_2. All other form factors coincide, except that we use lowercase letters. We have introduced the overall universality factor V_{cs} according to the Kobayashi–Maskawa matrix. (See reference 5 for notation and further references.)
3. LICHTENBERG, D. B. 1978. Unitary Symmetry and Elementary Particles, 2nd ed. Academic Press. New York.
4. DONOGHUE J. F. & B. R. HOLSTEIN. 1982. Phys. Rev. **D25:** 206.
5. GILMAN, F. 1986. *In* Review of Particle Properties, Particle Data Group, Phys. Lett. **170B:** 74.
6. CABIBBO, N. & L. MAIANI. 1978. Phys. Lett. **79B:** 109; CABIBBO, N., G. CORBO & L. MAIANI. 1979. Nucl. Phys. **B155:** 93, and references therein.
7. ALI, A. & E. PIETARINEN. 1979. Nucl. Phys. **B154:** 519.
8. CARSON, L. J., R. J. OAKES & C. R. WILLCOX. 1986. Phys. Rev. **D33:** 1356; KUBODERA, K., Y. KOHYAMA, K. OIKAWA & C. W. KIM. 1985. Nucl. Phys. **A439:** 695.
9. BRACKEN, P., A. FRENKEL & G. KARL. 1981. Phys. Rev. **D24:** 2984.

Status of *CP* Violation Experiments at Fermilab

G. J. BOCK

Fermilab
Batavia, Illinois 60510

INTRODUCTION

This paper presents two preliminary results from *CP* violation experiments at Fermilab. We concentrate on the ongoing experiment to search for direct *CP* violation in *K* decay. An experiment searching for *CP* violation in three-pion decay has preliminary limits that are also reported here.

Fermilab E731, a collaboration of physicists from five laboratories and universities (Chicago, Elmhurst, Fermilab, Princeton, and Saclay),[a] is an experiment intended to study *CP* violation through observations of the decays of neutral *K* mesons to two pions. More specifically the experiment addresses the issue of whether or not *CP* violation is confined to $\Delta S = 2$ interactions.

Neutral kaon phenomenology and the relation of *CP* violation to the Standard Model is well described in many review articles and in the papers presented at the First International Symposium for the Fourth-Family of Quarks and Leptons,[1] and only the highlights are reviewed here. Recall that the physical neutral kaon states ($K_{L,S}$) are almost, but not quite, the *CP* eigenstates ($K_1[CP = +1]$, $K_2[CP = -1]$). Specifically the K_L is: $K_L \sim K_2 + \epsilon K_1$, where $\epsilon \approx 0.002$. The *CP* mixture of the K^0 states results from a small *T* violation in the $\Delta S = 2$ process $K^0 \leftrightarrow \overline{K^0}$ allowed by second-order weak interactions. To date all observations of *CP* violation can be accounted for by the single parameter ϵ describing this imbalance. This is the prediction of the superweak model. There is also a parameter describing $\Delta S = 1$, or "milliweak" *CP* violation, ϵ', which is predicted to be small but nonzero in many models of the electroweak interactions. The usual definitions:

$$\eta_{+-} = \frac{\text{amp}(K_L \rightarrow \pi^+\pi^-)}{\text{amp}(K_S \rightarrow \pi^+\pi^-)}$$

$$\eta_{00} = \frac{\text{amp}(K_L \rightarrow \pi^0\pi^0)}{\text{amp}(K_S \rightarrow \pi^0\pi^0)}$$

expressed in terms of ϵ and ϵ' are:

$$\eta_{+-} = \epsilon + \epsilon', \qquad \eta_{00} = \epsilon - 2\,\epsilon'.$$

[a]The members of the E731 experiment are G. Bock, R. Coleman, R. Daudin, P. Debu, M. Dung, P. Cabezza, G. Gollin, G. Grazer, Y. Hsiung, P. Jarry, K. Nishikawa, J. Okamitsu, R. Patterson, B. Peyaud, E. Swallow, K. Stanfield, R. Stefanski, R. Turlay, B. Winstein, R. Winston, M. Woods, Y. Wah, H. Yamamoto, and T. Yamanaka.

Since the phases of ϵ and ϵ' are nearly equal,[1] we see that a test of the equality of $|\eta_{\pm}| = |\eta_{00}| = \epsilon$ is a test for $\Delta S = 1$ *CP* violation and we can write:

$$\frac{\epsilon'}{\epsilon} \approx \frac{1}{6}\left(1 - \left|\frac{\eta_{+-}}{\eta_{00}}\right|^2\right). \tag{1}$$

A number of recent calculations predict $\epsilon'/\epsilon > +0.002$.[1] Earlier measurements[2,3,4] of ϵ'/ϵ set lower limits about an order of magnitude higher. Motivated partially by the successes of the Standard Model and its natural incorporation of *CP* violation, two recently completed experiments[5,6] and two experiments underway[7,8] address the question of whether all the *CP* violation is confined to this imbalance or whether there may be some "milliweak" *CP* violation in the decay itself.

EXPERIMENTAL TECHNIQUE

This experiment, like its predecessor Fermilab E617 (Chicago/Saclay),[6] makes use of the double-beam technique: two distinct K_L beams are simultaneously incident on a spectrometer designed to detect two pion decays from the K_L. A thick boron carbide regenerator is located in one beam providing a source of K_S. Decays from this "regenerated" beam are used to normalize the *CP* violating decays detected in the neighboring vacuum beam. There are two modes of data collection. In the charged mode the detector records $K^0 \rightarrow \pi^+\pi^-$ decays; in the neutral mode, $K^0 \rightarrow \pi^0\pi^0$. In each mode the decay region and regenerator are the same and data are taken simultaneously from both the regenerated and vacuum beams. The charged mode is run separately from the neutral mode; there are slight changes to the apparatus depending on the mode. Four integrated decay rates are measured in this experiment and combined to produce the following quantity:

$$\left|\frac{\eta_{+-}}{\eta_{00}}\right|^2 = \left\{\left(\frac{N^v_{+-}}{N^r_{+-}}\right)\bigg/\left(\frac{N^v_{00}}{N^r_{00}}\right)\right\} \equiv \frac{R_{+-}}{R_{00}}, \tag{2}$$

where $N^{v(r)}_{+-}$ is simply the number of $\pi^+\pi^-$ events collected from the vacuum (regenerated) beam. The neutral mode data are defined similarly. Combining (1) and (2) we see that

$$\frac{\epsilon'}{\epsilon} = \frac{1}{6}\left(\frac{1 - R_{+-}}{R_{00}}\right).$$

To determine ϵ'/ϵ to 0.001 (statistically) would require roughly 75K events in each mode. In practice, of course, regenerated events are more easily obtained than vacuum events and charged data much easier than neutral, and these types of experiments are limited by the number of neutral, vacuum events.

SYSTEMATIC EFFECTS

Some potentially troublesome systematic effects are eliminated in the double-beam technique. Since both K_L and K_S decays are collected simultaneously in each mode, the measurement of R_{+-}/R_{00} does not directly depend on maintaining precise calibration over long periods of time. The most obvious potential problem might be the lead glass gains. However, while a gain change of 1–2% would change the acceptance for neutral events, to high order, the *ratio* of acceptances will not change. The technique is similarly insensitive to beam rate (accidentals) inefficiencies.

There are, however, several sources of systematic effects that do not cancel in this type of experiment. There is background under the mass peaks in both the charged and neutral mode data. This background results from semileptonic K decays in the charged mode data and $K \rightarrow 3\pi^0$ in the neutral mode sample. This correction is insignificant in the regenerated-beam sample because of the large ratio of K_S/K_L behind the regenerator, but must be made to each of the vacuum-beam samples.

The regenerated-beam samples require a different correction. The experiment is normalized to coherently (0 momentum transfer) regenerated events. There are two other processes that produce K_S in the boron carbide and their contributions to the sample must be subtracted. The largest of these two is the contribution from inelastically produced or regenerated events. Indeed this correction was the largest source of systematic error from E617. Although insignificant in E617, a small correction is necessary to account for elastic or diffractive regeneration, in which a K_L is transformed into a K_S during an elastic scatter in the regenerator.

There are geometrical acceptance differences between the vacuum and regenerated beam events that must be understood. This will amount to about a 4% correction (averaged over all momentum), and will require a good understanding of the spectrometer by Monte Carlo calculation. This acceptance difference arises because of the different longitudinal decay distributions of the K_L and K_S decays. We rely on the known distribution of the $K_L \rightarrow \pi^+\pi^-$ and $\pi^0\pi^0$ decays to check our understanding. Additionally we collected during actual data taking approximately 50 times more $K_L \rightarrow 3\pi^0$ than $K_L \rightarrow \pi^0\pi^0$. These events also have a known decay distribution and provide an invaluable check on the acceptance and performance of the spectrometer.

BEAMLINE AND DETECTOR

The result of E617 was $\epsilon'/\epsilon = -0.0046 \pm 0.0053 \pm 0.0024$, where the first error is statistical and the second is systematic. In order to achieve a precision of 0.001 it is clear that improvements had to be made in the systematics as well as statistics. A statistical limitation of the earlier experiment was the anomolously high flux of soft neutral (not kaons) particles in the beams that produced high singles rates throughout the spectrometer. For E731 a new long-lived neutral beam has been constructed with special attention paid to clean collimation and sweeping that significantly reduces the soft beam halos found in E617. A new primary proton beam for the experiment allows for variable production targeting angles about a nominal 5 mrad for optimizing the neutron/kaon content of the beams. The new beam also takes full advantage of the 800-GeV protons and high duty cycle of the Tevatron to provide a beam with five times more usable (100-GeV region) neutral kaon fluxes per hour than ever before at Fermilab. A typical spill during the last run had 1.0×10^{12} 800-GeV protons per 20-second pulse each minute and about 3.0×10^6 K_L/beam.

The spectrometer shown in FIGURE 1 has greater rate capability, a factor of 6 more geometrical acceptance, and much better background suppression capabilities than the E617 spectrometer. The 800-element lead glass calorimeter used in E617, was restacked in a circular array that, together with an increase in the gap of the analysis magnet and shortening the distance from the decay region to the lead glass, gave a factor of 6 increase in geometrical acceptance over E617. A new high rate 2000-wire drift chamber system (resolution better than 150 microns) was constructed to handle the higher fluxes and improve resolution needed for background suppression. The conversion hodoscope consisted of a 1-mm-thick scintillator plane, and a 0.1-radiation-length sheet of lead (for neutral mode running only), followed by another scintillator plane. The entire assembly was also in vacuum. The neutral trigger required the conversion of a single photon from the K decay. The electron pair was tracked back to the conversion plane to calculate the kaon's transverse momentum. This is then used to determine whether the event came from the vacuum or regenerated

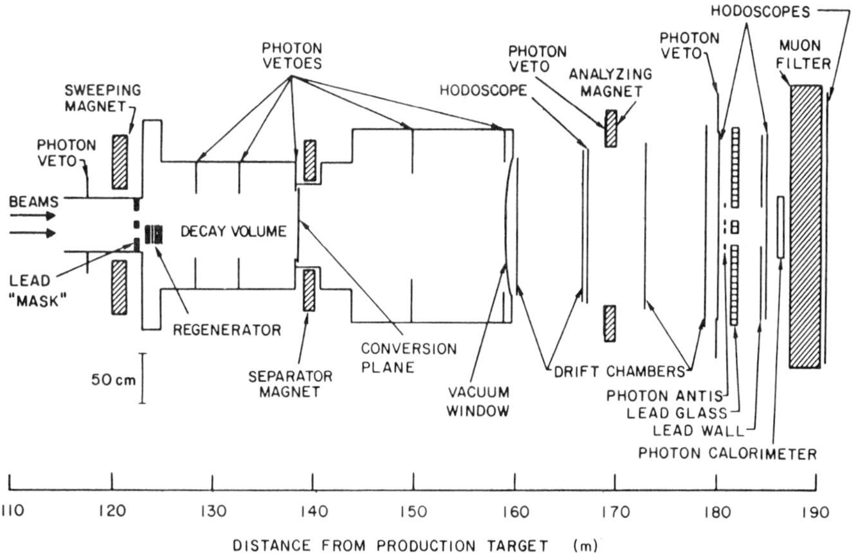

FIGURE 1. The E731 spectrometer. The twin K beams are shown entering from the left. Distances are given in meters from the production target.

beam. Occasionally, a neutron interaction in the lead sheet triggers the spectrometer. A π^0 from such an interaction can be detected in the lead glass calorimeter and serves as an added source of energy calibration during the data taking.

Background suppression was improved by including veto counters within and without the vacuum chamber, and a segmented, active regenerator, including hodoscopes for identifying the inelastic events in the K_S beam. The regenerated beam moves from one beam to the other on a pulse-by-pulse basis to cancel any small asymmetries in the apparatus.

1985 TEST RUN

During the 1985 run approximately 1500 $K_L \rightarrow 2\pi^0$ were collected per (good) week. That is over four times better than the previous experiment, even though there were significant deadtime losses. Approximately 8000 neutral mode and 35,000 charged mode K_L decays have been recorded along with about 24,000 neutral and 100,000 K_S decays. Additionally, we have almost 400,000 $K_L \rightarrow 3\pi^0$ decays taken along with the data for calibration purposes and acceptance checks.

Special momentum analyzed electron pair calibration data have also been taken during the run. Currently, the lead glass resolution calculated from these electron-pair data is 2.0% + $5\%/\sqrt{E}$, an improvement over the 1982 data.

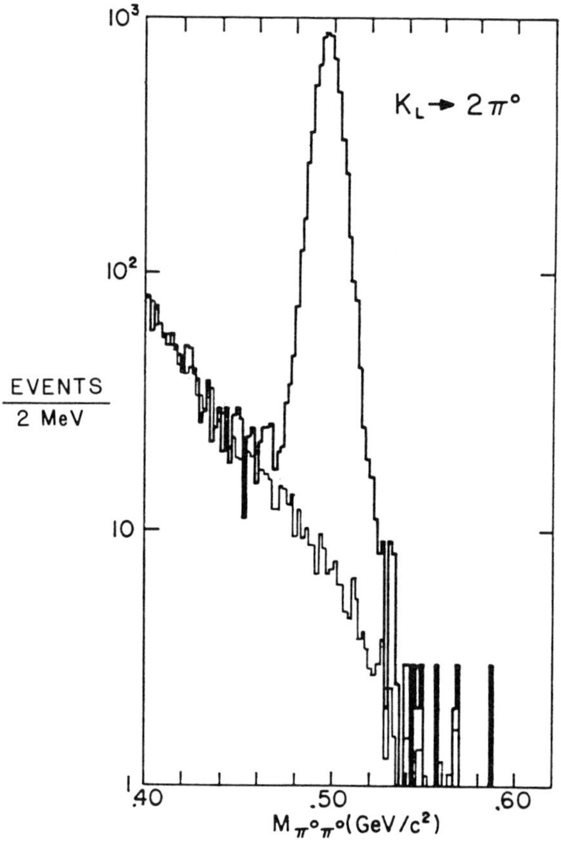

FIGURE 2. The invariant mass distribution for $K_L \rightarrow \pi^0\pi^0$. The 2% background is well described by Monte-Carlo-generated $K_L \rightarrow \pi^0\pi^0\pi^0$ events as shown here and described in the text.

FIGURES 2 and 3 show the K_L neutral and charged mode mass distributions from a sample of the 1985 data. Significant improvement in the background of the neutral mode data has been made using the new photon veto counters. These counters reduce the background from the copious $K_L \rightarrow 3\pi^0$ decays where two of the six photons escape (or one escapes and two fuse) the active area of the detector. The photon vetos are used in the off-line analysis of the data.

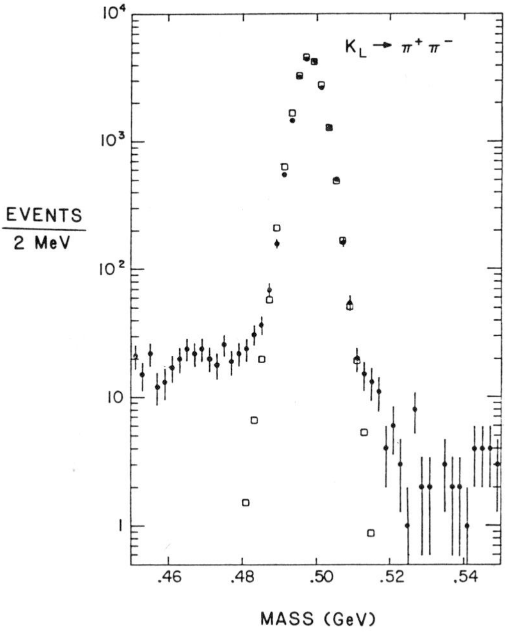

FIGURE 3. The mass distribution for $K_L \longrightarrow \pi^+\pi^-$ events. The open squares are the Monte Carlo events for this decay mode. The background is from misidentified semileptonic K decays.

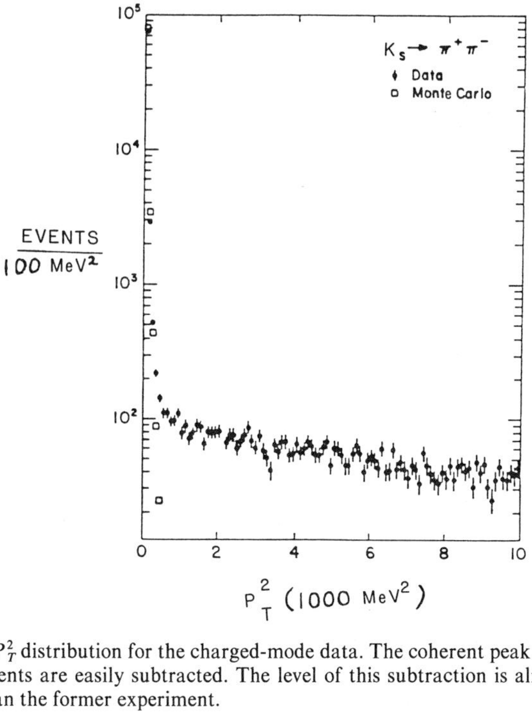

FIGURE 4. The P_T^2 distribution for the charged-mode data. The coherent peak is clearly visible. The diffractive events are easily subtracted. The level of this subtraction is almost an order of magnitude less than the former experiment.

The successful suppression of the inelastic correction in the K_S data by the active regenerator is shown in FIGURE 4. That subtraction was 15% in the E617 data and was the largest source of the systematic error. An order-of-magnitude improvement has been obtained using the active regenerator previously described. Some elastic (diffractive) regeneration remains and is easily subtracted.

PRELIMINARY RESULT

With the data from the 1985 test run, a preliminary result from E731 is

$$\frac{\epsilon'}{\epsilon} = +0.0035 \pm 0.003 \pm 0.002,$$

where the first error is statistical and the second is systematic. The systematic error at

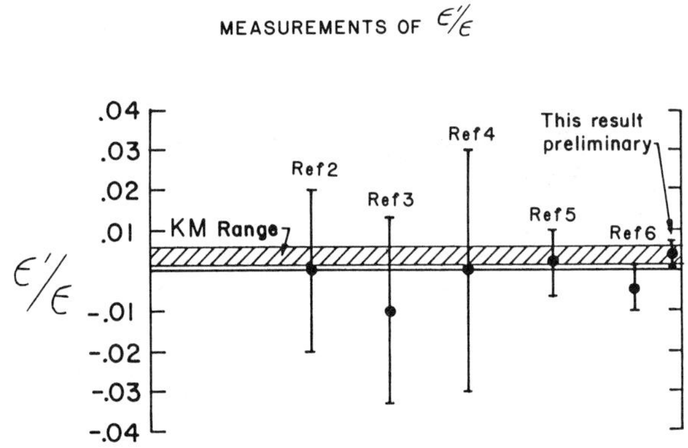

FIGURE 5. A summary of measurements of ϵ'/ϵ, including the preliminary result reported here.

the moment is dominated by uncertainties in the acceptance calculations previously described.

FUTURE WORK

While the analysis on the 1985 data continues, work is underway to ready the experiment for the final data run in June 1988. During that run approximately 100,000 neutral K_L events will be recorded. The order-of-magnitude increase in statistics from the test run will come primarily from three improvements. First, the experiment will run at design intensity ($\times 3$). A new segmented conversion hodoscope is being built to handle the increased rates. A trigger-level cluster finder and a completely new fastbus data acquisition system will eliminate the 50% deadtime problems of the 1985 run ($\times 2$). Finally, the 7000 events from the test run were collected during the final six weeks of the run. The run beginning in June will last for five months ($> \times 2$).

CP VIOLATION IN THREE PION DECAYS

In Fermilab experiment E621, a Rutgers, Minnesota, Michigan, and Wisconsin[9] collaboration is making a measurement of the parameter η_{+-0}, which in analogy to the two-pion case is defined as

$$\eta_{+-0} = \frac{\text{amp}\,(K_S \rightarrow \pi^+\pi^-\pi^0)}{\text{amp}\,(K_L \rightarrow \pi^+\pi^-\pi^0)}\,.$$

By collecting three pion decays in a *V* spectrometer close to a kaon production target, the experimenters look for a small interference between the K_L and K_S decays as a signal of *CP* violation in the $\pi^+\pi^-\pi^0$ decay mode. Acceptance corrections are crucial for the expected value of $\eta_{+-0} \cong \eta_{+-} \cong 0.002$, and the acceptance for K_S decays is monitored by simultaneously collecting the more copious *CP*-conserving $K_L \rightarrow \pi^+\pi^-\pi^0$ decays from a second target located upstream of the K_S target. Based on approximately 100,000 observed three-pion decays from a run in 1984, the group reports preliminary values of $|\eta_{+-0}| = 0.022 \pm 0.026$ and $\Phi_{+-0} = -39° \pm 80°$.[10] This is approximately one order of magnitude of improved sensitivity over previously published results. It is also one order of magnitude away from the expected signal. In hand and under analysis are approximately 3 million events from the 1985 run that should give a statistical improvement of one order of magnitude. It is quite possible that the proponents of E621 will be seeking more data in the future.

SUMMARY

We have presented two preliminary results from Fermilab experiments studying *CP* violation. Emphasis has been placed on the ongoing experiment to measure the ratio of the *CP* violation parameters ϵ'/ϵ to a precision of 0.001 as shown in FIGURE 5. The experimental technique and apparatus were described. We also discussed sources of systematic error and some preliminary data from the test run in 1985 was shown. A preliminary value is: $\epsilon'/\epsilon = +0.0035 \pm 0.003 \pm 0.002$. We also briefly described a second experiment that searched for evidence of *CP* violation in three pion decays—a measurement of the parameter η_{+-0}. We reported preliminary values of $|\eta_{+-0}| = 0.022 \pm 0.026$ and $\Phi_{+-0} = -39° \pm 80°$.

REFERENCES

1. CRONIN, J. 1981. Rev. Mod. Phys. **53:** 371, and references therein for a more complete discussion. See also GILMAN, F. J. 1987. Overview of the Standard Model. This volume.
2. HOLDER, M. *et al.* 1972. Phys. Lett. **B40:** 141.
3. BANNER, M. *et al.* 1968. Phys. Rev. Lett. **21:** 1103.
4. CHRISTENSON, J. *et al.* 1979. Phys. Rev. Lett. **43:** 1209.
5. BLACK, J. K. *et al.* 1985. Phys. Rev. Lett. **54:** 1628.
6. BERNSTEIN, R. *et al.* 1985. Phys. Rev. Lett. **54:** 1631.
7. HOLDER, M. CERN Proposal NA31. Unpublished; HOLDER, M. 1987. Status of *CP* violation experiments at CERN. This volume.
8. GOLLIN, G. *et al.* Fermilab Proposal E731. Unpublished.
9. THOMSON, G. *et al.* Fermilab Proposal E621. Unpublished.
10. THOMSON, G. Private communication.

Status of *CP* Violation Experiments at CERN

M. HOLDER
Fachbereich Physik
Universität Siegen
Siegen, West Germany

INTRODUCTION

We review the status of experiment NA31[1] on the determination of the parameter ϵ'/ϵ in K^0 decay. We then present the first measurement of the relative decay rate $\Gamma(K_S^0 \rightarrow 2\gamma)/\Gamma(K_L^0 \rightarrow 2\gamma) = 2.3 \pm 1.0 \pm 0.4$. Finally, we briefly discuss future plans for a measurement of the phase difference $\varphi_{+-} - \varphi_{00}$ by NA31 and on a K^0 experiment in $p\bar{p}$ annihilation at LEAR.

PRESENT STATUS OF THE *CP* VIOLATION PARAMETERS

The strong interaction eigenstates K^0 and $\bar{K}^0$ with different strangenesses are mixed by the strangeness-nonconserving weak interaction to form the eigenstates K_S^0 and K_L^0 with definite masses and lifetimes. The mixing can be described by two complex parameters, ϵ and δ, with well-defined symmetry properties. In the standard phase convention $CP(K^0) = \bar{K}^0$:

$$K_L^0 = N\{(1 + \epsilon + \delta)K^0 - (1 - \epsilon - \delta)\bar{K}^0\} \qquad \textbf{(1a)}$$

$$K_S^0 = N\{(1 + \epsilon - \delta)K^0 + (1 - \epsilon + \delta)\bar{K}^0\} \qquad \textbf{(1b)}$$

$$N = \frac{1}{\sqrt{2(1 + |\epsilon|^2)}} \qquad \text{for} \quad \delta = 0$$

$$\delta = 0, \qquad \text{if } CPT \text{ is conserved}$$

$$\epsilon = 0, \qquad \text{if } CP \text{ is conserved.}$$

Present experimental evidence is compatible with $\delta = 0$.

In addition to *CP* violation by the mixing of the K^0 and $\bar{K}^0$ states in an effective $\Delta S = 2$ interaction, described by the parameter ϵ, there may be *CP* violation in the $\Delta S = 1$ decay $K_L^0 \rightarrow 2\pi$, described by ϵ'. The K^0 decay amplitudes can be separated in components with definite isospin property of the final state

$$\langle 2\pi|_I H| K^0\rangle = A_I e^{i\delta_I}$$

$$\langle 2\pi|_I H| \bar{K}^0\rangle = A_I e^{i\delta_I}$$

$$I = 0, 2,$$

taking into account the strong interaction of the final state pions by the phase factor $e^{i\delta}$. If *CPT* is conserved, the transition amplitudes A and $\overline{A}$ are related by complex conjugation

$$\overline{A}_I = A_I^*.$$

CPT violation can be expressed by the parameters

$$\alpha_0 = \frac{A_0^* - \overline{A}_0}{A_0 + \overline{A}_0} \tag{2}$$

$$\alpha_2 = \frac{1}{\sqrt{2}} \frac{A_2^* - \overline{A}_2}{A_0 + \overline{A}_0} e^{i(\delta_2 - \delta_0)}. \tag{3}$$

If in addition to *CPT, CP* is also conserved, the amplitudes A_0 and A_2 have the same phase. The phase itself is not observable. The traditional convention[2] is A_0 = real. The phase difference is observable and related to the parameter ϵ':

$$\epsilon' = \frac{i}{\sqrt{2}} \frac{\operatorname{Im} A_2}{A_0} e^{i(\delta_2 - \delta_0)} \tag{4}$$

The measurable complex ratios of decay amplitudes are in the first order of the small quantities ϵ, δ, ϵ', α_0, α_2 given by

$$\eta_{+-} = \frac{\langle \pi^+\pi^- | H | K_L^0 \rangle}{\langle \pi^+\pi^- | H | K_S^0 \rangle} = \epsilon + \epsilon' - \delta + \alpha_0 + \alpha_2 \tag{5}$$

$$\eta_{00} = \frac{\langle \pi^0\pi^0 | H | K_L^0 \rangle}{\langle \pi^0\pi^0 | H | K_S^0 \rangle} = \epsilon - 2\epsilon' - \delta + \alpha_0 - 2\alpha_2 \tag{6}$$

$$\eta_{+-} - \eta_{00} = 3(\epsilon' + \alpha_2). \tag{7}$$

With $\delta = \alpha_0 = \alpha_2 = \epsilon' = 0$, the amplitudes η_{+-} and η_{00} are equal, and equal to ϵ. The phase of ϵ' is determined by the $\pi - \pi$ phase shifts δ_0, δ_2,[3]

$$\varphi_{\epsilon'} = (37 \pm 5)^0.$$

Also

$$\varphi_{\epsilon} = \arctan \frac{2\Delta m}{\Gamma_S} = (43.7 \pm 0.2)^0,$$

in the standard phase convention. The most precise method to determine ϵ'/ϵ is to measure the double ratio

$$\left| \frac{\eta_{00}}{\eta_{+-}} \right|^2 = \frac{[\Gamma(K_L^0 \rightarrow \pi^0\pi^0)]/[\Gamma(K_L^0 \rightarrow \pi^+\pi^-)]}{[\Gamma(K_S^0 \rightarrow \pi^0\pi^0)]/[\Gamma(K_S^0 \rightarrow \pi^+\pi^-)]}. \tag{8}$$

If at least two of the decays are observed at the same time, the measurement of this double ratio has the advantage that many systematic errors cancel. The present status of the *CP* violation parameters is

$$|\eta_{+-}| = (2.275 \pm 0.021)10^{-3} \qquad \text{(reference 4)}$$

$$\varphi_{+-} = (44.6 \pm 1.2)^0 \qquad \text{(reference 4)}$$

$$\varphi_{00} = (55.7 \pm 5.8)^0 \qquad \text{(reference 5)}$$

$$\text{Re}\,\epsilon = (0.33 \pm 0.2)\% \qquad \text{(reference 4)}$$

$$\left|\frac{\eta_{00}}{\eta_{+-}}\right|^2 = 1 - 6\,\text{Re}\,\frac{\epsilon'}{\epsilon} = 1.015 \pm 0.03. \qquad \text{(reference 6)}$$

The average of the quoted measurements has been taken for the last result.

A nonzero phase difference $\varphi_{+-} - \varphi_{00}$, at present a two-standard deviation effect, is related to a *CPT* violating amplitude α_2, as evidenced from equation 7.

CP violation may be accommodated within the framework of the standard electroweak theory by quark mixing, expressing the fact that the quark mass eigenstates do not coincide with the weak interaction eigenstates. In a mixing matrix of three equally charged quarks there is one free parameter relevant to *CP* violation; therefore, the parameters ϵ and ϵ' can be related in principle. Actual calculations are, however, complicated by the fact that the K^0 state is not easily accessible to perturbative QCD treatment, and by missing experimental information, such as the value of the top-quark mass. The theoretical situation was summarized by Gilman.[7] An estimate of the present uncertainty in the calculation of Re (ϵ'/ϵ) is also given in a review by Wolfenstein[8]

$$1.0 \cdot 10^{-3} < \text{Re}\,(\epsilon'/\epsilon) < 7 \cdot 10^{-3}.$$

A measurement of Re (ϵ'/ϵ) outside this range—for example, a negative value, in other words $|\eta_{00}| > |\eta_{+-}|$—would be an indication that our present understanding is incomplete and hint perhaps, but not necessarily, at a fourth generation of quarks.

THE STATUS OF ϵ'/ϵ MEASUREMENT IN EXPERIMENT NA31

The method used in the SPS experiment to determine the double ratio (equation 8) is to measure charged and neutral 2π decays simultaneously in either a K_S^0 or a K_L^0 beam. A layout of the apparatus and the beam line is shown in FIGURE 1. The experiment deviates in several aspects from previous approaches. The K_S^0 beam is not derived from a K_L^0 beam by regeneration, thus avoiding problems linked to neutron interactions and incoherent regeneration, but is taken directly from a target placed with a short collimator inside the decay volume. Moreover this K_S^0 target station, involving bending magnets and a beam dump, is moveable on a train, thus covering the entire decay fiducial volume (50 m) in K_L^0 decays with corresponding K_S^0 decays. Systematic errors in the dependance of acceptance on the decay point therefore cancel, except for a very small effect due to the lateral size of the beam (56 mm diameter in K_L^0 and 7 mm diameter in K_S^0 at the end of the final collimator). A further advantage of this approach is that neutrons or photons in the neutral beam can be kept inside a vacuum by providing a central hole for a beampipe in all equipment. The only background is then given by K^0 decays and by muons from the beam dump.

The experiment differs in two other respects from the standard, taking some advantages and also some risks. There is no converter for photons, which means that the lateral position of the decay point, and therefore the transverse momentum of the kaon, in $K^0 \longrightarrow 2\pi^0$ decays cannot be reconstructed. The longitudinal position of the vertex is reconstructed from the photon energies and impact points in the calorimeter using the K^0-mass as a constraint. In charged decays, $K^0 \longrightarrow \pi^+\pi^-$, the pion energies are not measured by a magnetic spectrometer but by calorimeters, with poorer resolution and therefore higher potential background from unidentified $K_L^0 \longrightarrow \pi^+\pi^-\pi^0$ and $K_L^0 \longrightarrow \pi e\nu$ decays. The trajectories of the charged pions are measured by just two wire

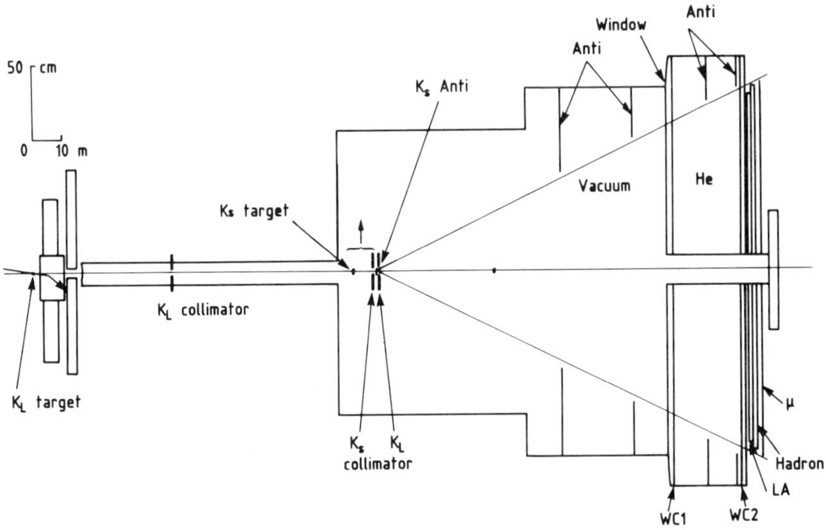

FIGURE 1. Layout of experiment NA31 in K_S^0 and K_L^0 beams.

chambers. Since the background in K_L^0 and K_S^0 runs may be very different, a different influence of accidentals on the detection of neutral and charged decays is dangerous and has to be investigated carefully.

The aim of the experiment is to measure ϵ'/ϵ with an accuracy of better than 0.1%. Therefore the individual decay rates that enter in the double ratio (equation 8) have to be measured with a relative systematic and statistical error of better than 0.3%. With proton intensities of $0.8 \cdot 10^{11}$ and $2 \cdot 10^7$ per burst ($\approx$2 sec) in K_L^0 and K_S^0 runs, respectively, about 800 events were collected per burst, among which an average 2(10) are $\pi^0\pi^0$ decays and 8(40) are $\pi^+\pi^-$ decays in K_L^0 (K_S^0) running. The total sample collected in 1986 consists of roughly 200,000 $K_L^0 \longrightarrow 2\pi^0$ decays with energies above 50 GeV. The accidental background was low. In 1% of random triggers a particle hit was reconstructed in the chambers; also in 1% of events accidental energy above 3 GeV appeared in the photon calorimeter, which was very similar for K_S^0 and K_L^0 runs.

A few words on the performance of the detector may be appropriate. The most important element is a lead–liquid-argon sampling calorimeter for photon and electron detection. This technique was chosen for its intrinsic linearity, ease of calibration, and the possibility of having fine granularity through a proper choice of the readout, even though better energy resolution may be obtained with lead glass Cerenkov detectors. Relatively thin lead plates (1.5 mm) and a narrow liquid-argon gap (2 mm) between lead plates and readout strips were chosen to optimize energy and spatial resolution (FIGURE 2). With readout strips of 1.25 cm width, and alternating between horizontal and vertical orientation in successive cells, a position resolution of better than 0.75 mm on electrons from K_{e3} decays was obtained. To facilitate ambiguity resolution, the readout of the calorimeter is independent in four quadrants centered on the beam axis. The energy resolution was calibrated with electrons between 10 GeV and 80 GeV and was better than $0.08/\sqrt{E/\text{GeV}}$ (root mean square). For electron rejection purposes the detector is read out in two longitudinal sections of 12 radiation lengths each. Between

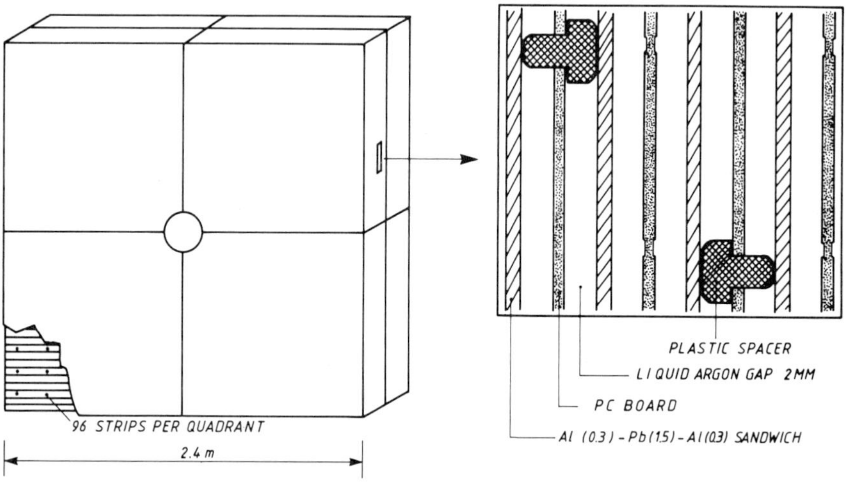

FIGURE 2. The liquid-argon detector.

these sections a layer of plastic scintillators (in liquid argon) was inserted to provide a fast trigger on photons. Accidents are monitored by recording—among other signals—the scintillator pulses with two different thresholds in 50-ns intervals for 1 μsec before and after each trigger. The absolute energy scale was calibrated with $K_S^0 \rightarrow 2\pi^0$ decays, using the K^0 mass and the location of the upstream end of the decay volume, marked by an anticounter preceded by 8 mm of lead. The energy scale is known to 0.1%. The liquid-argon response was found to drift by about 6% in two months for the final configuration. Runs with K_S^0 and K_L^0 beams were frequently altered, typically every two days. The energy scale for K_L^0 runs was then interpolated from the K_S^0 runs.

Resolution and background in $K_L^0 \rightarrow 2\pi^0$ decays may be judged from the distribution of the $\gamma\gamma$ invariant masses of the best combination to two π^0's in 4γ events (FIGURE 3). The background in K_L^0 decays is a strong function of the reconstructed decay-point position along the beam—an artifact of the reconstruction procedure for

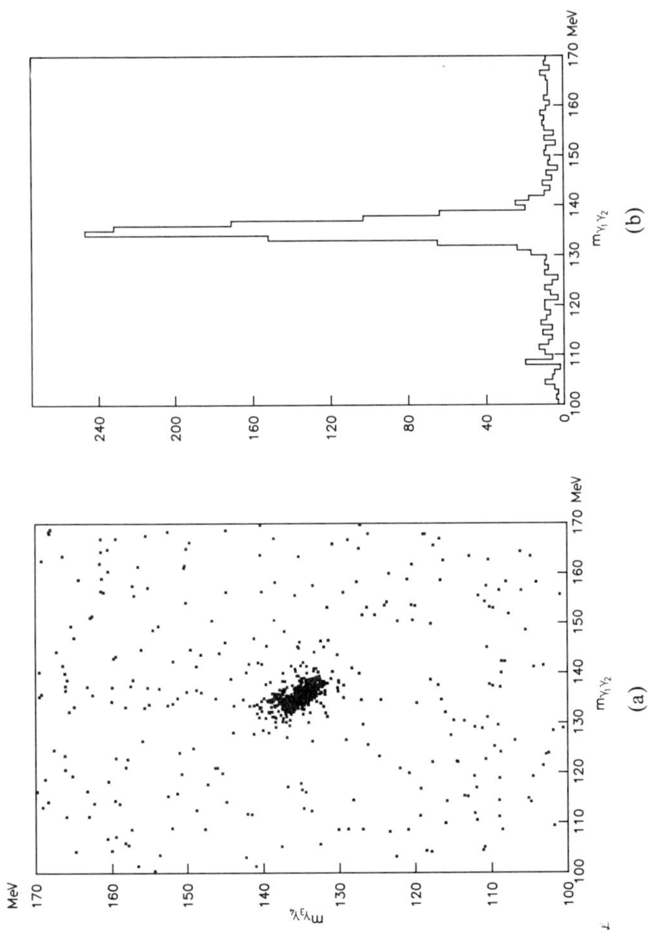

FIGURE 3. (a) Invariant masses of best combination of 4γ to $2\pi^0$ in K_L^0 decays. (b) 2γ invariant mass in $K_L^0 \longrightarrow 2\pi^0$ decay.

$3\pi^0$ decays in which two photons missed the detector and the anticounters. For a 45-m-long decay volume the average background is 4.5%; it can be determined with the desired precision of 0.3%, since it is uniformly distributed in the $(m_{\gamma_1\gamma_2}, m_{\gamma_3\gamma_4})$ space, as expected.

For charged pions the energy is measured by the combined responses of the liquid-argon calorimeter and a subsequent iron-scintillator sandwich calorimeter

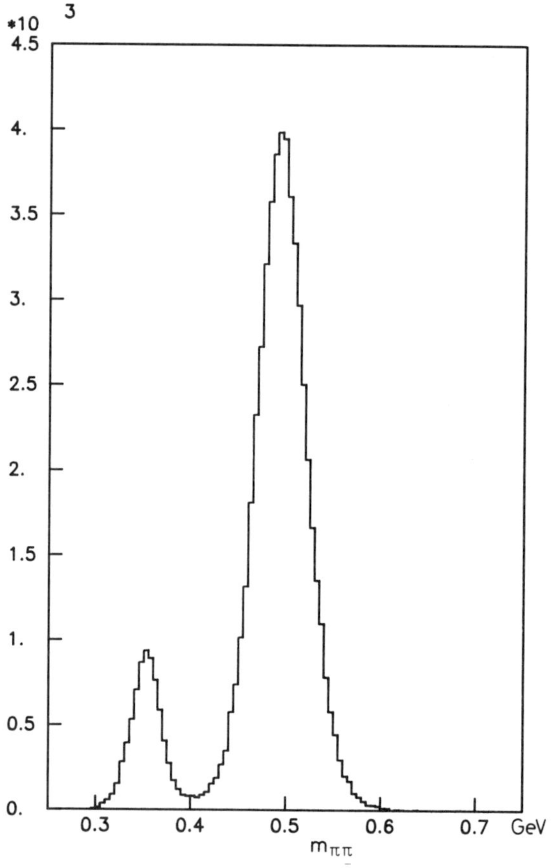

FIGURE 4. $\pi^+\pi^-$ invariant mass in $K_L^0 \rightarrow \pi^+\pi^-$ decay.

(2.5-cm-wide iron plates). With a weighing procedure a resolution σ/E of $0.7/\sqrt{E/\text{GeV}}$ for pion energies between 10 GeV and 80 GeV and a $K \rightarrow \pi\pi$ mass resolution of 25 MeV are obtained. The energy ratio of the observed pions is restricted to a maximum of 2.5 in order to avoid contamination by Λ decays in K_S^0 runs. About 1% of good events is lost by inefficiency or overflow in the chambers; 0.4% is rejected because of two large an error in the decay-point reconstruction. The background under the K^0

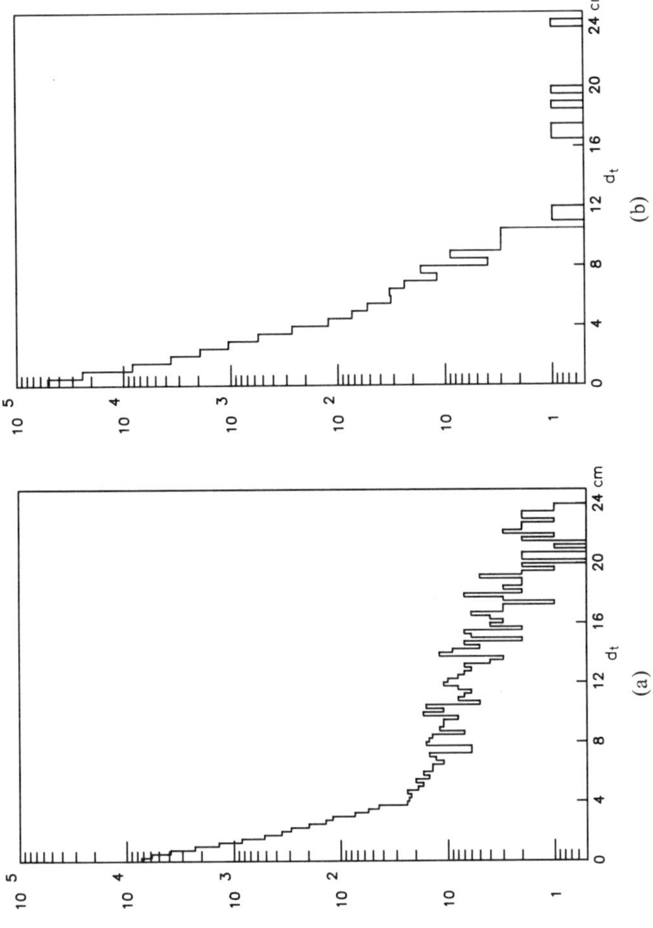

FIGURE 5. (a) Distance target – decay plane for $K_L^0 \rightarrow \pi^+\pi^-$ candidates with 430 MeV $< m_{\pi\pi} <$ 570 MeV. (b) Same as (a) for $K_S^0 \rightarrow \pi^+\pi^-$ decays.

mass peak (FIGURE 4) in K_L^0 decays can be estimated from the distribution in the distance d_t between proton target and decay plane. For two-body decays the target is in the decay plane. The accuracy of the d_t reconstruction can be obtained from K_S^0 runs. With the proper scale factor of about 2 for the difference in the distance between target and the average of the wire chamber positions, the d_t-distributions of K_S^0 and K_L^0 decays can be directly compared (FIGURE 5). There is a remaining background of approximately 1% from three-body decays under the K_L^0 mass peak. This background needs further study.

A MEASUREMENT OF THE RELATIVE DECAY RATE $\Gamma(K_S^0 \rightarrow 2\gamma)/\Gamma(K_L^0 \rightarrow 2\gamma)$

The $K_L^0 \rightarrow 2\gamma$ decays can be measured with very little (<1%) background, estimated from the center-of-gravity distribution of the photons. The background is mainly due to $2\pi^0$ decays in which two photons missed the detector. The relative branching ratio $K_L^0 \rightarrow 2\gamma/K_L^0 \rightarrow 2\pi^0$ was obtained from a sample of 109,058 two-photon and 61,313 two-pizero candidates collected during 1986. The result is

$$\frac{\Gamma(K_L^0 \rightarrow 2\gamma)}{\Gamma(K_L^0 \rightarrow 2\pi^0)} = 0.633 \pm 0.004 \pm 0.008,$$

where the first error is statistical and the second systematic, mainly caused by uncertainties in the acceptance calculation. The result agrees with a preliminary value[9] obtained from a short run in 1985.

The $K_S^0 \rightarrow 2\gamma$ decay is more difficult to observe, because the background from $2\pi^0$ decay is about 10^5 times stronger than in K_L^0 decays. A clean sample can, however, be obtained if one uses the observation that the transverse momentum p_t of photons in the $2\pi^0$ decay is restricted to $p_t < 228$ MeV/c, whereas the maximum p_t of a photon in the $K^0 \rightarrow 2\gamma$ decay is 249 MeV/c. When the K^0 mass is used as a constraint in the vertex reconstruction, a background event is shifted downstream by at least 8% of the decay-point distance to the photon detector due to this missing p_t. The interval

$$-0.04 < \delta < 0.06$$
$$\delta = \frac{z_v - z_{\text{anti}}}{z_{\text{LAC}} - z_{\text{anti}}},$$

where z_v, z_{anti}, and z_{LAC} are the positions along the beam of vertex, anticounter, and liquid-argon counter, was finally used as a signal region (see FIGURE 6). In this region $K_L^0 \rightarrow 2\gamma$ decays from the K_L^0 component in the K_S^0 beam provide a background that can be calculated from the observed $K_S^0 \rightarrow 2\pi^0$ rate. For 737,825 accepted $K_S^0 \rightarrow 2\pi^0$ decays in the momentum range 70 GeV to 170 GeV we find 60 $K^0 \rightarrow 2\gamma$ events, whereas 41 ± 2 events are expected from the $K_L^0 \rightarrow 2\gamma$ background. The remaining signal of 19 events above background is attributed to $K_S^0 \rightarrow 2\gamma$ decays and results in a branching ratio

$$\frac{\Gamma(K_S^0 \rightarrow 2\gamma)}{\Gamma(K_S^0 \rightarrow 2\pi^0)} = (7.6 \pm 3.3) \cdot 10^{-6}.$$

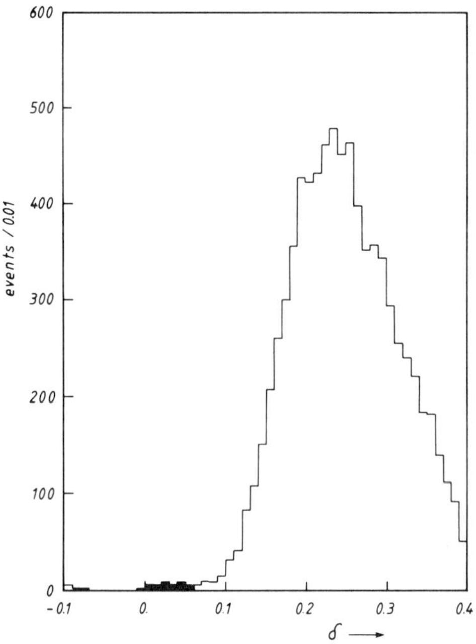

FIGURE 6. Normalized distance δ of decay point to collimator in K_S^0 2γ candidates (see text).

The probability that all observed events are due to a fluctuation of the $K_L^0 \to 2\gamma$ background is $5.6 \cdot 10^{-3}$. Using $|\eta_{00}|^2 = 5.2 \cdot 10^{-6}$, we then obtain

$$\frac{\Gamma(K_S^0 \to 2\gamma)}{\Gamma(K_L^0 \to 2\gamma)} = 2.3 \pm 1.0 \pm 0.4,$$

where the first error is statistical and the second systematic. Recent theoretical estimates[10] of this ratio are around 2.4.

FUTURE PLANS

Proposal[11] for a Measurement of $\varphi_{+-} - \varphi_{00}$ by NA31

The NA31 detector was used in 1987 to measure the phase difference between the *CP* violating amplitudes η_{+-} and η_{00}, by observing the interference pattern around 12

TABLE 1. Expected Precision of the LEAR Experiment

Quantity	Error
$\varphi_{+-} - \varphi_{00}$	2^0
$\|\epsilon'/\epsilon\|$	$2 \cdot 10^{-3}$
$\|\eta_{+-0}\|$	$<6 \cdot 10^{-4}$
$\|\eta_{000}\|$	$<8 \cdot 10^{-4}$

K_S^0 lifetimes behind the proton target. A new K_S^0 beam has been built for this purpose, with a long collimator (48 m or 34 m). Runs with two target positions are planned in order to have the interference pattern appear at two different locations in the decay volume. The statistical accuracy on the phase difference that can be obtained should be less than 1^0.

A LEAR Experiment[12] *on* CP *Violation in* K^0 *Decay*

At LEAR one can expect to have $2 \cdot 10^6\ \bar{p}$/sec stopping in a gaseous hydrogen target. With a branching ratio of $\approx 10^{-3}$, one finds the exclusive annihilation channels

$$
\begin{aligned}
p\bar{p} &\rightarrow K^+\pi^-\bar{K}^0 \\
&\rightarrow K^-\pi^+K^0 \\
&\rightarrow K^0\bar{K}^* \rightarrow K^0\pi^+K^- \\
&\rightarrow \bar{K}^0K^* \rightarrow \bar{K}^0\pi^-K^+,
\end{aligned}
$$

in which the strangeness of the neutral kaon is tagged by the charge of the charged kaon. By observing the partial rate difference between K^0 and $\bar{K}^0$ decays one can extract the *CP* violation parameters. Apart from the 2π decays it should also be possible to observe *CP* violating effects in three-body decays. The expected precision with a run of 10^{13} stopping $\bar{p}$'s is given in TABLE 1. The experiment is currently under construction, with some modifications of the initial proposal. A cylindrical track detector with subsequent time-of-flight and liquid Cerenkov counters for *K*-detection is arranged in a solenoid. Instead of a CsI counter array, a lead-streamer-tube sandwich was used as a photon detector.

REFERENCES

1. BURKHARDT, H., P. CLARKE, D. C. CUNDY, N. DOBLE, L. GATIGNON, R. HAGELBERG, G. KESSELER, J. VAN DER LANS, I. MANNELLI, T. MICZAIKA, H. G. SANDER, R. D. SCHAFFER, P. STEFFEN, J. STEINBERGER, H. TAUREG, H. WAHL, C. YOUNGMAN, G. DIETRICH, F. EISELE, W. HEINEN, D. J. CANDLIN, J. MUIR, K. J. PEACH, B. J. PIJLGROMS, I. P. SHIPSEY, B. STEPHENSON, H. BLÜMER, M. KASEMANN, K. KLEINKNECHT, B. RENK, E. AUGE, R. L. CHASE, M. CORTI, D. FOURNIER, P. HEUSSE, A. M. LUTZ, A. BIGI, M. CALVETTI, R. CAROSI, R. CASALI, C. CERRI, R. FANTECHI, A. NAPPI, G. PIERAZZINI, CH. BECKER, D. HEYLAND, M. HOLDER, G. QUAST, M. ROST, W. WEIHS, & G. ZECH. NA31 collaboration. CERN, Geneva, Switzerland.
2. WU, T. T. & C. N. YANG. 1964. Phys. Rev. Lett. **13:** 380.
3. LOSTY, M. F. *et al.* 1974. Nucl. Phys. **B69:** 185; ESTABROOKE, P. & A. D. MARTIN. 1974. Nucl. Phys. **D79:** 301.
4. PARTICLE DATA GROUP. 1986. Phys. Lett. **170B.**
5. CHRISTENSON, J. A. *et al.* 1979. Phys. Rev. Lett. **43:** 1209.
6. BLACK, J. K. *et al.* 1985. Phys. Rev. Lett. **54:** 1628; BERNSTEIN, R. H. *et al.* 1985. Phys. Rev. Lett. **54:** 1631.
7. GILMAN, F. J. 1987. Overview of the Standard Model. This volume.
8. WOLFENSTEIN, L. 1986. Annu. Rev. Nucl. Part. Sci. **36:** 137.
9. HOLDER, M. 1986. Presented at the 23rd Int. Conf. High Energy Physics, Berkeley, Calif.; SHIPSEY, I. P. J. 1986. Ph.D. Thesis, Univ. of Edinburgh, Edinburgh, Scotland.

10. CHAU, L. L. 1985. Phys. Rev. Lett. **54:** 1768; D'AMBROSIO, G. & D. ESPRIN. 1986. Phys. Lett. **175B:** 237.
11. Addendum to proposal P174. CERN/SPSC/86-6.
12. Basle-CERN-CEN Saclay-DAP Athens-Democritos-ETH-Fribourg-Liverpool-SIN-Stockholm-Thessaloniki collaboration; Proposal CERN/PSCC/85-6; Addendum CERN/PSCC/85-30.

Evidence for the Decay $\tau^- \rightarrow \pi^- \eta \nu_\tau$

J. REPOND

Argonne National Laboratory
Argonne, Illinois 60439

INTRODUCTION

Motivated by the 9% discrepancy between the one-prong topological branching ratio and the sum of the individual decay modes,[1-3] the High Resolution Spectrometer collaboration has searched for undetected decay channels of the τ-lepton. In particular, as the classification of one-prong decays indicates a higher rate than expected for the one-charged plus multiple-neutral channel,[4,5] we have searched for η-meson production in one-prong τ-decays.

The analysis[6] that gave evidence for the decay $\tau^- \rightarrow \pi^- \eta \nu_\tau$ is outlined. We then discuss the detailed checks that have been done, including the background from hadronic- and two-photon events, and the probability of missing photons. The latter plays an important role in determining the final state associated with η-mesons.

OBSERVATION OF η-MESONS IN τ-DECAYS AND EVIDENCE FOR THE DECAY $\tau^- \rightarrow \eta \pi^- \nu_\tau$

Applying standard τ-event selection cuts[7] yielded 4004 events in the 1–1 topology and 2553 events in the 1–3 topology.

The analysis depends on reconstructing $\eta \rightarrow \gamma\gamma$ decays via measurement of electromagnetic showers in the barrel shower counter. We show in FIGURE 1a the $\gamma\gamma$ mass spectrum for combinations where (1) the photons hit separate modules, (2) $E_\gamma >$ 1 GeV for both photons, and (3) no other photon shared the same module. A significant signal is seen both at the π^0 and at the η mass. The latter signal corresponds to a 3.4 standard deviation effect.

The data have been further divided into a low-energy and a high-energy sample with $E_{\gamma_1} + E_{\gamma_2}$ less than or greater than 4.5 GeV. The η signal persists in the $\gamma\gamma$ mass plot for the high-energy data of FIGURE 1c, but is less prominent in the low-energy spectrum of FIGURE 1b, thus favoring a low multiplicity final state.

The low branching ratio[7] for $\tau^- \rightarrow 5\pi^\pm\pi^0$ limits possible contributions from final states with multiple π^0's or η's, such as $\pi^-\eta\pi^0\pi^0\nu_\tau$ or $\pi^-\eta\eta\nu_\tau$, to a fraction of a percent, and so the two final states that could contribute to the signal of FIGURE 1a are $\pi^-\eta\nu_\tau$ and $\pi^-\eta\pi^0\nu_\tau$. The data of FIGURE 1a were, therefore, subdivided into 229 events with two and only two photons (FIGURE 1d), and 145 $\gamma\gamma$ combinations from 113 events with three or more neutral photons (FIGURE 1e). A minimum energy cut of 100 MeV was applied in defining a separate photon. The enhancement at the η mass in the inclusive data persists in the events with only two photons, whereas there is no significant signal for the $n_\gamma \geq 3$ selection. The data of FIGURE 1d provide the first evidence that the $\pi^-\eta\nu_\tau$ final state is being observed.

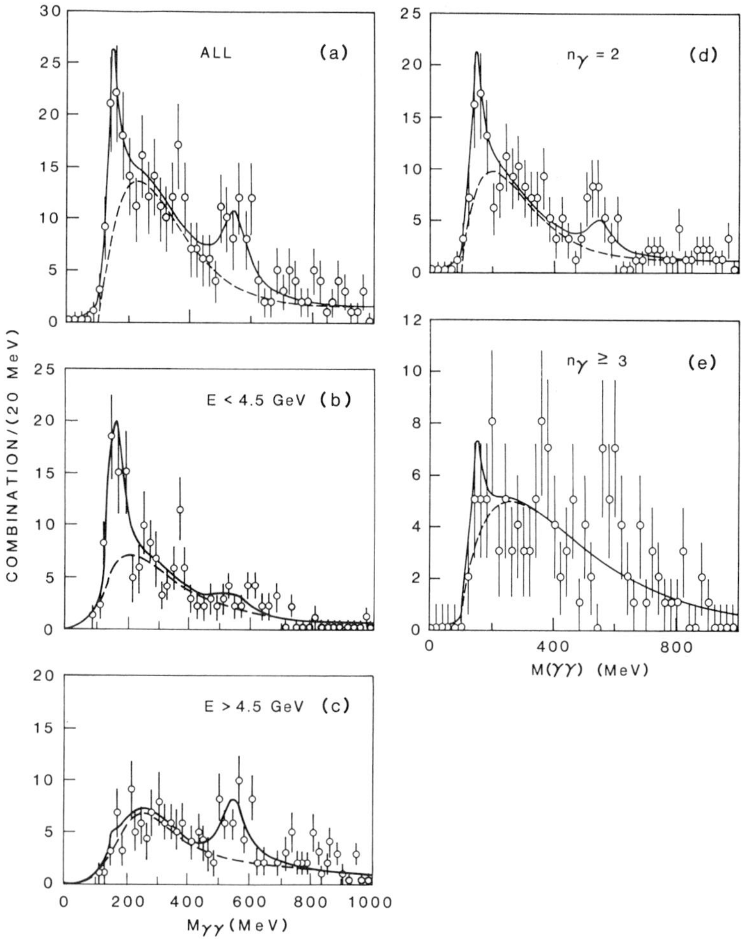

FIGURE 1. Effective mass of the $\gamma\gamma$ system for events with E_{γ_1}, $E_{\gamma_2} > 1$ GeV and having the photons in separate modules. (a) All combinations. (b) Combinations with $E_{\gamma_1} + E_{\gamma_2} \leq 4.5$ GeV. (c) Combinations with $E_{\gamma_1} + E_{\gamma_2} > 4.5$ GeV. (d) Events with two and only two photons. (e) Events with three or more photons. The lines show fits to the background function, plus contributions at the π^0 and η masses. In FIGURE 2e, the best fit requires no η contribution.

A Monte Carlo (MC) event generator with full detector simulation was used to determine the detection efficiencies. Because of the photon overlap problem, as well as the lower photon energy, the detection efficiency for the $\pi^-\eta\pi^0\nu_\tau$ final state is lower by roughly a factor of 10 than that for $\pi^-\eta\nu_\tau$. Associating the η-signal with the simplest decay, $\tau^- \rightarrow \pi^-\eta\nu_\tau$, leads to a branching ratio of $(5.1 \pm 1.0 \pm 1.2)\%$. This decay proceeds via a second-class current and violates the conserved vector current hypothesis. The assumption that the responsible decay is $\tau^- \rightarrow \pi^-\eta\pi^0\nu_\tau$ would give an unreasonably large branching ratio, although we cannot exclude the possibility that some of the observed η-mesons come from this decay.

CHECKS ON THE DATA AND ON THE ANALYSIS PROGRAM

All events with an entry in FIGURE 1a inside a band around the mass of the η (480 MeV to 620 MeV) were examined carefully, including all information from the barrel shower counter system. No unusual features were observed. The energies and the opening angles between the photons are distributed smoothly and show no unexpected enhancements.

The data were taken over a five-year period. TABLE 1 shows the distribution of events in the selected η-band by year and by topology. The numbers scale with the integrated luminosity obtained per year.

FIGURE 2a shows the $\gamma\gamma$ mass spectrum with the same cuts as used for FIGURE 1a, but this time selecting the three-prong hemisphere. No enhancement at the mass of the η is observed. This is also true if we analyze MC events of τ-decays that exclude final states containing η's.

TABLE 1. Distribution of Detected η-Mesons

Year	[pb^{-1}]	1–1	1–3	Total
1	20	6	1	7
2	90	11	5	16
3	74	10	7	17
4	72	11	5	16
5	44	5	3	8
Total	300	43	21	64

DISCUSSION OF HADRONIC BACKGROUND

Hadronic events ($e^+e^- \longrightarrow q\bar{q}$) as produced at $\sqrt{s} = 29$ GeV are distinguished from QED-events ($e^+e^- \longrightarrow \gamma\gamma, e^+e^-, \mu^+\mu^-, \tau^+\tau^-$) by their high track multiplicity. A Monte Carlo study[8] gives a ~5% hadronic background in the 1–3 topology and a negligible background in the 1–1 topology. Because of the shape of the hadronic multiplicity distributions,[9] if the events containing an η originate from a hadronic background, then their recoil against a single charged track would be suppressed by roughly a factor of 5 compared to their recoil against three charged tracks.

TABLE 1 shows that the ratio of 1–3 events to 1–1 events is $R_{3-1} = 0.42 \pm 0.06$, in good agreement with the MC prediction of $R_{3-1} = 0.49 \pm 0.02$. FIGURE 2b and c show the entries of FIGURE 1a split into the 1–1 and 1–3 topologies. The data suffer from limited statistics, but the enhancement at the η-mass is visible in both selections.

Hadronic events with a low charged track multiplicity are typically accompanied by a large number of photons originating from π^0-decays. The total invariant mass of all charged tracks (assumed to be pions) and photons in one hemisphere is usually much larger than the mass of the τ-lepton. We therefore rejected all τ candidates if the total invariant mass in either hemisphere exceeded the τ-mass. FIGURE 2d shows the mass of photon pairs under the same conditions as for FIGURE 1a, but with this additional cut. Most of the 46 events removed have a $\gamma\gamma$ mass above 700 MeV, and only two events with an entry near the η mass were rejected.

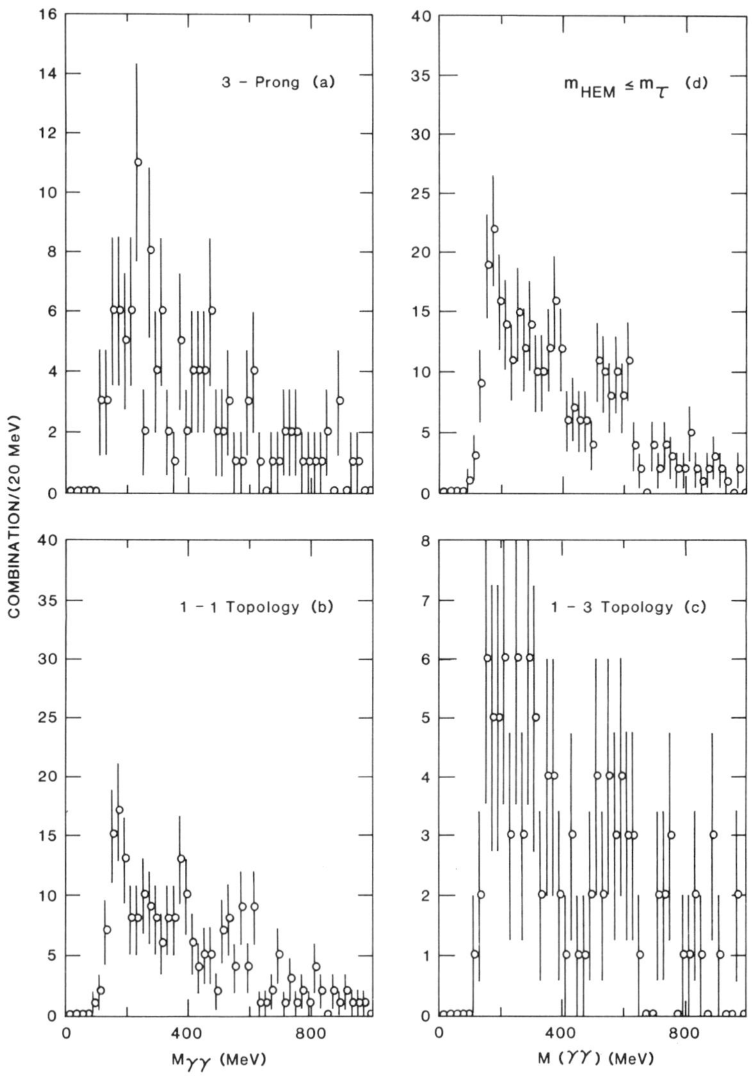

FIGURE 2. Effective mass of $\gamma\gamma$ system for events with $E_{\gamma_1}, E_{\gamma_2} > 1$ GeV and having the photons in separate modules. (a) All combinations (3-prong hemisphere). (b) Combinations (1-prong hemisphere) recoiling against 1-prongs. (c) Combinations (1-prong hemisphere) recoiling against 3-prongs. (d) All combinations (1-prong hemisphere); events with total invariant mass above the τ-mass rejected.

Finally, the track momentum spectrum and the hemisphere invariant mass of the charged tracks, plus all photons recoiling against a detected η, were compared with the total data. No significant shifts to lower momenta or higher masses are observed for the η events.

Similarly, the invariant mass on the three-prong side of events recoiling against a detected η on the one-prong side is similar to the inclusive data sample.

We conclude that the η-meson production is not associated with hadronic annihilations.

BACKGROUND FROM TWO-PHOTON EVENTS

Two-photon events, where the two photons produce an η-meson via the reaction $e^+e^- \rightarrow e^+e^-\eta$ could represent a major background in the 1–1 topology. In FIGURE 3 we show the distribution of the dip angle (defined relative to the e^+-beam direction) for the charged tracks associated with a reconstructed η in the 1–1 topology. The shadowed area corresponds to the positive tracks only. Two-photon events would characteristically show a strong enhancement at small (large) angles for positive (negative) tracks. This effect is not observed. Furthermore, both charged tracks in the event would have to be electrons. The η-events in the 1–1 topology contain at most one track in the hemisphere of the η and 10 tracks in the opposite hemisphere with E/p ratios compatible with being electrons. Using $B(\tau^- \rightarrow e^-\nu_e\nu_t) = 17.9\%$, we expect nine tracks recoiling against an η to be electrons. Thus no excess of electrons is observed.

From these considerations, we conclude that the η-mesons observed are not due to background from two-photon events.

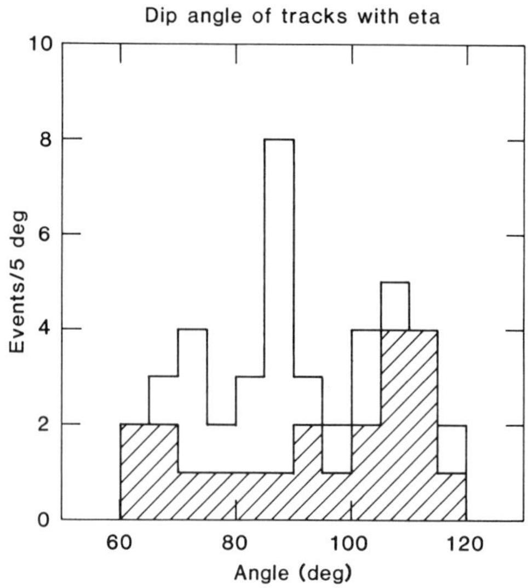

FIGURE 3. Dip angle of tracks with reconstructed η. Shadowed area corresponds to positive tracks only.

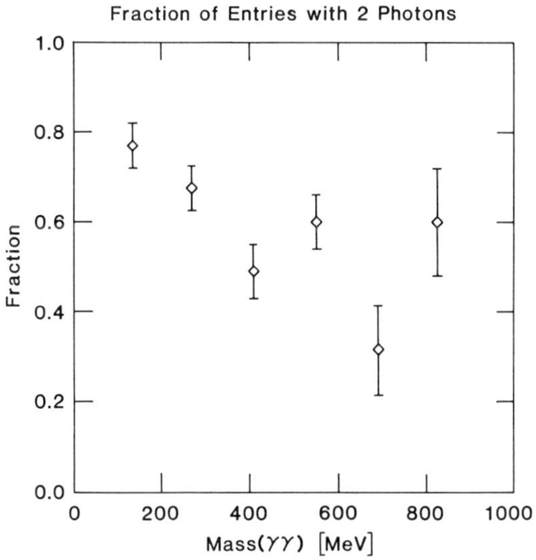

FIGURE 4. Fraction of entries in FIGURE 1a with two and only two γ's as a function of the mass of the two γ's.

PROBABILITY OF MISSING PHOTONS

The probability of missing photons from the π^0 decay is a crucial consideration in distinguishing the forbidden decay ($\tau^- \rightarrow \pi^-\eta\nu_\tau$) and the allowed decay ($\tau^- \rightarrow \pi^-\eta\pi^0\nu_\tau$).

Let us assume that the observed η's originate from the allowed decay and try to understand how we could have missed the two γ's from the decay of the additional π^0. The first possibility is that the γ's could be missed because they hit the cracks between the modules of the barrel shower counter (~3% of the width of one module) or because they overlap with the charged track. Monte Carlo and other studies[10] indicate that such effects are small and, furthermore, the energy depositions of the charged tracks are consistent with being from charged π's with no overlapping photons.

Second, the γ's from the π^0 decay could overlap with γ's from the η-decay. This clearly leads to a wrong energy assignment for at least one γ from the η-decay, thus shifting the mass outside of the η peak region.

Third, π's could be missed because, due to an unknown mechanism, they are emitted with low momenta, and their decay γ's would fall under the 100 MeV lower energy limit. In order to study this possibility, the cut used to split the data in a subset of two and only two γ's per hemisphere (see FIGURE 1d) was lowered to 50 MeV and raised to 200 MeV. These changes rejected 7 or added 6 events, which were distributed uniformly over the $\gamma\gamma$ mass range.

Finally, we show in FIGURE 4 the fraction of entries in FIGURE 1a that have two and only two γ's as a function of the mass of the two γ's. Even though the shape of the distribution is not known, there is some enhancement of events with only two photons in the η-meson mass region.

CONCLUSIONS

We report the observation of η production in one-prong τ-decays and give evidence for the decay $\tau^- \rightarrow \pi^- \eta \nu_\tau$. This decay proceeds via a second-class current and violates the conserved vector current hypothesis. The branching ratio of 5.1% gives a 3.6% contribution to the one-prong topology, since the branching ratio of η to neutrals is 71%.

The data have been thoroughly checked and no unusual features have been observed. Background from hadronic events or from two-photon events were excluded as possible sources for the η-signal. The probability of missing γ's and thus confusing the two considered decays, $\tau^- \rightarrow \pi^- \eta \nu_\tau$ and $\tau^- \rightarrow \pi^- \eta \pi^0 \nu_\tau$, is small.

REFERENCES

1. GILMAN, F. J. & S. H. RHIE. 1985. Phys. Rev. **D31:** 1066.
2. BURCHAT, P. R. 1986. Presented at the 23rd Int. Conf. High Energy Physics. Berkeley, Calif.
3. GILMAN, F. J. 1987. SLAC-PUB-4265.
4. AIHARA, H. *et al.* 1986. Phys. Rev. Lett. **57:** 1836.
5. BURCHAT, P. R. *et al.* 1987. Phys. Rev. **D35:** 27.
6. DERRICK, M. *et al.* 1987. ANL-HEP-PR-86-106. Phys. Lett.
7. AKERLOF, C. *et al.* 1985. Phys. Rev. Lett. **55:** 570.
8. BYLSMA, B. *et al.* 1987. Phys. Rev. **D35:** 2269.
9. DERRICK, M. *et al.* 1986. Phys. Rev. **D34:** 3304.
10. AKERLOF, C. *et al.* 1985. Phys. Lett. **156B:** 271.

Overview of the Standard Model[a]

FREDERICK J. GILMAN

Stanford Linear Accelerator Center
Stanford University
Stanford, California 94305

INTRODUCTION

The Standard Model, with the gauge group $SU(3)_C \times SU(2) \times U(1)$, assignment of fermions to left-handed doublets and right-handed singlets under the electroweak $SU(2)$, and 18 or more parameters, is in excellent shape. Experiment has again and again confirmed its predictions to whatever accuracy they can be predicted and measured in a given situation. With the beginning of experiments at the Z factories in the near future, the one-loop corrections to the electroweak portion of the Standard Model will be tested to high accuracy.

Instead of reviewing these successes, we concentrate on a few possible problem areas: the Kobayashi–Maskawa matrix, *CP* violation in the neutral kaon system as expressed through the parameters ϵ and ϵ', $B - \bar{B}$ mixing, and τ-decays.

THE KOBAYASHI–MASKAWA MATRIX

The Kobayashi–Maskawa (KM) matrix[1] is defined as the matrix transformation that takes us from the mass eigenstates of the d, s, and b quarks to the weak eigenstates, d', s', and b', the partners in weak doublets of the u, c, and t quarks, respectively, which by convention are unmixed:

$$\begin{pmatrix} d' \\ s' \\ b' \end{pmatrix} = \begin{pmatrix} V_{ud} & V_{us} & V_{ub} \\ V_{cd} & V_{cs} & V_{cb} \\ V_{td} & V_{ts} & V_{tb} \end{pmatrix} \begin{pmatrix} d \\ s \\ b \end{pmatrix}. \tag{1}$$

The 1986 Review of Particle Properties[2] gave the following results for the magnitudes of those matrix elements that can be measured up to the present time.

(1) Nuclear beta decay, when compared to muon decay, gave

$$|V_{ud}| = 0.9729 \pm 0.0012,$$

which included refinements[3] that had lowered $|V_{ud}|$ by 0.13%.

(2) Analysis[4] of hyperon and K_{e3} decays yielded

$$|V_{us}| = 0.221 \pm 0.002.$$

[a]This work was supported by the Department of Energy under Contract DE-AC03-76SF00515.

(3) From ν and $\bar{\nu}$ production of charm, the CDHS group[5] had deduced

$$|V_{cd}| = 0.24 \pm 0.03.$$

(4) By comparing the experimental value[6] for $\Gamma(D \rightarrow \bar{K}e^+\nu_e)$ with the expression that follows from the standard weak interaction amplitude, one derives

$$|f_+^D(0)|^2 |V_{cs}|^2 = 0.51 \pm 0.07,$$

where f_+^D is the form factor for D_{l3} decay, which is the analogue of f_+ for K_{l3} decay. With the conservative assumption that $|f_+(0)| < 1$,

$$|V_{cs}| > 0.66.$$

(5) The ratio $|V_{ub}{}^* V_{cb}|$ is obtained from the semileptonic decay of B-mesons by fitting to the lepton energy spectrum as a sum of contributions involving $b \rightarrow u$ and $b \rightarrow c$. As more data had accumulated, the inadequacy of previous parametrizations of the lepton spectrum became clear.[6] Using only the lepton momentum region beyond the end-point for $b \rightarrow cl\bar{\nu}_l$ resulted in[6]

$$\frac{\Gamma(b \rightarrow ul\bar{\nu}_l)}{\Gamma(b \rightarrow cl\bar{\nu}_l)} < 0.08,$$

which translates to

$$\left|\frac{V_{ub}}{V_{cb}}\right| < 0.19.$$

(6) The magnitude of V_{cb} itself can be determined if the measured semileptonic bottom hadron partial width is assumed to be that of a b-quark decaying through the usual $V - A$ interaction (which from result 5 is $BR(b \rightarrow cl\bar{\nu}_l)$ to within 8%):

$$0.037 < |V_{cb}| < 0.053.$$

One cannot prove there are three generations from these data, but only show consistency with that as a hypothesis. The crucial test comes from the constraint that unitarity of the 3×3 matrix imposes on the first row:

$$(0.9729 \pm 0.0012)^2 + (0.221 \pm 0.002)^2 + (<0.01)^2 = 0.9954 \pm 0.0025, \qquad (2)$$

with a couple of standard deviations being nothing to make a fuss about.

Since the Review of Particle Properties went to press, there have been some small shifts in the central values of some of the matrix elements due to reanalysis and/or new data, such as incorporating newer charm semileptonic branching ratios in extracting[7] $|V_{cs}|$. More importantly, a change[8] in the order $Z\alpha^2$ coulomb corrections brought different experiments into better agreement and raised the value of $|V_{ud}|$:

$$|V_{ud}| = 0.9747 \pm 0.0010,$$

to be compared with a very recent result[9]

$$|V_{ud}| = 0.9755 \pm 0.0017,$$

which also improves on previous analyses of this quantity, primarily in terms of the electron screening correction.[b] The unitarity sum for the first row is now 0.9989 ± 0.0021 or 1.0004 ± 0.0035, depending on which new result one uses. One couldn't ask for better agreement with three generations. Turning this around, and using unitarity to restrict the coupling between the u-quark and a new charge $-1/3$ quark results in

$$|V_{ub'}| \leq 0.06. \tag{3}$$

This is not very restrictive and, looking at its primary origin in the error bar on $|V_{ud}|$, it seems unlikely that there will be a very significant improvement upon it in the future.

CP VIOLATION

CP violation has still only been observed in the neutral K system. There it is conveniently parametrized in terms of ϵ, which characterizes *CP* violation in the kaon mass matrix, and ϵ', which is nonzero only due to *CP* violation in the neutral kaon decay amplitude.

For ϵ one takes the short distance contribution (corresponding to the box diagram with W's and quarks). For $m_q^2 \ll M_W^2$ this leads to the expression:[10]

$$\epsilon \approx \frac{e^{i\pi/4}}{\sqrt{2}} \frac{BG_F^2 f_K^2 m_K}{6\pi^2 \Delta M_K}$$

$$\times\; s_1^2 s_2 s_3\, s_\delta \left[-\eta_1 m_c^2 + \eta_2 s_2 (s_2 + s_3 c_\delta) m_t^2 + \eta_3 m_c^2 \ln\left(\frac{m_t^2}{m_c^2}\right) \right], \tag{4}$$

where the s_i are the sines of the KM angles θ_i, $i = 1, 2, 3$ (these are known to be small so that the approximation $c_i = \cos\theta_i = 1$ has been used in equation 4). A nonzero value of the angle δ is indicative of *CP* violation in the KM parametrization.[1] The factors η_1, η_2, and η_3 are due to strong interaction (QCD) corrections and have the values 0.7, 0.6, and 0.4, respectively, with usual quark and Wboson masses.[11] The infamous parameter B is the ratio of the actual value of the matrix element between the K^0 and $\bar{K}^0$ states of the operator composed of the product of two $V - A$ neutral, strangeness-changing currents divided by the value of the same matrix element obtained by inserting the vacuum between the two currents.

One can see that there can be a potential problem in getting the right-hand side of equation 4 to reproduce the experimental value of $|\epsilon| = 2.27 \times 10^{-3}$ if the combination of the KM angles, m_t, and B are not large enough. To get some idea of where the experimental situation places us, take $B = 1/3$, a b-quark lifetime of 1 picosecond, and $b \to u/b \to c < 0.04$. Then for $m_t \lesssim 60$ GeV we would have trouble satisfying equation 4.

This is not far from the situation that existed a couple of years ago. Since then there has been a retreat from confrontation. On the experimental side, as noted in the previous section, the upper limit on $b \to u/b \to c$ has become less stringent. This allows s_3 to be larger, relieving some of the pressure on the right-hand side of equation 4.

[b]We thank Bill Marciano for a clarifying discussion of the comparison of the two new results.

On the theoretical side, the argument[12] that $B \approx 1/3$, was found to have potential 100% corrections from the next order (in chiral $SU(3)$ breaking) and is therefore unreliable.[13] A flurry of papers[14] on the subject followed with values of B ranging from 0.3 to 1.5 or so. Finally, there is the value of m_t, which theoretical prejudice somehow kept near its ever-increasing lower bound or at some nominal "discovered" value. Higher values of m_t lead to less of a problem in explaining the value of ϵ; they may now even be demanded by the size of $B - \bar{B}$ mixing (see the next section).

The situation for ϵ' has also become more open. By inserting experimentally measured quantities, the contribution to ϵ' from the "penguin" operator contribution to $K \rightarrow \pi\pi$ can be written[15]

$$\frac{\epsilon'}{\epsilon} = 6.0\, s_2 s_3\, s_\delta \left(\frac{\text{Im}\ \hat{C}_6}{-0.1}\right)\left(\frac{\langle \pi\pi\, | Q_6 |\, K^0 \rangle}{1.0\ \text{GeV}^3}\right), \quad (5)$$

where Q_6 is the "penguin" operator in the short distance expansion of the strangeness-changing weak Hamiltonian[16] and Im $\tilde{C}_6$ is the imaginary part of the corresponding Wilson coefficient with the Kobayashi–Maskawa factor taken out.

The value of -0.1 for this last quantity is relatively stable from calculation to calculation, as the imaginary part depends on momentum scales from m_c to m_t where the short distance expansion is well justified. The value of the matrix element of Q_6 is much less certain. If it is large enough to explain the magnitude of $A(K \rightarrow \pi\pi)$, then, combined with the value of $s_2 s_3 s_\delta$ needed to fit ϵ (see the preceding), it yields the prediction $\epsilon'/\epsilon \approx +10^{-2}$. This was the basic observation in reference 17: If the "penguin" operator is to be an explanation of the $\Delta I = 1/2$ rule and the magnitude of $A(K \rightarrow \pi\pi)$, then ϵ'/ϵ should be at the 1% level.

We note first that larger values of m_t or of B indirectly make the prediction for ϵ' decrease, for they allow a smaller KM factor from fitting ϵ. Early calculations of the matrix element of Q_6 gave numbers of order 1 GeV^3, to which we have normalized in equation 5. Later calculations incorporating current algebra constraints in a correct manner gave much smaller numbers,[17] but they do not allow one to understand the magnitude of the overall $K \rightarrow \pi\pi$ amplitude. Recent results from lattice gauge theory[18] (still in its infancy in regard to this kind of calculation) and from a $1/N$ expansion approach to weak matrix elements[19] give hope of understanding the magnitude of the K-decay amplitude while indicating somewhat smaller values of ϵ'/ϵ. There have also been calculations of the electromagnetic "penguin" contributions that have already undergone a rather tortuous history from significantly suppressing the magnitude of ϵ/ϵ',[20] to finding two sign errors that led to a cancellation and a small suppression,[21] to uncovering another error and a net small-to-large enhancement.[22]

By comparison, the most recent experiments obtain[23–25]

$$\frac{\epsilon'}{\epsilon} = (-0.46 \pm 0.53 \pm 0.24) \times 10^{-2} \qquad \text{(reference 23)}$$

$$\frac{\epsilon'}{\epsilon} = (+0.17 \pm 0.82) \times 10^{-2} \qquad \text{(reference 24)}$$

$$\frac{\epsilon'}{\epsilon} = (+0.35 \pm 0.30 \pm 0.20) \times 10^{-2}. \qquad \text{(reference 25)}$$

In light of this, very divergent possibilities are still open. It is still possible that "penguin" contributions are "large," that CP violation originates in the KM phase, and that we will still end up with a value of $|\epsilon'/\epsilon|$ of order 1%. The opposite alternative, that "penguin" contributions are small, is open as well. In either of these cases it is possible that there is a fourth generation, thereby allowing the severing of the connection between "large penguin" contributions and "large" (of order 1%) ϵ'/ϵ values.[26] Finally, it is possible that CP violation does not stem from the KM phase and has its origin in physics at a much higher mass scale.[27]

$B^0 - \bar{B}^0$ MIXING

As in the neutral K system, the neutral B system is capable of exhibiting mixing between an initial $B_d^0(\bar{b}d)$, for example, and its charge conjugate state, $\bar{B}_d^0(\bar{d}b)$. A typical signature (but not the only one) arises from the ensuing semileptonic decay involving a negatively charged lepton instead of the positively charged one that would come from a B_d^0. If we examine the 2×2 mass matrix of the $B^0 - \bar{B}^0$ system, we expect that Γ_{12}/M_{12} will be of order m_b^2/m_t^2, and that it is the mass difference between the two neutral B eigenstates, $|\Delta M| = \frac{1}{2}|M_{12}|$, that is the primary source of the mixing. This should be short-distance dominated, with the box diagram involving t-quarks. It is straightforward to rewrite the K result for B_d:[28]

$$|\Delta M| = \eta_{\text{QCD}} \frac{G_F^2 f_B^2 B_B m_B}{6\pi^2} |-e^{i\delta} \cdot s_1 s_2|^2 m_t^2, \tag{7}$$

where the QCD correction factor,[29] $\eta_{\text{QCD}} = 0.85$, and the factors of $-e^{i\delta} \approx V_{tb}$ and $s_1 s_2 \approx V_{td}^*$ are KM angle factors that arise at the $t - b$ and $t - d$ vertices, respectively. We have again used the form of the answer valid when $m_t^2/m_W^2 \ll 1$, although it is simple enough to copy the longer, exact expression.[10] Moreover, for the $B_s^0(\bar{b}s)$ meson, we need simply make the replacement, $V_{td}^* \rightarrow V_{ts}^* \approx s_2 + s_3 e^{-i\delta}$, to get the appropriate expression for ΔM_{B_s}.

The relationship to experiment is made through the quantity

$$r = \frac{(\Delta M)^2 + (\Delta\Gamma/2)^2}{2\Gamma^2 + (\Delta M)^2 - (\Delta\Gamma/2)^2} \approx \frac{(\Delta M/\Gamma)^2}{2 + (\Delta M/\Gamma)^2}. \tag{8}$$

For the case where the initial B is tagged as being a B^0 rather than $\bar{B}^0$, $r = l^-/l^+$, the number of "wrong" to "right" sign leptons in its semileptonic decay. For uncorrelated $B^0 + \bar{B}^0$ pairs it follows that $2r/(1 + r^2) = l^\pm l^\pm/l^+ l^-$, but for correlated pairs produced at the $\Upsilon(4S)$, $r = l^\pm l^\pm/l^+ l^-$.

Experimentally, Mart II has an upper limit[30] on the mixing for uncorrelated B's at PEP, as does the CLEO experiment[31] for correlated B's at the $\Upsilon(4S)$ that corresponds to $r_d < 0.24$. UA1, on the other hand, claims[32] a several standard deviation signal in same-sign dimuons, but has attributed it to B_s. The Argus collaboration[33] has recently claimed for the first time the observation of a nonzero value of $r_d \approx 0.20$ using three different methods from data taken at the $\Upsilon(4S)$. Moreover, they claim a 90% C.L. lower bound of $\Delta M/\Gamma > 0.44$. This is similar to a K_S^0, that is, $\Delta M_K/\Gamma_{K_S} \approx 0.5$, but for a particle about 10 times heavier and 100 times shorter lived!

How does this match with theory? For B_d, the quantity of merit for the mixing is

$$\left(\frac{\Delta M}{\Gamma}\right)^2_{B_d} \approx 0.06 \left(\frac{f_{B_d}}{f_K}\right)^4 B^2_{B_d} \left(\frac{s_2}{0.1}\right)^4 \left(\frac{m_t}{35\ \text{GeV}}\right)^4, \tag{9}$$

where we have reproduced an old result,[34] written with the parameters normalized to "standard" values of a few years ago (0.1 was then, and is still roughly now, the maximum allowed value of s_2). If we stretch the matrix element to the large end of a "reasonable" range ($f_B = f_K$ and $B_B = 1$) and do the same for the KM factor ($s_2 = 0.1$), then $r_d = 0.2$ is reproduced for $m_t \approx 60$ GeV. If we reduce the matrix element and KM factor ($B_B = 0.7$ and $s_2 = 0.07$), then m_t is pushed above ≈ 100 GeV.[35]

Note that for the B_s system, the relevant KM factor is $|V_{ts}|$, which in the (very good) approximation of small KM angles, θ_i, is equal to $|V_{cb}|$. The latter is fixed by the B lifetime, so there is no room to maneuver with respect to the KM angles in the expression for $\Delta M/\Gamma$, which becomes

$$\left(\frac{\Delta M}{\Gamma}\right)^2_{B_s} \approx 1.0 \left(\frac{f_{B_s}}{f_K}\right)^4 B^2_{B_s} \left(\frac{m_t}{35\ \text{GeV}}\right)^4. \tag{10}$$

It is impossible to avoid large mixing of B_s for any reasonable set of parameters that are in accord with moderate mixing for B_d.

These new data do not then necessitate new physics beyond the standard model with three generations, although the matrix element, KM factor, and m_t are being squeezed upward, and some or all of them must have larger values than was assumed over the last few years. This is the most important short-term consequence. Over the longer term, the most important implication of reasonably large B_d mixing may well be as engineering information for studies of *CP* violation in *B*-decay at a *B* factory or the SSC. The new results have given a boost to the chances of doing this important but difficult physics in the future.

TAU-DECAY

As is now known, the properties of the tau-lepton and the tau-neutrino are generally in excellent accord with those expected in the Standard Model for the third-generation charged and neutral leptons, respectively. The decays $\tau \to \nu_\tau e \bar{\nu}_e$, $\tau \to \nu_\tau \mu \bar{\nu}_\mu$, $\tau \to \nu_\tau \pi$, $\tau \to \nu_\tau 2\pi$, $\tau \to \nu_\tau 3\pi$, and $\tau \to \nu_\tau 4\pi$, occur[36] at the expected rates and account for almost 90% of tau-decays.

However, there is a nagging problem in accounting for all tau-decays.[32] It shows up in doing the bookkeeping for the sum of the exclusive modes that result in one charged prong and comparing the result with the corresponding inclusive topological branching ratio[36] of 86.3 ± 0.3%. Taking all the modes previously listed plus Cabibbo-suppressed modes into account, there remains about 7% of tau-decays that are not assignable to known exclusive channels.

The least disturbing explanation for this discrepancy—that the measurements of the major one-prong exclusive channels needed to be scaled up—became less likely as new and more accurate measurements became available.[38] By the time of the Kyoto Conference in 1985, the world-average branching ratios had significantly reduced errors and the discrepancy remained.[39]

However, data published in the past year seemed to point toward a resolution of the problem in a still relatively conventional manner through decays of the tau involving eta-mesons. The Mark II[40] and TPC[41] collaborations each used slightly different tagging techniques and, constraining the sum of all the branching fractions to be unity, found that tau-decays into a neutrino, charged pion, and two or more neutral hadrons were well in excess of the theoretical expectations[48] for the sum of just $\tau^- \rightarrow \nu_\tau \pi^- 2\pi^0$ and $\tau^- \rightarrow \nu_\tau \pi^- 3\pi^0$. As other modes involving only pions and kaons had been shown to make very small contributions to the one-prong plus multineutrals topology,[48] by process of elimination decays involving the eta-meson in the "multineutrals" became prime suspects.[42]

The finger was put more directly on the eta by the Crystal Ball[43] and the HRS[44] collaborations, both of which claimed evidence for an eta-signal in tau-decays. HRS reported a branching fraction for $\tau^- \rightarrow \nu_\tau \pi^- \eta^+$ anything of $5.0 \pm 1.0 \pm 1.2\%$. Thus, by the Berkeley Conference the discrepancy seemed to be moving toward a resolution in terms of tau-decays to an eta-meson.[36,45]

Even though no one expected such large branching ratios involving the eta, everything seemed to be quieting down to a quite conventional result. Now, the situation has come alive again with the HRS collaboration claiming that the decays involving the eta occur at the 5% branching-ratio level in exactly the mode, $\tau^- \rightarrow \nu_\tau \pi^- \eta$, that is not expected in the Standard Model.[46] The $\eta\pi$ system, which is G-odd, has natural spin-parity and in the Standard Model it must come from the vector current, which is G-even; we have by definition a process that involves a second-class current. Within the Standard Model this should happen at a level of roughly α^2 in the rate when compared to processes arising through the usual first-class currents. Such a decay would be completely negligible.[47] Other experiments should soon be able to confirm or refute this result.[c]

If we stay within the confines of the Standard Model, we can still ask the question of whether tau-decays involving the eta can account for the remaining one-prong modes. The most obvious candidate that is allowed in the Standard Model is $\tau \rightarrow \nu_\tau \eta 2\pi$. This proceeds through the vector current and can be related by CVC to a weighted integral $\sigma_{e^+e^- \rightarrow \eta\pi^+\pi^-}$. The data that have recently become available on the latter process allow one to place the bound[48] $BR(\tau^- \rightarrow \nu_\tau \eta \pi^- \pi^0) < 0.24\%$. Examination of all the other tau modes involving eta's shows[48] that a generous upper limit on the sum of their branching ratios is 2%. Within the Standard Model, the "missing" exclusive modes have to come from somewhere else, unless production of etas is a major part of the e^+e^- total cross section at low energy and has been completely missed. The puzzle has only become deeper as the possibilities have been further narrowed.

[c]After the 1st Int. Symp. for the Fourth Family of Quarks and Leptons, the Argus collaboration set an upper limit of 1.5% at 95% C.L. on the branching fraction for $\tau^- \rightarrow \nu_\tau \pi^- \eta$ (talk at the 22nd Rencontre de Moriond, Les Arcs, France, March 8–16, 1987) and the TPC collaboration set an upper limit on the branching fraction of the $SU(3)$ related decay, $\tau^- \rightarrow \nu_\tau K^- K^0$, of 0.26% (H. Aihara *et al.*, LBL Preprint LBL-23176, 1987. Unpublished).

REFERENCES

1. KOBAYASHI, M. & T. MASKAWA. 1973. Prog. Theor. Phys. **49:** 652.
2. PARTICLE DATA GROUP. 1986. Phys. Lett. **170B:** 74.
3. MARCIANO, W. J. & A. SIRLIN. 1986. Phys. Rev. Lett. **56:** 22.
4. LEUTWYLER, H. & M. ROOS. 1984. Z. Phys. **C25:** 91.
5. ABRAMOWICZ, H. *et al.* 1982. Z. Phys. **C15:** 19.
6. The *c* and *b* decay data at that time were reviewed by THORNDIKE, E. 1986. *In* Proc. 1985 Int. Symp. Lepton and Photon Interactions at High Energies. M. Konuma and K. Takahashi, Eds., 406. Kyoto Univ., Kyoto, Japan. For a review of the experimental situation as of the summer of 1986, see GILCHRIESE, M. G. D. 1986. Rapporteur talk at the 23rd Int. Conf. on High Energy Physics, Berkeley, Calif. (Cornell Preprint CLNS-86/754. Unpublished.)
7. KLEINKNECHT, K. & B. RENK. 1986. Mainz Preprint MZ-ETAP/86-4. Unpublished.
8. SIRLIN, A. & R. ZUCCHINI. 1986. Phys. Rev. Lett. **57:** 1994.
9. DRUKAREV, E. G. & M. I. STRIKMAN. 1986. Leningrad Institute of Nuclear Physics Preprint 1221. Unpublished.
10. INAMI, T. & C. S. LIM. 1981. Progr. Theor. Phys. **65:** 297. Contains the full expression for all values of m_q/M_W.
11. GILMAN, F. J. & M. B. WISE. 1980. Phys. Lett. **93B:** 129; GILMAN, F. J. & M. B. WISE. 1983. Phys. Rev. **D27:** 1128. Note that the QCD correction calculation has only been done for $m_q^2 \ll M_W^2$; it needs to be redone for the case where $m_t \gtrsim M_W$.
12. DONOGHUE, J. F., E. GOLOWICH & B. R. HOLSTEIN. 1982. Phys. Lett. **119B:** 412; APPELQUIST, T., J. D. BJORKEN & M. S. CHANOWITZ. 1973. Phys. Rev. **D7:** 2225.
13. BIJNENS, J., H. SONODA & M. B. WISE. 1984. Phys. Rev. Lett. **53:** 2367.
14. For a sample, see Table 6 in LANGACKER, P. 1986. *In* Proc. 1985 Int. Symp. Lepton and Photon Interactions at High Energies. M. Konuma and K. Takahashi, Eds., 186. Kyoto Univ., Kyoto, Japan.
15. GILMAN, F. J. & J. S. HAGELIN. 1983. Phys. Lett. **133B:** 443. We are neglecting corrections to equation 5 from $\eta - \eta'$ mixing and "electromagnetic penguins" (see the later discussion).
16. GILMAN, F. J. & M. B. WISE. 1979. Phys. Lett. **83B:** 83; GILMAN, F. J. & M. B. WISE. 1979. Phys. Rev. **D20:** 2392.
17. PHAM, T. N. 1984. Phys. Lett. **145B:** 113; PHAM, T. N. & Y. DUPONT. 1984. Phys. Rev. **D29:** 1368; DONOGHUE, J. F. 1984. Phys. Rev. **D30:** 1499.
18. See, for example, SHARPE, S., R. GUPTA, G. GURALNIK, G. W. KILCUP & A. PATEL. 1987. SLAC Preprint SLAC-PUB-4236. Unpublished.
19. BURAS, A. J., W. A. BARDEEN & J. M. GERARD. 1980. Phys. Lett. **180B:** 133; Max Planck Institute Preprints MPI-PAE/Pth-45/86 and MPI-PAE/Pth-55/86. Unpublished.
20. DONOGHUE, J. F., E. GOLOWICH, B. R. HOLSTEIN & J. TRAMPETIC. 1986. Phys. Lett. **179B:** 361; Erratum, 1987. Unpublished.
21. BURAS, A. J. & J. M. GERARD. 1987. Max Planck Institute Preprint MPI-PAE/Pth-7/87. Unpublished.
22. SHARPE, S. 1987. SLAC Preprint SLAC-PUB-4296. Unpublished.
23. BERNSTEIN, R. H. *et al.* 1985. Phys. Rev. Lett. **54:** 1631.
24. BLACK, J. K. *et al.* 1985. Phys. Rev. Lett. **54:** 1628.
25. WINSTEIN, B. 1987. Invited talk at the Division of Particles and Fields Meeting, Salt Lake City, Utah.
26. GRONAU, M. & J. SCHECHTER. 1985. Phys. Rev. **D31:** 1668; DECKER, R., E. A. PASCHOS, U. TURKE & H. USLER. 1985. Nucl. Phys. **B258:** 313; PASCHOS, E. A. 1987. Possible Signatures of the Fourth Generation at Present Energies. This volume.
27. For other recent reviews, see LANGACKER, P. Reference 14; WOLFENSTEIN, L. 1985. Commun. Nucl. Part. Phys. **14:** 135; Carnegie-Mellon Preprint CMU-HEP87-01.
28. ELLIS, J., M. K. GAILLARD, D. V. NANOPOULOS & S. RUDAZ. 1977. Nucl. Phys. **B131:** 285.
29. HAGELIN, J. S. 1981. Nucl. Phys. **B193:** 123.

30. SCHAAD, T. *et al.* 1985. Phys. Lett. **160B:** 188.
31. BEAN, A. *et al.* 1987. Phys. Rev. Lett. **58:** 183.
32. ALBAJAR, C. *et al.* 1986. CERN Preprint CERN-EP/86-209. Unpublished.
33. ALBRECHT, H. *et al.* 1987. Phys. Lett. **192B:** 245.
34. GILMAN, F. J. & J. S. HAGELIN. 1983. Phys. Lett. **133B:** 443.
35. For other comments on this situations, see BIGI, I. 1987. Indirect Searches for Very Heavy Quarks. This volume; PASCHOS, E. 1987. Possible Signatures of the Fourth Generation Present Energies. This volume; after the meeting, other directly related papers appeared: ELLIS, J., J. S. HAGELIN & S. RUDAZ. 1987. CERN Preprint CERN-TH.4679/87. Unpublished; CHAU L.-L. & W.-Y. KEUNG. 1987. UC Davis Preprint UCD-87-02. Unpublished.
36. GILCHRIESE, M. G. D. Reference 6.
37. GILMAN, F. J. & S. H. RHIE. 1985. Phys. Rev. **D31:** 1066.
38. BALTRUSAITIS, R. M. *et al.* 1985. Phys. Rev. Lett. **55:** 1842; ASH, W. W. *et al.* 1985. Phys. Rev. Lett. **55:** 2118.
39. THORNDIKE, E. Reference 6.
40. BURCHAT, P. R. *et al.* 1987. Phys. Rev. **D35:**27.
41. AIHARA, H. *et al.* 1986. Phys. Rev. Lett. **57:** 1836.
42. GAN, K. K. *et al.* 1987. SLAC Preprint SLAC-PUB-4110. Unpublished. Very recently the Mark II collaboration has been able to measure directly the branching ratio for $\tau^- \rightarrow \nu_\tau \pi^- 2\pi^0$ and found it to be in accord with these expectations, thereby tightening the argument further. If these authors assume that the additional tau-decay mode is $\tau^- \rightarrow \nu_\tau \eta \pi^- \pi^0$, then they find a corresponding branching ratio of $4.2^{+0.7}_{-1.2} \pm 1.6\%$.
43. KEH, S. *et al.* 1986. Contribution to the Berkeley Conf. quoted in reference 6.
44. ABACHI, S. *et al.* 1986. Contribution to the Berkeley Conf. quoted in reference 6.
45. BURCHAT, P. R. 1986. Invited talk at the 23rd Int. Conf. High Energy Physics, Berkeley, Calif. UC Santa Cruz Preprint SCIPP 86/72. Unpublished.
46. DERRICK, M. *et al.* 1987. Argonne Preprint ANL-HEP-PR-86-106. Unpublished.
47. BERGER, E. L. & H. J. LIPKIN. 1987. Phys. Lett. **189B:** 226; MEURICE, Y. & C. K. ZACHOS. 1987. Argonne Preprint ANL-HEP-PR-87-09. Unpublished.
48. GILMAN, F. J. 1987. SLAC Preprint SLAC-PUB-4265; GILMAN, F. J. Phys. Rev. **D35.** In press.

Theoretical Expectations for Mass Scales of the Fourth Generation and Higgs Bosons

CHRISTOPHER T. HILL

Fermi National Accelerator Laboratory
P.O. Box 500
Batavia, Illinois 60510

INTRODUCTION

A major problem of modern particle physics is to understand the origin and to compute the values of the quark and lepton masses and mixing angles. It is not clear that this will be possible in short order since, if considerations from grand unified theories (GUTs) are relevant, the details at inaccessibly high energy scales may be involved. Furthermore, the mechanism that breaks the electroweak symmetries is fundamentally distinct from the mechanism that produces the masses (certainly this is true in the Standard Model, since the Higgs–Yukawa couplings are parameters independent of the gauge couplings and Higgs potential; in many models this distinction is maintained).

Nature may oblige us with helpful new information at the TEV scale that sheds light on *The Theory*. However, there is another interesting possibility that, with a fourth generation, becomes somewhat likely: heavy quark and lepton masses are *universal*. That is, perhaps the details of the origin of the Higgs–Yukawa couplings are buried at the GUT scale, but are *irrelevant* in determining the masses measured at low energies due to the evolution of the parameters by the renormalization group. The low energy values may essentially be fixed points and can be computed by a knowledge of (1) the renormalization group equations up to the GUT scale, (2) the existence of a desert (this is really a simplifying assumption), (3) the existence of sufficiently heavy fermions that are governed by these fixed points. This idea was first proposed by Pendleton and Ross[1] and subsequently refined and applied to the fourth generation in reference 2. We review the application of fixed points to fermion masses, mixing angles, and the Higgs system for which the ideas apply and suggest natural scales.[3] We emphasize that the natural scale for Higgs–Yukawa couplings in superstring inspired models is already large ($\approx g_{\text{gauge}}$) and this possibility is quite plausible there.

There are many other theories of mass and mixing angles, including horizontal unification and extended technicolor. We do not discuss these at present, but rather propose a novel idea based upon the essential teachings of superstrings and Kaluza–Klein theories on the problem of fermion masses: Fermions emerge as zero-modes on the compact manifold associated with dimensional reduction, and the number of families is associated with a topological index. Little has been done to understand the resulting systematics of masses and mixings. We show how a geometric heirarchy might emerge from a novel realization of these ideas.

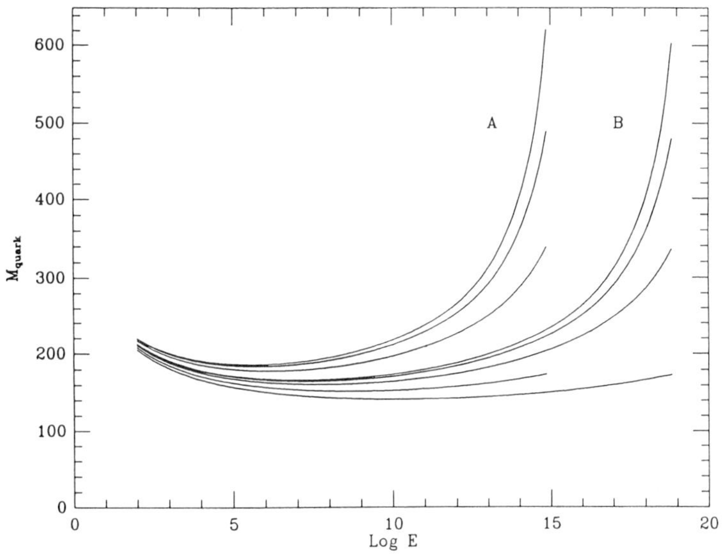

FIGURE 1. Evolution from (A) 10^{15} GeV and (B) 10^{19} GeV of effective mass of single heavy $+2/3$ charge quark.

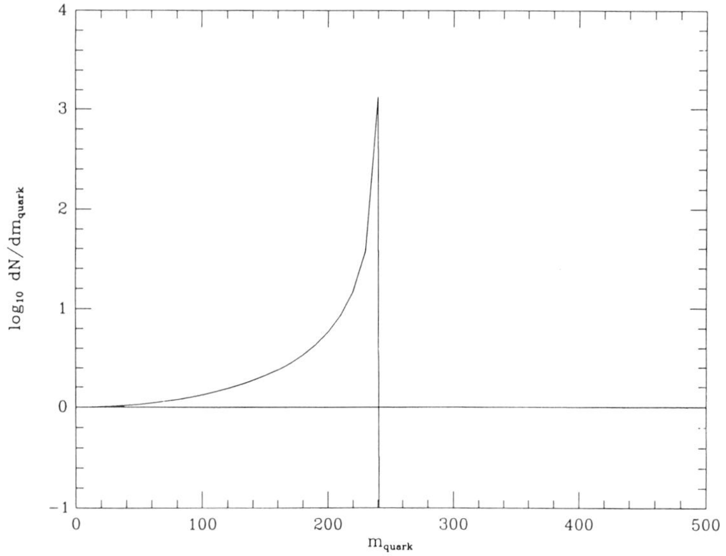

FIGURE 2. Probability distribution of masses of heavy quark with equal likelihood of any mass as defined at M_X.

RENORMALIZATION GROUP FIXED POINTS

Consider the evolution of the Higgs–Yukawa coupling constant of a Standard Model +2/3 charged quark. The renormalization group equation is

$$D \ln g_{+2/3} = \frac{3}{2} g_{+2/3}^2 - \frac{3}{2} g_{-1/3}^2 + \sum_{\text{all flavors}} g_f^2 - 8 g_3^2 - \frac{9}{4} g_2^2 - \frac{17}{12} g_1^2, \tag{1}$$

where $16\pi^2 D = d/d \ln E$. We may translate this into an effective "running mass" by multiplying the coupling constant by the Higgs VEV of 175 GeV. The resulting evolution given by equation 1 from a large scale, presumeably the GUT scale, for various choices of initial value is seen in FIGURE 1. Here we consider two candidate GUT scales, $M_X \approx 10^{15}$ GeV and $M_X \approx 10^{19}$ GeV. We notice that for a wide range of initial values the resulting low-energy top-quark mass is universal, that is, it is insensitive to the choice of initial value and it is insensitive to the choice of GUT mass.

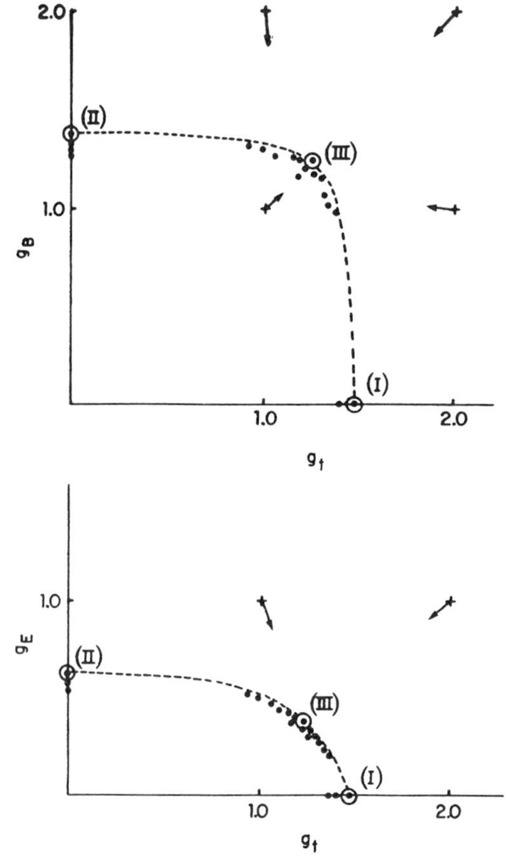

FIGURE 3. Scatter plots of g_t, g_b, g_e from initial array at M_X (integer initial values).

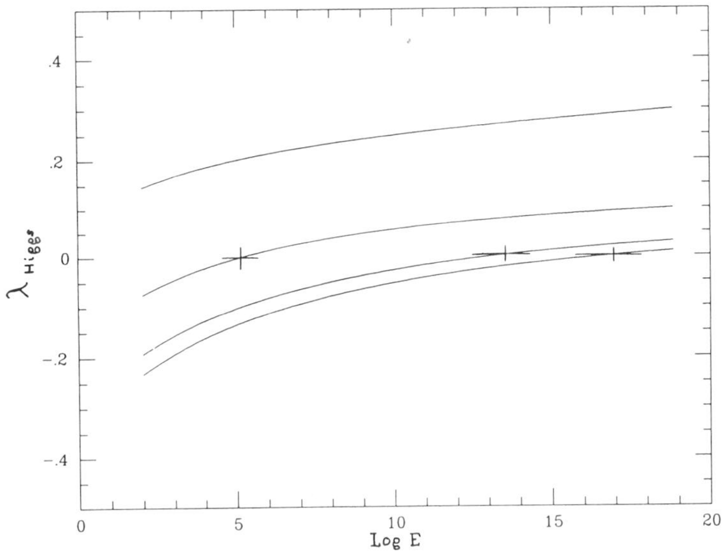

FIGURE 4. Higgs-quartic coupling integrated from 10^{19} GeV. Coleman–Weinberg breaking effectively occurs at zero crossing.

We would predict at one loop $m_{\text{quark}} \approx 240$ GeV. Fischler and Oliensis[4] have given a comprehensive two-loop analysis and obtain $224 < m_{\text{quark}} < 232$ GeV with $\Lambda_{\overline{MS}} = 0.16$ GeV. Also, to one loop we may compare supersymmetric $SU(5)$ that yields $m_{\text{quark}} \approx 205$ GeV.[5] Such a prediction for the top-quark mass cannot yet be ruled out by experiment, though it may be in conflict with the best estimates of the weak isospin splitting permitted by the ρ parameter.[6] Moreover, some other considerations in standard GUTology may be relevant.[7]

This may be stated another way: Given no a priori knowledge of the value of the quark mass at M_X, that is, a flat distribution of initial values at M_X, we determine the probability distribution at low energies as in FIGURE 2. We see it is peaked at the fixed point, which corresponds to an upper limit as first discussed in reference 8.

We turn to the fourth generation and consider the evolution equations (assuming a massless neutrino):[2]

$$D \ln (g_T) = \frac{9}{2} g_T^2 + \frac{3}{2} g_B^2 + g_E^2 - 8g_3^2 - \frac{9}{4} g_2^2 - \frac{17}{12} g_1^2 \tag{2}$$

$$D \ln (g_B) = \frac{9}{2} g_B^2 + \frac{3}{2} g_T^2 + g_E^2 - 8g_3^2 - \frac{9}{4} g_2^2 - \frac{5}{12} g_1^2 \tag{3}$$

$$D \ln (g_E) = \frac{5}{2} g_E^2 + 3g_B^2 + 3g_T^2 - \frac{9}{4} g_2^2 - \frac{15}{4} g_1^2. \tag{4}$$

We have here set all Cabibbo angles to zero. The results of a numerical integration are best presented as scatter plots for various initial points, as shown in FIGURE 3. It should be noted that no particular point on the fixed 2-surface in the 3-space of the three coupling constants is particularly favored, though the point up ~220 (204); down ~215 (198); lepton ~60 (68) GeV, appears to be an attractor (we give some two-loop results from reference 4 in parentheses). There is insufficient running time to significantly attract the three couplings to this point. However, the surface is definitely nonspherical; yet one can simply quote a sum rule.[9] Knowledge, however, of two masses determines the third.

APPLICATION TO THE HIGGS SYSTEM

It is quite interesting to apply these ideas to the Higgs system.[3] First we consider a single doublet and then generalize to the case of the two-doublet model.

First we assume the standard potential:

$$V(\phi) = -\frac{m_H^2}{2}|\phi^\dagger\phi| + \frac{\lambda}{4}|\phi^\dagger\phi|^2 \tag{5}$$

whence $v^2 = m_H^2/\lambda = (175\ \mathrm{GeV})^2$ and the Higgs boson has mass $m^2 = v^2\lambda$. Hence, a knowledge of λ determines the Higgs boson mass; λ is governed by a logarithmic renormalization group (RG). We find for λ the RG equation

$$\begin{aligned} D\lambda = {} & 12\lambda^2 - 3\lambda(g_1^2 + 3g_2^2) + \frac{3g_2^4}{2} + \frac{3(g_1^2 + g_2^2)^4}{4} \\ & + 4\,\lambda(3g_t^2 + 3g_b^2 + g_l^2) - 4(3g_t^4 + 3g_b^4 + g_l^4), \end{aligned} \tag{6}$$

where we have included the contributions from a single heavy generation.

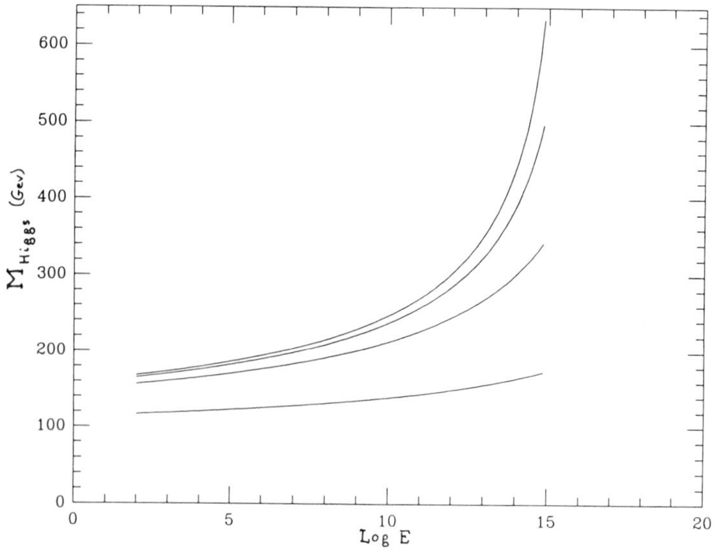

FIGURE 5. Evolution of Higgs effective mass from 10^{15} GeV. No apparent fixed point occurs in absence of heavy fermions.

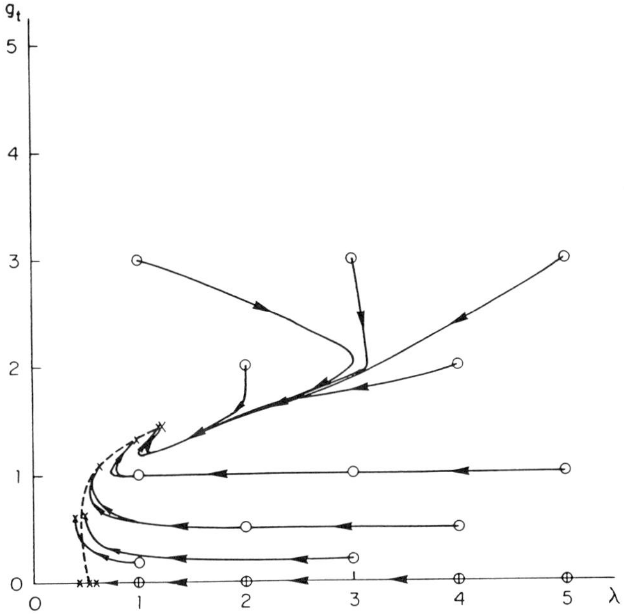

FIGURE 6. Joint evolution of large Higgs-Yukawa coupling and quartic Higgs coupling.

Curiously, we find that if we set the fermion contributions to zero (all fermions assumed light), then $D\lambda > 0$, and since we are descending in energy from a presumed M_X, small initial λ values can be driven negative. Mass scales at which λ becomes negative are essentially Coleman–Weinberg symmetry-breaking scales. This is shown in FIGURE 4. Moreover, there is no real fixed point as previously described, but there is a practical upper limit to the low-energy value of λ, and thus an upper limit on the Higgs mass. This is no longer universal, FIGURE 5.

Allowing for a single heavy quark we may study the joint evolution of the Higgs–Yukawa coupling, equation 1 (there is no backreaction of the Higgs quartic coupling onto the Higgs–Yukawa coupling). This is indicated in FIGURE 6. Thus, at the fixed line we may construct a relationship between the Higgs mass and the quark mass valid for $m_{\text{quark}} > 100$ GeV, as shown in FIGURE 7.

We turn now to the two-doublet model that has the general potential

$$V(\phi_1,\phi_2) = \mu_1^2\phi_1^\dagger\phi_1 + \mu_2^2\phi_2^\dagger\phi_2 + \frac{\lambda_1}{2}|\phi_1^\dagger\phi_1|^2 + \frac{\lambda_2}{2}|\gamma_2^\dagger\phi_2|^2$$
$$+ \lambda_3\phi_1^\dagger\phi_1\phi_2^\dagger\phi_2 + \lambda_4|\phi_1^\dagger\phi_2|^2 + \frac{\lambda_5}{2}e^{i\epsilon}(\phi_1^\dagger\phi_2)^2 + \text{h.c.} \quad \textbf{(7)}$$

There are obvious quartic stability constraints:

$$\lambda_1, \lambda_2 > 0 \quad \textbf{(8)}$$

and several others involving $\sqrt{\lambda_1\lambda_2}$ for various ranges of the other parameters.[3] Finally, there are ten ways of sewing the two doublets together with a generation of quarks and leptons (modulo $\phi_1 \leftrightarrow \phi_2$), such as (this is a fairly obvious notation; schemes **I.** through **X.** are listed in reference 3; no coupling may be viewed as coupling to light fermions and we always assume a light neutrino):

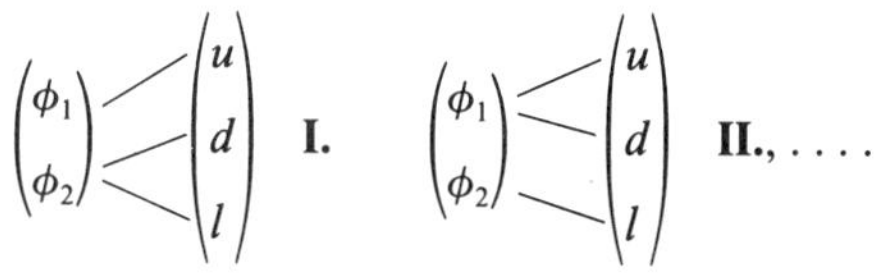

(9)

Since we have nothing to say about the evolution of the masses, we use as input parameters the known combination, $v_1^2 + v_2^2 = (175 \text{ GeV})^2$ and an unknown vacuum angle, $\tan\theta = v_1/v_2$.

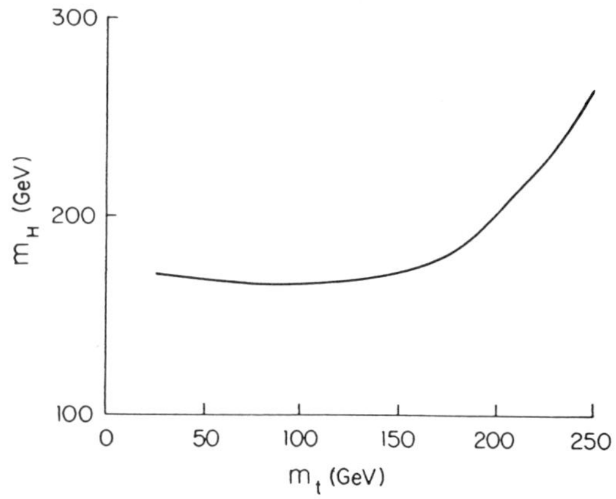

FIGURE 7. Fixed line relationship between a heavy "t-quark" and Higgs boson masses.

Now we must face several issues with the two-doublet model:

(1) Why is the breaking ferromagnetic $\uparrow\uparrow$ as opposed to $\uparrow\downarrow$, so as to preserve the $U(1)_{EM}$? This requires $\lambda_4 < 0$, so why should such a value be naturally selected?

(2) Are the quartic stability constraints naturally respected by evolution across the desert? (If not we get Coleman–Weinberg at the appropriate scale.)

(3) The model admits an axion in the ferromagnetic breaking when $\lambda_5 = 0$. Is there anything special, or natural, about that value?

(4) Does the model admit infrared RG fixed points?

In fact we find that the RG behavior of the theory is interesting only in the presence of a heavy quark or lepton (or both), and appears to develop quartic instabilities otherwise. For example, the evolution of the quartic coupling constants is plotted in

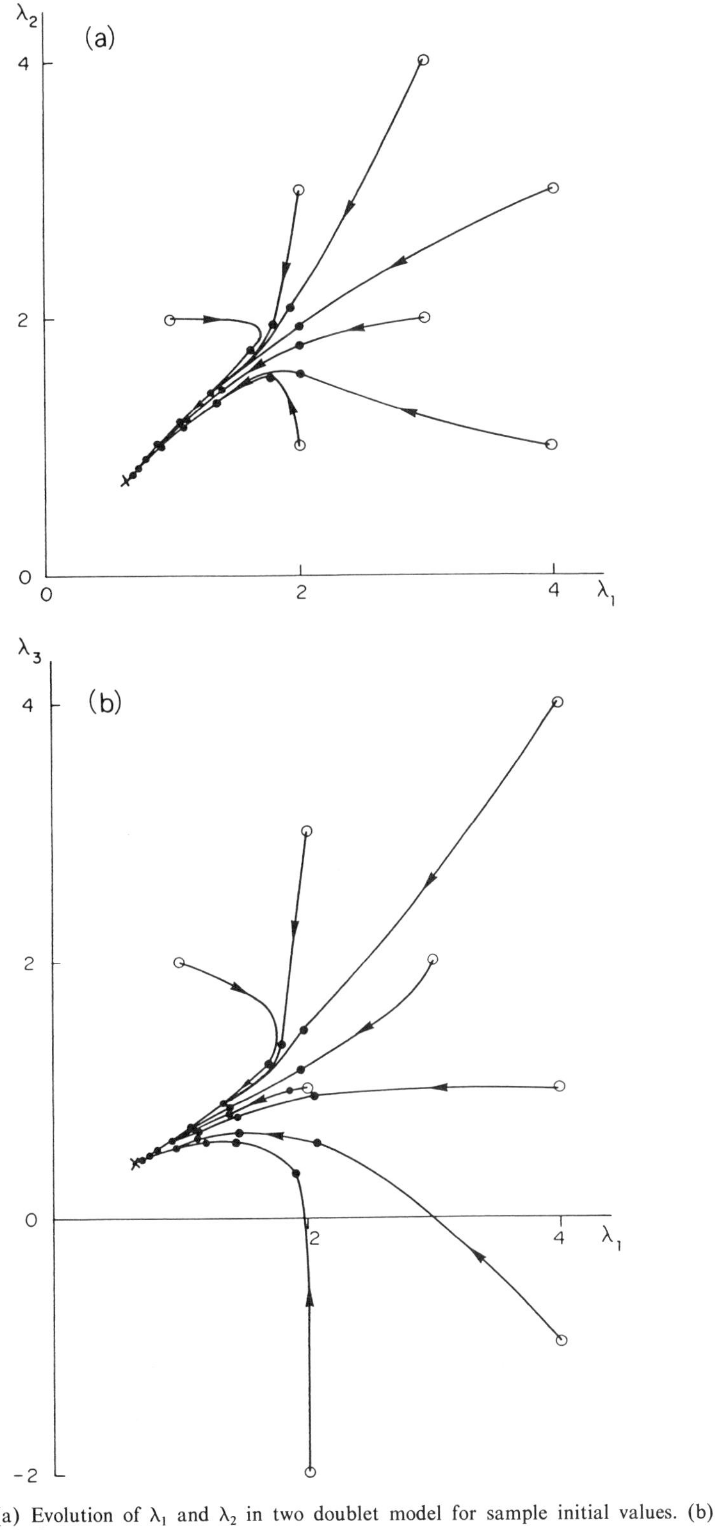

FIGURE 8. (a) Evolution of λ_1 and λ_2 in two doublet model for sample initial values. (b) Evolution of λ_1 and λ_3.

FIGURES 8a and 8b in scheme **I**. Remarkably, $\lambda_4 < 0$ occurs naturally as a fixed-point result (FIGURE 8c). Also, though there is insufficient evolution time to produce a bonafide low-mass axion, the coupling λ_5 is driven small and we get a neutral pseudoscalar as the lightest Higgs boson. Hence, $\lambda_5 = 0$ is a stable IR fixed point. The heavy neutrals have masses at the fixed point determined up to v_1/v_2, FIGURE 9. In

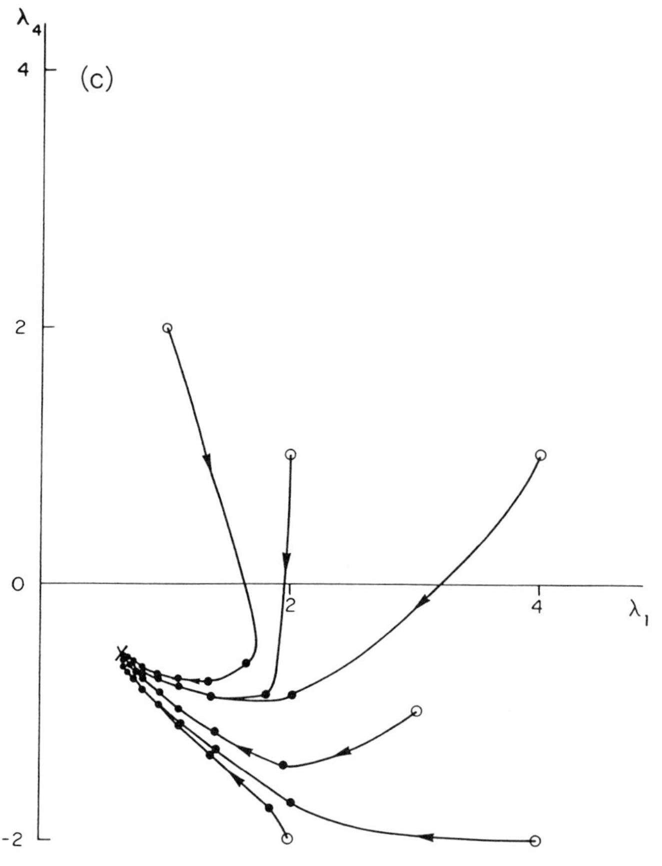

FIGURE 8. (c) Evolution of λ_1 and λ_4; note ferromagnetic ($\lambda_4 < 0$) fixed point occurs.

TABLE 1 we give the other masses at this fixed point, again for scheme **I.** Here, in addition to a heavy fourth generation, we include the effects of the top quark.

For comparison, if there are no heavy >100-GeV fermions, the evolution of the parameters is unstable.[3] In terms of a reasonable fundamental theory in the presence of the desert the two Higgs doublet scheme (and presumeably multi-Higgs) seems to make sense only in presence of a fourth generation.

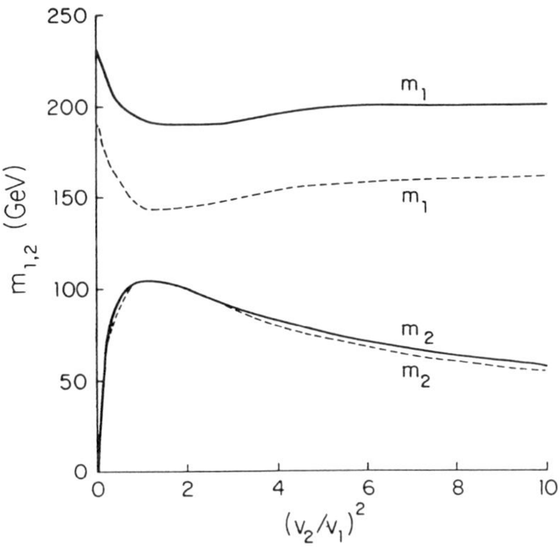

FIGURE 9. Scheme **I.** masses of neutral Higgs bosons in the two-doublet model.

TABLE 1. Higgs Masses (GeV) at the Fixed Point (Scheme **I**.)

m_{quark}	$m_{\pm}$	m_1^0	m_2^0	m_{axion}
0	188	190–230	0–105	60
50	186	188–225	0–105	60
172	136	164–216	0–131	25

TOPOLOGICAL THEORY OF MASSES AND MIXING ANGLES

One of the major lessons of compactified higher dimensional unified theories, such as Kaluza–Klein and superstring models, is that the low-energy laboratory fermions may be (1) zero-modes on a compactified manifold, and generally (2) the number of generations is determined by an index theorem on the compactified manifold. This latter point is important since the parent higher dimensional theory may have only one fermionic field that "spawns" the larger observed number of generations. There has been, however, little understanding to date as to how the fermion mass matrix arises, how it may lead to an approximate geometrical mass heirarchy, and how the texture of the CKM matrix emerges. Here we give some preliminary sketches of how these questions may be answered. Furthermore, we choose to study a modified problem that may either be viewed as a laboratory for the study of compactified higher dimensional theories or as a fundamentally new approach to dimensional reduction *without employing compactification.*

Spontaneous symmetry breaking of gauge theories can give rise to objects such as solitons, vortices, and monopoles. In D space dimensions the generalization of the flux tube is to a $D - 2$-dimensional object, and the solutions go over directly. If fermions obtain a mass by coupling to the complex scalar field of an abelian vortex, massless states localized on the vortex, known as Jackiw–Rossi zero-modes, can exist. Here a single fermion field in higher dimensions spawns n zero-modes if the vorticity of the flux tube is n. Presumably higher spin local bosonic condensates can yield vector and spin-2 zero-modes, candidate effective Yang–Mills, and gravitational fields, in a manner analogous to Witten's bosonic superconducting strings.[10]

An intriguing possibility then arises. If the universe is actually an infinite $1 + D$ spacetime, there might occur such objects in such a universe. The scale of symmetry breaking forming the soliton, M_s may be quite large, $M_s \gg M_W$ and probably $M_s \sim M_{\text{Planck}}$. The trapped zero-mode fermions would then be described by an effective $1 + D - d$ theory, and we choose $D - d = 3$ to correspond to the observed physical universe. Thus, *we have dimensionally reduced the world without compactifying the extra dimensions.*

The vortex is a nontrivial axially symmetric configuration of a complex scalar field ϕ and an optional gauge field A that we set to zero. The scalar field will be taken as a fixed c-number of the form: $\phi(r) = e^{in\theta} f(r)$, where, (r, θ, z) are cylindrical coordinates normal to the string. For definiteness, we will take $n > 0$. The asymptotic forms of the fields are $f(r) \rightarrow 0$ for $r \rightarrow 0$ and $f(r) \rightarrow f_0$ for $r \rightarrow \infty$.

The simplest way to exhibit the heirarchy phenomenon is to write the Lagrangian for the fermions:

$$\begin{aligned} L = {} & i\psi^\dagger \partial\!\!\!/ \psi + i\chi^\dagger \partial\!\!\!/ \chi + i\lambda^\dagger \partial\!\!\!/ \lambda + i\delta^\dagger \partial\!\!\!/ \delta \\ & + ig\phi(r, \theta)\psi^T \epsilon \chi - ig\phi^*(r, \theta)\lambda^T \epsilon \delta \\ & + h\sigma(\psi^T \epsilon \delta + \chi^T \epsilon \lambda) + \text{h.c.} + L(\sigma), \end{aligned} \qquad \textbf{(10)}$$

where ψ, χ, λ, and δ are 2-component (left-handed Weyl) spinors. Here σ is a scalar field that develops a VEV localized on the string, $\langle \sigma \rangle = \sigma_0 \exp(-\kappa r)$. This is analogous to Witten's bosonic superconducting string[10] and can easily be engineered by dynamics. It is presented here to simplify the arguments to come and is not essential for the present effect (we can set $g = g'$ with the σ condensate; alternatively we can generate the geometric hierarchy with σ = constant and $g' \gg g$).

Far from the flux tube the Higgs field may be regarded as having a constant VEV. Then we see that the pairs of Weyl spinors, (ψ, χ) and (λ, δ) form two four-component massive Dirac fields. On the vortex there occur Jackiw–Rossi zero-modes, and (ψ, χ) produce n chiral left-movers, while (λ, δ) produce n chiral right-movers for a vorticity n.

For any vorticity n we have to a good approximation the sequence of normalized zero-modes from $p = 0$ to $p = \pm(n - 1)$: $\psi_p = \beta_p(r)\alpha(z,t)e^{ip\theta}$, and one finds that the $\beta_p(r)$ have the behaviors

$$\beta_p(r) \rightarrow c_p r^{|p|} \qquad r \rightarrow 0; \qquad \beta_p(r) \rightarrow e^{-|gf_0|r} \qquad r \rightarrow \infty. \qquad \textbf{(11)}$$

We shall assume $\kappa \gg |gf_0|$, that is, the radial profile of the σ condensate varies rapidly compared to the fermionic zero-modes. This simply requires a choice of small g.

Let us now consider the effects of the mass terms. We may now compute the mass matrix for the fermionic zero-modes by integrating the transverse wave-function radial profile as given in equation 11:

$$M_{pq} = \int r\, dr\, d\theta \beta_p \beta_q^* h\sigma_0 e^{-\kappa r} e^{i(p-q)\theta}. \qquad \textbf{(12)}$$

Clearly, owing to the angular dependence of the modes, the mass matrix is diagonal; that is, generation number is topologically conserved at this level and no flavor mixing occurs. We see, however, that a geometrical heirarchy emerges:

$$M_{pq} \approx 2\pi c_p^2 \delta_{pq} \int r\, dr\, r^{|2p|} \sigma_0 e^{-\kappa r} \approx \delta_{pq} K_p \epsilon^{|2p+1|}, \tag{13}$$

where $\epsilon \sim |gf_0|/\kappa \ll 1$ and $K_p \sim 1$. (We have numerically solved the transverse Dirac equation and verified that this heirarchy does indeed occur; the K_p are interesting for making model predictions and must be determined numerically.[11])

Mixing can occur by introducing higher dimension operators into the Lagrangian as fermionic mass terms, but this also involves the vortex Higgs. For example, we might add terms involving $\psi^\dagger\Phi^2\psi$ and $\psi^\dagger\Phi^3\psi$ to pairs of Weyl spinors with the appropriate charges (this requires introducing different charged fermions). Such terms will naturally be suppressed by $1/M_{\text{heavy}}$, $1/M^2_{\text{heavy}}$, etc. These terms will lead to mixing between families and a rudimentary KM matrix emerges.

Of course, the principal difficulty is to give a convincing explanation of gravity and Yang–Mills fields trapped on the flux tube. These issues will be addressed in a forthcoming paper.[11]

SUMMARY

We have discussed two classes of theories that have something to say about fourth-generation quark and lepton masses (as well as Higgs). One is based upon universality of low-energy parameters as determined by the renormalization group: infrared fixed points. The other is a toy model of geometric heirarchy that may arise in a novel scheme of dimensional reduction without compactification.

REFERENCES

1. PENDLETON, B. & G. G. ROSS. 1981. Phys. Lett. **98B:** 291.
2. HILL, C. T. 1981. Phys. Rev. **D24:** 691.
3. HILL, C. T., C. N. LEUNG & S. RAO. 1985. Nucl. Phys. **B262:** 517.
4. FISCHLER, M. & J. OLIENSIS. 1983. Phys. Rev. **D28:** 194; FISCHLER, M. & J. OLIENSIS. 1982. Phys. Lett. **119B:** 385.
5. BAGGER, J., S. DIMOPOULOS & E. MASSÓ. 1985. Phys. Rev. Lett. **55:** 920; GOLDBERG, H. 1985. Phys. Lett. **165B:** 292.
6. MARCIANO, W. J. 1987. Constraints and speculations on fourth-generation physics. This volume.
7. FISCHLER, M. & C. T. HILL. 1981. Nucl. Phys. **B193:** 53; BAGGER, J., S. DIMOPOULOS & E. MASSO. 1985. Phys. Lett. **145B:** 211.
8. CABIBBO, N. *et al.* 1979. Nucl. Phys. **B158:** 295.
9. BAGGER, J., S. DIMOPOULOS & E. MASSÓ. 1985. Phys. Rev. Lett. BAGGER, J. S. DIMOPOULOS & E. MASSÓ. 1985. Nucl. Phys. **B253:** 397.
10. WITTEN, E. 1985. Nucl. Phys. **B249:** 557.
11. HILL, C. T. In preparation.

Constraints and Speculations on Fourth-Generation Physics[a]

W. J. MARCIANO

Physics Department
Brookhaven National Laboratory
Upton, New York 11973

INTRODUCTION

A fourth generation of quarks and leptons, which we refer to as (t', b', L, ν_L), is neither predicted nor disallowed by the Standard Model. In our opinion, it is the most conservative guess one could make as to what new physics lies ahead; so, we should keep an open mind regarding its existence.

In this paper, we describe some low-energy constraints on the properties of a fourth generation and point out some implications it might have for neutrino oscillations. To begin with, we give some results of a recent global analysis[1] of weak neutral current phenomenology that provide constraints on fourth-generation masses. Then we survey bounds on fourth-generation quark mixing from charged current unitarity and point out that $K^+ \longrightarrow \pi^+\nu\bar{\nu}$ provides an important probe of fourth-generation physics. Finally, we speculate about some properties a fourth-generation neutrino might exhibit, with emphasis on neutrino oscillations in matter.

NEUTRAL CURRENT CONSTRAINTS

A global analysis of all existing weak neutral current data combined with m_W and m_Z measurements has been recently completed.[1] One of the results of that work was to obtain from a 2-parameter fit

$$\rho = 0.998 \pm 0.0086 \tag{1a}$$

$$\sin^2\theta_W = 0.229 \pm 0.0064, \tag{1b}$$

where the central values correspond to $m_t = 45$ GeV and $m_{\text{Higgs}} = 100$ GeV, but the errors allow m_t to be as large as 100 GeV and $m_{\text{Higgs}} \leq 1$ TeV. A fourth generation of quarks and leptons with large mass splittings between isodoublet partners would have shown up in that analysis by leading to $\rho > 1$. (They enter through loop corrections.[2]) For $|V_{t'b'}| \simeq 1$, the quarks produce a shift

$$\Delta\rho = \frac{3\alpha}{16\pi\sin^2\theta_W}\left[\left(\frac{m_{t'} - m_{b'}}{m_W}\right)^2 + \frac{2m_{t'}m_{b'}}{m_W^2}\left(1 + \frac{m_{t'}m_{b'}}{m_{t'}^2 - m_{b'}^2}\ln\frac{m_{b'}^2}{m_{t'}^2}\right)\right], \tag{2}$$

[a]This work was supported by the U.S. Department of Energy under Contract DE-AC02-76CH00016.

while the lepton contribution is down by a 1/3, such that for $m_{\nu_L} \simeq 0$

$$\Delta\rho = \frac{\alpha}{16\pi \sin^2\theta_W} [m_L^2/m_W^2]. \tag{3}$$

Assuming that those shifts are not canceled by some other new effects, such as Higgs triplets, one finds[1] from equations 1–3

$$|m_{t'} - m_{b'}| \leq 180 \text{ GeV} \qquad 90\% \text{ C.L.}$$
$$m_L \leq 310 \text{ GeV} \qquad 90\% \text{ C.L.} \tag{4}$$

The constraint on $|m_{t'} - m_{b'}|$ is quite restrictive. Combined with the UA1 bound[3]

$$m_L \geq 41 \text{ GeV} \tag{5}$$

it implies that if a fourth generation exists, the lepton and quark masses are likely to be less hierarchial than we have become accustomed to. For example, based on previous charged lepton–quark mass ratios, one might have inferred from equation 5 that $m_{b'} \gtrsim$ 100 GeV; however, equation 4 then limits $m_{t'}$ such that $m_{t'}/m_{b'} \lesssim 2.8$. That ratio is to be compared with $m_c/m_s \simeq 8$ and $m_t/m_b > 6$.

If the t' and b' mix with the lighter generations and we parametrize that mixing by V_{ij}, $i = u, c, t, t'$, $j = d, s, b, b'$, then the ρ constraint also implies

$$\sum_j |V_{t'j}|^2 \frac{|m_{t'} - m_j|^2}{m_W^2} \lesssim 5 \tag{6a}$$

and

$$\sum_j |V_{ib'}|^2 \frac{|m_{b'} - m_i|^2}{m_W^2} \lesssim 5. \tag{6b}$$

That confirms one's intuition that very heavy quarks should have small mixing with light quarks.

CHARGED-CURRENT UNITARITY

In the three-generation standard model, the charged-current quark mixing matrix

$$V = \begin{pmatrix} V_{ud} & V_{us} & V_{ub} \\ V_{cd} & V_{cs} & V_{cb} \\ V_{td} & V_{ts} & V_{tb} \end{pmatrix} \tag{7}$$

is constrained by unitarity to satisfy

$$\sum_i V_{ij}^* V_{ik} = \sum_i V_{ji}^* V_{ki} = \delta_{jk}. \tag{8}$$

The presence of a fourth generation could lead to an apparent deviation from three-generation unitarity via its mixing. The current experimental values[4]

$$|V_{ud}| = 0.9744 \pm 0.0010 \tag{9a}$$

$$|V_{us}| = 0.220 \pm 0.002 \tag{9b}$$

$$|V_{ub}| < 0.012 \tag{9c}$$

$$|V_{cd}| = 0.207 \pm 0.024 \tag{9d}$$

$$|V_{cs}| = 0.95 \pm 0.14 \tag{9e}$$

$$|V_{cb}| = 0.043^{+0.006}_{-0.008} \tag{9f}$$

are completely consistent with three-generation unitarity. For example,

$$|V_{ud}|^2 + |V_{us}|^2 + |V_{ub}|^2 = 0.9979 \pm 0.0021. \tag{10}$$

From such results one finds the 90% C.L. constraints

$$|V_{ub'}| \leq 0.07$$

$$|V_{cb'}| \leq 0.63$$

$$|V_{t'd}| \leq 0.14$$

$$|V_{t's}| \leq 0.62. \tag{11}$$

Those constraints are not very stringent. Indeed, if a fourth generation exists, we would normally expect its mixing with the first and second generations to be much less than the values in equation 11.

A fourth generation could also show up via flavor-changing loop effects. One can use $K_L \longrightarrow \mu^+\mu^-$, $K_L - K_S$ mass difference and CP violation parameters to bound the masses and mixings of t'. The announced observation of $B_d^0 - \bar{B}_d^0$ oscillations by the Argus collaboration[5] may be the first harbinger of fourth-generation effects. Alternately, it may imply that m_t and/or $|V_{td}|$ are larger than anticipated. A fourth generation can also enhance the $K^+ \longrightarrow \pi^+\nu\bar{\nu}$ decay rate.[6,7] In fact, a branching ratio of $O(10^{-9})$ would very likely point to a fourth generation. An experimental measurement of that branching ratio with a sensitivity of 2×10^{-10} is currently underway at Brookhaven National Laboratory (BNL).

NEUTRINO OSCILLATIONS

A fourth-generation neutrino, ν_L, would influence neutrino oscillation phenomena. Oscillations are dependent on neutrino-mass-squared differences $\Delta m_{ij}^2 = m_i^2 - m_j^2$, mixing angles, and (in the case of propagation through matter) indices of refraction n_{ν_l}. It seems plausible to assume a neutrino-mass hierarchy

$$m_4^2 \gg m_3^2 \gg m_2^2 \gg m_1^2 \tag{12}$$

and if the quark section is a proper guide, one should anticipate small mixing. Indeed, the grand unified theory (GUT) seesaw mechanism[8] often leads to $m_\nu^2 = m_l^4/M^2$ or m_q^4/M^2, where M is a very large mass scale $10^{10} - 10^{18}$ GeV, while m_l and m_q are lepton and quark masses. Using the former with $m_L \simeq 60$ GeV gives

$$m_4^2 : m_3^2 : m_2^2 : m_1^2 :: 10^{20} : 10^{14} : 10^9 : 1, \qquad (13)$$

but the overall scale is very sensitive to the value of M. If we assume that the MSW effect[9] is the solution to the solar neutrino problem, then some $\Delta m_{i1}^2 \simeq 10^{-7} \sim 10^{-4}\,\mathrm{eV}^2$. In a four-generation scenario, it may be more likely that Δm_{31}^2 or Δm_{41}^2 is in that mass range.[10] The first case corresponds to $M \simeq 10^{12}$ GeV, while in the second, $M \simeq 10^{15}$ GeV. In the latter scenario one might eventually sort the mass and mixing out using solar neutrino experiments, but it will be very hard. In the event that $\Delta m_{31}^2 \simeq 10^{-7} \sim 10^{-4}\,\mathrm{eV}^2$, then one might expect from equation 13

$$\Delta m_{4j}^2 \simeq 10^{-2} \sim 10^2\,\mathrm{eV}^2. \qquad (14)$$

That range is currently being explored at reactor and accelerator facilities where $\bar{\nu}_e$ and ν_μ oscillations are searched for. Oscillations into $\overset{(-)}{\nu}_L$ would lead to an effective disappearance of flux in charged-current reactions. Unfortunately, if the fourth-generation mixing with the first two generations is small (as anticipated), the signal will probably be undetectable, since flux normalization uncertainties generally limit such disappearance experiments to 10% or more conversion.

If a source of very high-energy neutrinos $\gtrsim 2$ TeV were available, then in principle oscillations into $\overset{(-)}{\nu}_L$ would be directly observable via $\nu_L + N \rightarrow L + X$ (i.e., E_ν must be above L threshold). The cross section would in fact be quite large $\gtrsim 10^{-35}\,\mathrm{cm}^2$. Unfortunately, there again small mixing and very long vacuum oscillation lengths $(\Delta m^2/E_\nu)^{-1}$ would render such measurements impossible. One can inquire as to whether matter effects can be used to enhance high-energy oscillations. We know from the MSW effect that strong enhancements can occur if certain resonance conditions are met. We close this section by describing what conditions might render oscillations into $\overset{(-)}{\nu}_L$ large.

Given the properties previously discussed, two types of resonance phenomena involving $\overset{(-)}{\nu}_L$ may occur

$$\nu_e \longrightarrow \nu_L \qquad (15a)$$

$$\bar{\nu}_\mu \quad \text{or} \quad \bar{\nu}_\tau \longrightarrow \bar{\nu}_L \qquad (15b)$$

Assuming $\Delta m_{41}^2 \simeq 10^{-2} - 10^2\,\mathrm{eV}^2$, the first case, which is controlled by[9]

$$p_\nu(n_{\nu_e} - n_{\nu_L}) \simeq -\sqrt{2}G_\mu N_e, \qquad (16)$$

where $G_\mu = 1.16636 \times 10^{-5}\,\mathrm{GeV}^{2}$ and N_e = electron density, leads to resonance for solar densities if $E_\nu \gtrsim 5$ TeV (which is probably above L production threshold). The second scenario results from higher order loop corrections. At the one-loop level, one finds for a neutral isoscalar medium[11]

$$\begin{aligned} p_\nu(n_{\nu_L} - n_{\nu_\mu}) &\simeq p_\nu(n_{\nu_L} - n_{\nu_\tau}) \\ &\simeq -p_\nu(n_{\nu_e} - n_{\nu_L})\frac{\alpha}{16\pi \sin^2\theta_W} \\ &\quad \cdot \left(\frac{20X_L + X_L^2}{1 - X_L} + \frac{24X_L - 3X_L^2}{(1 - X_L)^2}\ln X_L\right) \\ X_L &= m_L^2/m_W^2. \end{aligned} \qquad (17)$$

So, for $m_L \simeq 60$ GeV,

$$p_\nu(n_{\nu_L} - n_{\nu_\mu}) \simeq p_\nu(n_{\nu_L} - n_{\nu_\tau}) \simeq 7 \times 10^{-3} p_\nu(n_{\nu_e} - n_{\nu_L}) \qquad \mathbf{(18)}$$

and one expects resonant conversion of $\bar{\nu}_\mu$ or $\bar{\nu}_\tau$ into $\bar{\nu}_L$ (for solar densities) for $E_\nu >$ several hundred teraelectronvolts. (Of course, the preceding estimates are very rough.)

A candidate source for high-energy $\overset{(-)}{\nu}_L$ via matter-enhanced oscillations may be cosmic accelerators,[12] such as Cygnus X-3, which is a neutron star-ordinary star binary. It is believed that very high-energy ν_e, $\bar{\nu}_e$, ν_μ, and $\bar{\nu}_\mu$ may be produced by that source. As they propagate through the dense companion of the neutron star, some high-energy ν_e and $\bar{\nu}_\mu$ may be converted to ν_L and $\bar{\nu}_L$ by the just-mentioned matter-enhanced oscillation mechanism. They may be observed on earth via $\overset{(-)}{\nu} + N \rightarrow L^\pm + X$ with the subsequent decay $L^\pm \rightarrow \mu^\pm \nu\bar{\nu}$. The signal would be a high-energy $\mu^\pm$ coming from the direction of the cosmic source, but smeared by relatively high p_T due to the mass of the decaying $L^\pm$.

CONCLUSION

Low-energy phenomenology is starting to provide bounds on the masses and mixings of a potential fourth generation of quarks and leptons. During the next few years those constraints should tighten from the results of $K^+ \rightarrow \pi^+\nu\bar{\nu}$ and $B^0 - \bar{B}^0$ oscillation experiments or an indirect signal for a fourth generation may be observed. Of course, Z-width measurements and searches for $e^+e^- \rightarrow \nu\bar{\nu}\gamma$ at high energies should be directly sensitive to the presence of a light fourth-generation neutrino.

If a fourth generation does exist, its neutrino may further enrich neutrino oscillation phenomena. One interesting possibility, matter-enhanced $\nu_e \rightarrow \nu_L$ and $\bar{\nu}_\mu \rightarrow \bar{\nu}_L$ oscillations, was speculated on in this paper. A potential source of such a phenomenon might be cosmic accelerators such as Cygnus X-3. The signal, production of a massive $L^\pm$ with subsequent decay into a fairly high p_T $\mu^\pm$ may be observable in terrestrial detectors and should be searched for.

SUMMARY

We have surveyed some constraints from low-energy phenomenology on fourth-generation fermion masses and mixings. We then speculated on properties of matter-enhanced high-energy fourth-generation neutrino oscillations.

REFERENCES

1. Amaldi, U., A. Böhm, L. S. Durkin, P. Langacker, A. Mann, W. Marciano, A. Sirlin & H. H. Williams. 1987. Phys. Rev. **D36:** 1385.
2. Veltman, M. 1977. Nucl. Phys. **B123:** 89.
3. Cline D. & M. Mohammadi. 1986. *In* Proc. Workshop on SSC Simulation. Madison, Wis.
4. Marciano, W. J. & A. Sirlin. 1987. *In* Proc. DPF Meeting. Salt Lake City, Utah. 172.
5. Albrecht, H. 1987. Phys. Lett. **B192:** 245.

6. MARCIANO, W. & Z. PARSA. 1986. Annu. Rev. Nucl. Part. Sci. **36:** 171; MARCIANO, W. 1986. *In* Proc. 2nd Conf. on the Intersections Between Particle and Nuclear Physics. Lake Louise, Alta., Canada. 596.
7. TÜRKE, U. 1986. Phys. Lett. **B168:** 296.
8. GELL-MANN, M., P. RAMOND & R. SLANSKY. *In* Supergravity. D. Freedman & P. van Nieuwenhuizen, Eds. North-Holland, Amsterdam.
9. MIKHEYEV, S. P. & A. Yu. SMIRNOV. 1978. Nuovo Cimento **9C:** 17; WOLFENSTEIN, L. 1978. Phys. Rev. **D17:** 2369.
10. LANGACKER, P., S. PETCOV, G. STEIGMAN & S. TOSHEV. 1987. Nucl. Phys. **B282:** 589.
11. BOTELLA, F. J., C. S. LIM & W. J. MARCIANO. 1987. Phys. Rev. **D35:** 896.
12. EICHLER, D. 1978. Nature **275:** 725; VESTRAND W. & D. EICHLER. 1982. Astrophys. J. **261:** 251.

Patterns of Fermion Masses, Mixings, and *CP* Violation[a]

M. GRONAU

Department of Physics
Technion–Israel Institute of Technology
Haifa, Israel

INTRODUCTION

There are two mutually related well-known facts about the standard $SU(3) \times SU(2) \times U(1)$ gauge theory, which illustrate the crucial role played by fermion masses in this theory. We first elaborate somewhat lengthily on these two points to advance the viewpoint taken in this paper.[1]

(1) The model without fermion masses is considerably prettier than the real world. (One of the few ironies of nature.) It has a huge global symmetry of $U(N) \times U(N) \times U(N) \times U(N) \times U(N)$, where N = number of families. The unitary symmetries in the product group are family symmetries that act on N left-handed weak doublet quarks, on N charge 2/3 ("up") right-handed singlet quarks, on N charge $-1/3$ ("down") right-handed singlet quarks, on N left-handed doublet leptons, and on N right-handed charged leptons. This enormous symmetry is broken in the massive theory.

In the grand unified theories (GUTs), which relate quarks and leptons, the massless theory has a more restricted global symmetry. A massive GUT uses up some of the reduced symmetry to relate quark and lepton masses. Thus massless $SU(5)$ has a global $U(N) \times U(N)$ symmetry (for N families of $\bar{5}$ and 10 representations) and massless $SO(10)$ has merely a $U(N)$ symmetry (for N families of 16 representations). The massive versions of these theories have the potential of relating charged lepton and down-quark masses of a given family [$SU(5)$[2]], and can even establish a relationship among the masses of all the fermion species of one family, including those of massive neutrinos [$SO(10)$[3]].

Since mass generation breaks the previously mentioned global symmetries, the problem of fermion masses is tightly linked to the following theoretical questions:

- What determines the pattern of symmetry breaking?
- Which symmetries are broken explicitly?
- Which symmetries are broken spontaneously?
- Which symmetries are part of a family unifying gauge symmetry?
- Are there some fundamental discrete symmetries?

These and many other interesting questions cannot be unambiguously answered at present. Different models may have different answers. A somewhat pragmatic way to

[a]This work was supported in part by a U.S.-Israel Binational Science Foundation (BSF) grant, and in part by Jubiläumsfonds der Österreichischen Nationalbank Project 2765.

proceed seems to be to search for patterns in the data that may have some clues to the answers. The clues should not be looked for merely in the fermion mass spectrum, as will be explained in the next point.

(2) The rather old puzzle of "who ordered the muon"[b] has turned the standard $SU(3) \times SU(2) \times U(1)$ gauge theory into a long list of very important questions to which the theory provides no answer. To mention a few:

- Why do fermions (quarks and leptons) appear in families?
- What determines the overall number of these families N?
- What fixes the actual values of fermion masses?
- What determines flavor mixing by the weak interactions?
- What is the origin of *CP* violation, which in the standard 3-family model is treated just as another phase-mixing parameter?

These questions about the various observed phenomena are closely related to the purely theoretical questions listed under point 1. The Standard Model parameterizes the

TABLE 1. Fermion Masses (in GeV)

	First Family			Second Family			Third Family
ν_e	<0.00000002		ν_μ	<0.00025		ν_τ	<0.07
e	0.000511	205 →	μ	0.105	17 →	τ	1.78
d	0.009 ± 0.003	20 →	s	0.17 ± 0.05	20–40 →	b	5.3 ± 0.1
u	0.005 ± 0.002	200–500 →	c	1.4 ± 0.1	>25 →	t	>40

phenomena related to these questions in terms of a large number (about 20) of arbitrary parameters without providing a quantitative explanation for their magnitudes. In the model the masses are generated from arbitrary Yukawa couplings between two fermions and a spinless field. The answers to the preceding questions are all given by the yet unknown structure of the fundamental (weak eigenstate) fermion-mass matrix generated when the scalar field gets a vacuum expectation value. Systematic studies of the preceding measured quantities (the number of families, masses, mixing, and *CP* phases) may be used to investigate this structure, or alternatively the structure of the Higgs system. If the latter is an effective low-energy description of a more fundamental mechanism of fermion-mass generation, then such studies may provide the first window into this new layer of physics. Again, at this point it seems premature to answer the preceding questions with an a priori theory. The modest task one may undertake is to study systematics in masses, mixing, and the *CP* phase, looking for clues for the structure of the fermion-mass matrix. We adopt this viewpoint in the following discussion.

[b] This phrase is attributed to I. Rabi.

THE DATA

A summary of the present existing information relevant to the question of fermion masses is given in TABLE 1.

In the table the numbers on the horizontal arrows describe mass ratios of adjacent families. The running quark masses are all given at a scale of 1 GeV.[4] Note that the light-quark-mass ratios are known from current algebra to a much higher accuracy than the absolute values of these masses.

The most obvious patterns that emerge from the table of masses are the following.

(1) *A family hierarchy:* Similar fermions of adjacent families have masses that differ by factors of a couple of tens or a few hundreds (see next point). This may indicate a successive cascade process, in which the masses of a given family are radiatively generated from the masses of the heavier family.[5]

(2) *A charge-dependent gap:* Up-like quarks are heavier than down-like quarks roughly by an order of magnitude.[c] This feature is not shared by the first family, in which one may attribute the reverse order to a small additive perturbation on the tiny double-cascade value.

(3) Neutrinos are extremely light (or even massless), exceedingly lighter than the corresponding charged leptons.[d]

Quark Mixing

The quark mixing matrix for three families is defined in terms of three angles and a phase. A convenient convention,[e] both phenomenologically and theoretically, is to use the magnitudes of the three upper off-diagonal elements as the sines of the three angles that mix the corresponding families. The experimental values of these magnitudes are given below.[6]

$$V = \begin{pmatrix} & 0.220 \pm 0.002 & 0 - 0.012 \\ & & 0.043^{+0.006}_{-0.008} \\ & & \end{pmatrix} \begin{matrix} u \\ c \, . \\ t \end{matrix} \qquad (1)$$
$$\quad d \qquad\qquad s \qquad\qquad b$$

One notes an obvious pattern in flavor mixing. The mixing of two adjacent families

[c]The t to b mass ratio is at least 7. For upper limits on this ratio, see N. G. Deshpande, Neutrino Counting from $\bar{p}p$ and e^+e^- Machines. This volume; also W. Marciano, reference 6.

[d]For rather tight upper limits (astrophysical and others) on the muon- and tau-neutrino masses, see S. Nussinov, Mixing of Fourth-Generation Neutrinos. This volume.

[e]The advantage of this convention over the original KM parametrization has been noted by many. The order in which the three subsequent 2-flavor mixing should be carried out, the generalization to more families, and a convenient way of introducing CP phases are discussed with references to previous work by M. Gronau.[1]

becomes weaker the higher up the family ladder the families are located. Yet a smaller mixing characterizes the coupling of the first and third families. This may be approximately parametrized in terms of increasing powers of the Cabibbo angle:[7]

$$V_{us} \approx \sin\theta_c \qquad V_{cb} \approx \sin^2\theta_c \qquad V_{ub} < \sin^3\theta_c. \tag{2}$$

We refer to this pattern as the family hierarchy of quark mixing.

The CP *Phase*

Having chosen the three mixing angles to be given by the three upper off-diagonal elements, the fourth parameter that defines the mixing matrix is the phase-convention-free *CP* phase ϕ = phase of the product $P = V_{cs}^* V_{us} V_{cb} V_{ub}^*$.[8] The first three elements in this product may be chosen to be real by convention such that the phase is restricted to V_{ub},[7] the smallest of the four elements for which there exists only an upper limit.

The phase $\phi = \arg P$ may in principle be determined from the measured *CP* impurity (ϵ) in the neutral kaon system. However, in practice this is rather difficult. As any other *CP* violating quantity, ϵ contains the universal factor[9] $J = \mathrm{Im}\, P = |P|\sin\phi$, which is proportional to $|V_{ub}|$. It also depends on the t-quark mass and on the value of the B parameter,[6] for which the theoretical uncertainty is large. Assuming $m_t < 66$ GeV (motivated by the neutrino-counting analysis of W/Z production in $\bar{p}p$ colliders[10]) and $B < 1$ (indicated by lattice QCD calculations) one finds $\phi > \theta_c$. This lower bound on the *CP* phase increases with decreasing values of m_t, B, and $|V_{ub}|$. The phase approaches $\pi/2$ with, for example, $B = 0.7$, $|V_{ub}| = 0.004$, and $m_t = 60$ GeV, which are not unreasonable values.

The Number of Families

The number of families is limited by $3 < N < 4$ or 5, where the upper value decreases with increasing t-quark mass.[10]

ARE MASSES, MIXING ANGLES, AND THE PHASE RELATED?

The hierarchial structure exhibited both by the quark masses and mixings suggests that they may be related to each other. The idea that masses and mixings are related is an old one. Such a relationship may be argued for heuristically by demanding that the physics of very light quarks cannot be strongly affected by the existence of very heavy quarks.[f] This implies that the mixing V_{ij} between two families i, j vanishes in the limit in which the two-family mass ratio vanishes:

$$V_{ij} \to 0 \qquad \text{when} \quad \frac{m_i}{m_j} \to 0. \tag{3}$$

[f] We heard this argument first advocated by H. Harari.

A 19-year-old relation that satisfies this criterion[11]

$$V_{12} = \sin\theta_c = \sqrt{m_d/m_s} \tag{4}$$

is an example that works pretty well. Its most straightforward extension to the mixing between the second and third families does not hold, however

$$\sin^2\theta_c \approx V_{23} \neq \sqrt{\frac{m_s}{m_b}} = (1-2)^{-1/2}\sin\theta_c. \tag{5}$$

Here we are using the following approximations for mass ratios

$$m_d{:}m_s{:}m_b = \theta_c^4{:}\theta_c^2{:}1-2 \qquad m_u{:}m_c{:}m_t = (1-2)\,\theta_c^6{:}\theta_c^2{:}{>}1. \tag{6}$$

This should not disturb us too much. To conform with equation 3, equation 4 may also contain on the right-hand side the ratio m_u/m_c. There are two simple ways of including this ratio without spoiling the success of equation 4:

$$V_{12} = \sqrt{\frac{m_d}{m_s} - \frac{m_u}{m_c}} \quad \text{or} \quad V_{12} = \sqrt{\frac{m_d}{m_s}} - i\sqrt{\frac{m_u}{m_c}}. \tag{7}$$

In the second relation the imaginary relative phase was chosen to minimize the effect of the second term. Equation 7 may be extended to the mixing between the second and third families

$$V_{23} = \sqrt{\frac{m_s}{m_b} - \frac{m_c}{m_t}} \quad \text{or} \quad V_{23} = \sqrt{\frac{m_s}{m_b}} - \sqrt{\frac{m_c}{m_t}}. \tag{8}$$

The real relative phase in the second relation is chosen to allow for a large destructive effect. The value $V_{cb} \approx \sin^2\theta_c$ requires cancellation between the leading terms of equations 8. This means that within some approximation $m_c/m_t = m_s/m_b$. Equations 7 and 8 may be derived in specific schemes to be studied in the following section.

One may further raise the question whether a similar relation applies to the *CP* phase ϕ. Here the situation is different when one tries to follow the preceding heuristic argument. *CP* violation requires the existence of three families and is proportional to the sines of the three mixing angles and to $\sin\phi$. In the limit of vanishing two-family mass ratio the corresponding mixing angle vanishes (by equation 3), and *CP* violation disappears while ϕ remains arbitrarily large.

There is obviously the interesting possibility that $\phi \approx \pi/2$, which is favored by low values of m_t, B, and V_{ub} (see the previous section). At first sight this seems to have no particular significance, since the *CP* phase was an arbitrary choice. (For instance, we could have based its definition on the product of the lower left-hand-side elements of the mixing matrix.) In the next section we comment on the possibility that $\phi \approx \pi/2$, and we interpret it as an indication for a tight (one-to-one) connection between flavor mixing and *CP* violation.

THE STRUCTURE OF THE FERMION MASS MATRIX

The patterns described in the second section and some of the relations between masses and mixings given in the last section are exhibited by specific structures of the quark mass matrix. Let us briefly describe a few of the most appealing forms and comment on their successes as well as their limitations.[g]

The Fritzsch Mass Matrix[12]

It is assumed that only the heaviest quark has a diagonal mass matrix element, whereas the lighter masses arise (radiatively) through nearest neighbor mixing

$$M_i = \begin{pmatrix} 0 & A_i & 0 \\ A_i' & 0 & B_i \\ 0 & B_i' & C_i \end{pmatrix}, \qquad i = u, d; \qquad |A_i| = |A_i'|; \qquad |B_i| = |B_i'|. \tag{9}$$

In addition to the six magnitudes of the parameters A_i, B_i, C_i there are two physically significant phases, σ and τ. Given the family mass hierarchy, one finds for the mixing elements the following approximate expressions:

$$|V_{us}| \approx \left|\sqrt{\frac{m_d}{m_s}} - e^{i\sigma}\sqrt{\frac{m_u}{m_c}}\right|, \qquad |V_{cb}| \approx \left|\sqrt{\frac{m_s}{m_b}} - e^{i(\tau-\sigma)}\sqrt{\frac{m_c}{m_t}}\right|,$$
$$|V_{ub}| \approx \sqrt{\frac{m_u}{m_c}}\left|V_{cb} + e^{-i\sigma}\frac{m_s}{m_b}\right|. \tag{10}$$

Comparing the first two equations with the second of equations 7 and 8, we see that the data are consistent with equations 10 and seem to prefer the phases $\sigma = \tau = \pi/2$. This was further discussed by Shin in his paper (footnote g). The measured value of V_{cb} may be used to obtain an upper bound of 90 GeV on the t-quark mass. The prediction for V_{ub} is $|V_{ub}| = (2\text{–}5)\ 10^{-3}$.

The Stech Form[13]

The up and down mass matrices are assumed to be Hermitian. Both contain a symmetric term (S), whereas only the second matrix also has an antisymmetric part (A) (s = real parameter).

$$M_u = S \qquad M_d = sS + A. \tag{11}$$

In the absence of the small perturbation A the masses of the three up- and three down-quarks become proportional and flavor mixing disappears, which is a zeroth-order approximation to observations. Assuming the family hierarchy of quark masses this leads to the following relations, true to lowest order in mass ratios:

$$|V_{us}| \approx \sqrt{\frac{m_d}{m_s} - \frac{m_u}{m_c}} \qquad |V_{cb}| \approx \sqrt{\frac{m_s}{m_b} - \frac{m_c}{m_t}}. \tag{12}$$

[g] See also contributions by H. Fritzsch (reference 5) and M. Shin.

This is in accord with the first of equations 7 and 8. The measured value of V_{cb} imposes a tight upper limit on the t-quark mass $m_t < 50$ GeV.

A Hybrid Form[14]

Noticing that the Fritzsch and Stech forms are complementary to each other (when the phases in the first mass matrices are left free), we have suggested a scheme that imposes the two structures simultaneously. Namely, we assume equation 11 in which both S and A have a Fritzsch texture. This leads to a complete determination of the mixing matrix in terms of quark-mass ratios. One obtains equations 12 as well as the last of equations 10, where σ is given in terms of mass ratios. Furthermore, the CP phase ϕ is found to have a value close to $\pi/2$ as a result of the anomalously light u-quark.

When this scheme is generalized to any number of families all the mixing parameters are predicted in terms of the quark masses. Hence this model may be used as a guide to the physics of a possible fourth family.[15] Among its predictions let us mention the relatively low $m_{b'}$ ($m_{b'} < 55$ GeV) and the possibly large tb' mixing, which may give rise to somewhat enhanced $B - \overline{B}$ mixing effects.

At this point we comment on a possible interpretation[16] of $\phi \approx \pi/2$ if borne out by future experiments. As indicated by the last example, a maximal value of the CP phase in the mixing matrix seems to be related to a maximal phase between the diagonal and off-diagonal elements of the mass matrix. This can be directly shown when assuming the mass matrix to be of the following Hermitian structure

$$M_u = \text{(real diagonal)}; \qquad m_d = \text{(real diagonal)} + \text{(imaginary off-diagonal)}. \quad \textbf{(13)}$$

This is a nine-parameter description and therefore has a single prediction for the mixing matrix in terms of the quark masses. Using the family hierarchies in masses and mixings, equations 6 and 2, one finds $\phi \approx \pi/2$. This result also holds, to leading order in mass ratios, when M_u contains the same imaginary off-diagonal terms as M_d.

The previous discussion shows that the significance of $\phi \approx \pi/2$, if supported by future measurements, may be to indicate at a very strong link between flavor mixing and CP violation. As the preceding scheme illustrates, the entire Lagrangian may be divided into two parts: A CP-even family-diagonal term and a CP-odd family-mixing term:

$$\mathcal{L} = \mathcal{L}_1\,(CP = +, \Delta F = 0) + \mathcal{L}_2\,(CP = -, \Delta F \neq 0). \quad \textbf{(14)}$$

Flavor mixing is due solely to the CP violating term, namely mixing disappears when CP violation is switched off. This feature, which definitely goes beyond the Standard Model, may become a useful clue to the origin of fermion masses.

It was claimed by Jarlskog[17] that the scheme of equation 13, a particular example of equation 14, can have no physical significance. This argument was based on the physical equivalence of different weak eigenstate bases related to each other by some of the $U(N)$ transformations discussed in the introduction. If this were true, this argument would invalidate the physical significance of any specific form of the mass matrix, such as, for example, the Fritzsch texture. Since this argument was repeated on various occasions[18] we wish to briefly clarify this point. The preceding equivalence is restricted to the massless or to the Standard Model, in which the mass matrix remains arbitrary, of course. In any model that attempts to explain fermion masses the preceding huge symmetry must be broken. The pattern of symmetry breaking determines which of the previous transformations are no more symmetries when fermions acquire their masses. The structure of equation 13 or the Fritzsch form are determined by this pattern and cannot simply be "rotated" to different forms without affecting the physics. In short, the invariant functions of the mass matrix introduced by Jarlskog are blind to physics beyond the Standard Model and cannot be applied to the structure of the fermion mass matrix.

In spite of the phenomenological successes of the previous three schemes, one cannot overlook some of their shortcomings:

(1) Family hierarchy in masses must be assumed before deriving a similar hierarchy between mixings. There exist a few scenarios about the origin of the mass hierarchy;[5] however, no particular scheme is illuminating.
(2) To obtain a hierarchy between V_{us} and V_{cb} one requires a fine-tuned cancellation between mass ratios. We are not familiar with any simple argument for such an approximation, besides the assumption of certain small couplings.
(3) The charge-dependent gap seems to be unrelated to the other large mass ratios.
(4) To obtain these structures within specific models one must typically adopt in ones model as many assumptions as the number of predictions one is aiming at.

With this last critical remark let us focus for a moment on $SO(10)$ GUT upon which the Stech scheme was originally based.[19] This will also illustrate the role of leptons in the study of the structure of the fermion mass matrix.[20] In $SO(10)$ fermion masses may arise from Yukawa coupling to scalars belonging to 10, 126, and 120 representations. The first two give rise to symmetric mass matrices (S and S', respectively), whereas the 120 is antisymmetric in flavor (A). These matrices may be made Hermitian by imposing a suitable discrete (CP) symmetry on the Lagrangian. Quark and lepton masses, including Dirac and Majorana neutrino masses (left, $\Delta I_W = 1$ and right, $\Delta I_W = 0$) are all expressed in terms of these three matrices and ratios of VEVs (s, s', etc). The fermion mass matrices, evaluated at 1 GeV, also contain the renormalization group factors ($r_1 \sim r_2 \sim 3$) that distinguish between the ways that quark and lepton masses evolve:

$$\begin{aligned} M_u &= S + S' + a_1 A & M_d &= sS + s'S' + A \\ r_1 M_e &= sS - 3s'S' + a_2 A & r_2 M_{\nu_D} &= S - 3S' + a_3 A \\ r_3 M_{\nu_L} &= s'_L S' & r_4 M_{\nu_R} &= s'_R S'. \end{aligned} \tag{15}$$

Clearly the Stech model, equations 11, is an approximation, in which some of the VEVs are assumed to be smaller than others. The need for the three terms in equations 15 can be briefly summarized as follows:

- S: This is the dominant term, which yields proportional up-quark, down-quark, and charged lepton masses ($m_b/m_\tau \sim 3$), and zero mixing angles.
- A: This term is responsible for fermion mixing and CP violation. It is less important for the larger up-quark masses than for the smaller down-quark

masses. Therefore it was neglected by Stech in M_u. If the Stech limit $m_t < 50$ GeV is violated by future experiments, then A must also be included in M_u (with a coefficient that may be as large as m_t/m_b).
- S': This term is needed for lepton masses and in particular for the seesaw mechanism of neutrino masses.[21]

When imposing a Fritzsch texture on each of the matrices of equation 15 these three terms yield a consistent description of all charged fermion masses and mixings. Furthermore, with $m(N_R) \sim 10^{13}$ GeV, they may accommodate the following neutrino masses and mixings: $m(\nu_e) \sim 10^{-7}$ eV, $m(\nu_\mu) \sim 10^{-4}$ eV, $m(\nu_\tau) \sim 10^{-2}$ eV, $\sin^2 \theta_{e\tau} \sim 10^{-3} - 10^{-2}$, which are within the MSW solution of the solar neutrino problem.[22]

In summary, fermion masses and mixings exhibit certain hierarchy patterns that may be traced back to the structure of the fermion mass matrix. Much more progress is needed to understand the physics that underlies this structure. Crucial to such progress may be improvements in measurements of m_t, V_{ub}, N, and—in theoretical (lattice?) calculations—of m_s and B.

SUMMARY

In this paper the problem of fermion masses, which remains unsolved in the standard $SU(3) \times SU(2) \times U(1)$ gauge theory, was approached from a rather phenomenological point of view. The family hierarchy patterns observed both in masses and mixings seek an explanation in a more fundamental theory, in which the form of the fermion mass matrix is predictable. Without embarking on a specific theory we reviewed a few phenomenologically suggestive forms of the mass matrix, proposed some time ago by Fritzsch, by Stech, and—in a slightly more predictive scheme—by Johnson, Schechter, and myself. The role of $SO(10)$ GUT in deriving these forms was briefly discussed. To agree with experiment, a fine-tuned cancellation of mass ratios in expressions of mixing angles was required in these schemes. This means that the t-quark mass is restricted to lying in a rather well-defined region given by the other quark masses. The implication of a different t-quark mass measurement was noted and the possible embedding of a fourth family of fermions in these schemes was briefly commented on.

ACKNOWLEDGMENT

We would like to thank V. Gupta, R. Johnson, and J. Schechter for an enjoyable collaboration.

REFERENCES

1. For previous reviews on the question of fermion masses, which contain further references, see: WISE, M. 1984. Paper presented at the Physics in Collision IV Conf.; RAMOND, P. 1986. Commun. Nucl. Part. Phys. **16:** 1; GRONAU, M. 1986. *In* Proc. Int. Symp. on the Production and Decay of Heavy Hadrons, Heidelberg. K. R. Schubert and R. Waldi, Eds.: 379.

2. BURAS, A. J., J. ELLIS, M. K. GALLARD & D. V. NANOPOULOS. 1978. Nucl. Phys. **B135:** 66; GEORGI, H. & C. JARLSKOG. 1979. Phys. Lett. **86B:** 297.
3. See, for example, BARR, S. 1981. Phys. Rev. **D24:** 1895, and references therein.
4. GASSER, J. & H. LEUTWYLER. 1982. Phys. Rep. **87:** 77.
5. FRITZSCH, H. 1987. Paper presented at the 1st Int. Symp. for the Fourth Family of Quarks and Leptons. Santa Monica, Calif. For a somewhat different approach, see DIMOPOULOS, S. 1983. Phys. Lett. **129B:** 417; BIJNENS, J. & C. WETTERICH. 1987. Nucl. Phys. **B283:** 237.
6. GILMAN, F. J. 1987. Overview of the Standard Model. This volume; MARCIANO, W. J. 1987. Constraints and Speculations on Fourth-Generation Physics. This volume.
7. WOLFENSTEIN, L. 1984. Phys. Rev. Lett. **51:** 1945.
8. GRONAU, M. & J. SCHECHTER. 1985. Phys. Rev. Lett. **54:** 385. This invariant phase is practically identical to the phase defined by MAIANI, L. 1977. *In* Proc. Int. Symp. on Lepton and Photon Interactions at High Energies. Hamburg. 867.
9. CHAU, L. L. & W. K. KEUNG. 1984. Phys. Rev. Lett. **53:** 1802; JARLSKOG, C. 1985. Phys. Rev. Lett. **55:** 1039.
10. DESHPANDE, N. Footnote *c*.
11. GATTO, R., G. SARTORI & M. TONIN. 1968. Phys. Lett. **28B:** 128. This numerical relation was also noticed by R. Oakes.
12. FRITZSCH, H. 1978. Phys. Lett. **73B:** 317.
13. STECH, B. 1983. Phys. Lett. **130B:** 189.
14. GRONAU, M., R. JOHNSON & J. SCHECHTER. 1985. Phys. Rev. Lett. **54:** 2176.
15. JOHNSON, R., J. SCHECHTER & M. GRONAU. 1986. Phys. Rev. **D33:** 2641; KANG, K. & M. SHIN. 1985. Phys. Lett. **165B:** 383.
16. GRONAU, M., V. GUPTA, R. JOHNSON & J. SCHECHTER. 1986. Phys. Rev. **D33:** 3368.
17. JARLSKOG, C. 1987. Invariant Functions of Mass Matrices: A New Approach to Family Problems. This volume.
18. JARLSKOG, C. 1986. *In* Proc. Int. Symp. on the Production and Decay of Heavy Hadrons. Heidelberg, 331; Univ. of Stockholm Preprint 16, Oct. 1986.
19. Similar mass matrices may be obtained in $SU(2)_L \times SU(2)_R \times U(1)$ models: ECKER, G., W. GRIMUS & H. NEUFELD. 1984. Nucl. Phys. **B247:** 70; GRONAU, M. & R. N. MOHAPATRA. 1986. Phys. Lett. **168B:** 248; MOHAPATRA, R. N. 1987. Theoretical Origin of Quark Mass Matrices. This volume.
20. The following discussion is due to: JOHNSON, R., S. RANFONE & J. SCHECHTER. 1987. Phys. Rev. **D35:** 282; also see BOTTINO, A., C. W. KIM, H. NISHIURA & W. K. SZE. 1986. Phys. Rev. **D34:** 862.
21. GELL-MANN, M., P. RAMOND & R. SLANSKY. 1979. *In* Supergravity. North-Holland, Amsterdam; YANAGIDA, T. 1979. *In* Proc. Workshop on Unified Theory and the Baryon Number of the Universe. KEK.
22. ROSEN, S. P. & J. M. GELB. 1986. Phys. Rev. **D34:** 969, and references therein.

Mixing of Fourth-Generation Neutrinos

S. NUSSINOV

School of Physics and Astronomy
Raymond and Beverly Sackler Faculty of Exact Science
Tel Aviv University,
Tel-Aviv 69978, Israel

INTRODUCTION

We briefly review some of the contraints on the mixing of massive decaying neutrinos. Some of the possible implications for neutrino physics of the recent supernova, and in particular the apparent overabundance of neutrino energy, are discussed as well.

DISCUSSION

The UA1 experiments on the W and Z branchings, ASP-type experiments ($e^+e^- \rightarrow \gamma$ + nothing), and light elements of Big-Bang nucleosynthesis[1] limit the number of neutrino generations—hopefully allowing for one more generation.

No direct experimental evidence for neutrino mixings exists and large domains in the $\sin^2 2\theta$ versus $\Delta(m^2)$ planes (for the various $\nu_i - \nu_j$ pairs) are excluded by reactor and accelerator experiments.[2] Various authors[3] have applied Fritzsch and Stech mass matrices, in conjunction with the seesaw mechanism, to neutrino masses and mixings. This suggests an hierarchical pattern of mixing, so that in particular $U_{e\tau} < U_{e\mu}$ or $U_{\mu\tau}$—an important fact, since the $e^-\tau$ mixing is least constrained ($\sin^2\theta < 0.1$) by direct experiments.[2]

Two classes of general considerations favor, however, nonvanishing neutrino masses and mixings: The solar neutrino puzzle[4] and neutrinos as dark matter candidates.

Particle physics attempts to explain that the suppressed counting in the chlorine solar neutrino experiment involving vacuum[5] and matter-enhanced[6] oscillations or decays of one species into others[7] all require ν masses and mixings. A maximal suppression by a $1/N_\nu$ factor of an initial pure population of any neutrino species (and not just of the experimentally relevant ν_e) is possible—unless maximal CP violation is invoked—only if $N_\nu = 2^k$.[8] Thus, if $k = 2$, than $N_\nu = 4$, the subject of the First International Symposium for the Fourth Family of Quarks and Leptons.

Light neutrinos are not ideal candidates for dark matter. They freely stream until a very late stage and cluster only on very large scales (10^{15} solar masses for a 30-eV neutrino mass). One may attempt avoiding these difficulties by assuming earlier clustering of heavier neutrinos that then decay.

A main theme of the present paper is the limits on decays of heavy neutrinos: The "electromagnetic" decays $\nu_H \rightarrow \nu_e + e^+e^-$ and $\nu_H \rightarrow \nu_i + \gamma$ or the invisible decays $\nu_H \rightarrow \nu_i + X^0$ with X^0 an unobservable Majoron, or decays into three neutrinos.[a]

[a] H. Harari and Y. Nir in reference 3 and references therein.

The bounds come from (a) laboratory experiments, (b) cosmological considerations, and (c) supernovas.

(a) We can search for neutrino mixings in K or π decaying into $\nu_H + e$ (or μ)[9] or via beam dump experiments[10] looking directly for the νee decay with rate

$$\tau^{-1} = 0.5\ 10^{-4}(m_{\nu H}/\text{MeV})^5 U_{\text{He}}^2\ \text{sec}^{-1}.$$

In the latter experiments the proton beam is driven into a fixed target followed by a 10^3-m dump and a 30-m-long decay chamber. The preceding mode has a very clear signature—two colinear e^+e^- pointing to and pulsed with the initial beam. This yields very good direct bounds on the ν_τ, which is copiously produced in the pp collisions via the associated F^+F^- production. For a tentative fourth-generation neutrino the analog-associated production of the heavy fourth-generation quarks is below threshold. Therefore only the π- and K-decay experiments yield direct bounds in this case. (The neutrino ν_4 could be produced also in the decay $\tau^- \longrightarrow \nu_4 + \bar{u} + d$, but in this case another unknown mixing angle θ_{34} is involved.)

(b) By demanding that the neutrino energy density will not overclose the universe we find the famous bound on masses of stable neutrinos $m_\nu < 60$ eV or $m_\nu > 3$ GeV. Even when neutrinos decay we need to verify that the decay products redshift enough to reduce their energy density to an acceptable value, yielding $m_\nu^2 \tau_\nu < 2.10^{20}\ \text{eV}^2\ \text{sec}$.[11]

More elaborate considerations include speeding up of evolution and shortening the age of the universe and entropy production diluting the baryon-to-photon ratios required for the light element nucleosynthesis. Late electromagnetic decays will produce energetic gammas that will not equilibrate with, and consequently distort, the 3-degree background radiation. Finally, an electromagnetic decay of a neutrino heavier than a few megaelectronvolts will produce,[12] directly or via inverse comptonization, γ's energetic enough to dissociate deuterium to a level below the currently observed abundance.

Together with the direct experimental bounds these considerations practically rule out electromagnetically decaying neutrinos of mass > 60 eV. This is not the case for the fourth-generation neutrino precisely because of the comment made at the end of part (a).

(c) We next focus on the bounds for the decay $\nu_H \longrightarrow e^+ + e^- + \nu_e$ from supernovas. These are in many ways much more direct than the cosmological bounds, and certainly bolster our belief that no massive unstable neutrinos exist.[13] Supernovas are the world's best factories for neutrinos of mass < 50 MeV. In the typical type II supernova $0.2M_\odot c^2 = 3.6\ 10^{53}$ ergs of gravitational energy is gained in forming the neutron star core of a size equal to 10 km = 10^6 cm. All neutron star masses measured to date are consistent with a canonical value of $M = 1.4\ M_\odot$[14] and the preceding energy release.

Only one to three percent of the energy is in the kinetic energy of the blown mantle and in electromagnetic radiation. The bulk of the energy is emitted via neutrinos. High-energy ν_e are initially generated in electron capture $e^- + p \longrightarrow \nu_e + n$. These neutrinos have large cross sections and are trapped in the dense core. Eventually the neutrinos thermalize with the photons, electron–positron pairs, and all the other neutrino and antineutrino species, and finally diffuse out of the core. Since the mu- and tau-neutrinos have lower cross sections they are emitted from slightly deeper in the star than the electron neutrinos and reflect the higher temperature of the inner layer.

Detailed numerical simulations[15] yield $T_{\bar{\nu}(e)} = 5$ MeV and $T_{\bar{\nu}(\mu)}$ [or $\bar{\nu}(\tau)$] = 10 MeV. (The latter would apply also to any fourth-generation neutrino.) The average neutrino energies are 3 T, that is, 15 and 30 MeV, respectively, and almost 10^{58} neutrinos of each species are emitted. (Note that the Boltzmann suppression factors $f_B = 2.8$ $(m_H/\mathrm{T})^{3/2} \exp(-m_H/\mathrm{T})$ are fairly mild—not less than 0.01 even for $m_H = 100$ MeV.) Consider next a neutrino decay experiment on astrophysical scales[13] in which supernova ↔ accelerator, red giant ↔ beam dump, and galaxy ↔ decay chamber.

On average, a supernova explosion occurs in our galaxy every 50 years. For ν_H lifetime, 10^{11} sec $> \tau > 10^3$ sec, the decay occurs in the galaxy but outside the star. Positrons are thus injected into the galaxy at a rate of $\approx 10^{49} f_B$/sec, trapped in the galactic magnetic fields, and then slowed down—over a period of about 10^5 years—and annihilate at rest. In the steady-state situation established this yields a constant flux $\Phi = f_B 10^3$ cm^2/sec of the 511-keV annihilation line, which vastly exceeds the $\sim 10^{-3}$ cm^2/sec background.

If $\tau < 10^3$ sec, the ν decays inside the star depositing the energy in the envelope. If, furthermore, f_B exceeds 0.1, too much energy will be deposited as compared to what is inferred from the speed of the receding envelope.

We have assumed all along that the e^+e^- mode dominates over $\nu_H \rightarrow \nu + \gamma$. This is not the case if the mixing of a heavy neutrino with the τ-neutrino vastly exceeds its mixing with the electron neutrino. Since the photons from a supernova are not captured in the galaxy, considerations based on establishing a steady-state situation do not apply.

How will all the preceding be modified if a supernova would have just now (from the point of view of our telescopes!) exploded at a distance of 50 kiloparsecs (kpc)? This is precisely what happened three days before the Symposium (February 26–28, 1987), and we utilize the rest of the space to discuss this SN1987a.

Taking 0.3 10^{58} electron antineutrinos emitted at a distance $L = 1.5\ 10^{53}$ cm, we have a flux of $\Phi = 10^{10}\,\bar{\nu}_e/\mathrm{cm}^2$. Even if we consider only interactions on the free protons (of which there are $N_p \approx 10^{32}$ in the ~2-kiloton volume used in Kamiokande II), then the expected number of neutrino interactions is

$$N_{\text{events}} = \Phi \cdot \sigma \cdot N_p = 10,$$

where σ is the cross section for the ν_e capture:

$$\sigma(\nu_e + p \rightarrow e^+ + n) \simeq 10^{-41}\,(E/10\ \mathrm{MeV})^2\mathrm{cm}^2.$$

Coincident neutrino pulses from 1987(a) have been observed in the Kamiokande II[16] and IMB[17] detectors. The number of the neutrinos (11 and 8, respectively) and the few seconds pulse durations agree with the general theoretical expectations. The long transit time, t(transit) = 5.10^{12} sec, and high magnetic fields encountered by the neutrinos inside and near the forming neutron star are much larger than anything obtainable in terrestrial experiments. This suggests utilizing these few events to constrain new neutrino physics, such as nonvanishing neutrino masses, neutrino mixing (in vacuum and in matter), neutrinos decays, nonvanishing neutrino magnetic moments, and unusual new interactions of neutrinos.

(1) If the electron neutrino had a mass $m(\nu_e)$, then a neutrino of energy E is delayed with respect to that of a simultaneously emitted but much more energetic ν

by

$$\Delta t = (m^2/2\,E^2)t(\text{transit}).$$

This yields new 10 eV upper bounds on the electron neutrino mass.[18] In obtaining this bound and in much of the other ensuing analysis, it is assumed that anti-ν_e capture on protons dominates. If the reaction on oxygen is important, then the initial ν energies could be twice as high and the mass bound correspondingly weaker.

The cross sections for detecting $\nu(\mu)$ and $\nu(\tau)$ via neutral current interactions on electrons are much smaller and we cannot put similar bounds on the masses of $\nu(\tau)$ and $\nu(\mu)$.

Vacuum neutrino oscillations can occur en route from the supernova. The relevant range of parameters: $\Delta(m^2) > 10\ \text{eV}^2$ and $\sin(\theta)^2 = 0.1$, necessary for an observable delayed, $\nu(e)$ pulse, is allowed[2] only for $\nu_e \leftrightarrow \nu_\tau$.

The 7-sec gap between the first nine and last three Kamioka events is a bit puzzling. These three events have also an average energy that is about half of the average energy of the remaining events, and thus could perhaps be attributed to background. These events may reflect, however, the joint effect of neutrino vacuum oscillations and different trapping times for the $\nu(e)$ and $\nu(\tau)$.

(2) The observation of a $\nu(e)$ signal implies that the lifetime of 10 MeV $\nu(e)$ exceeds the transit time of 5.10^{12} sec and excludes the decay of such neutrinos en route from the sun as a possible explanation of the solar neutrino puzzle.[7]

In practically all schemes for neutrino masses one is led to expect a strong hierarchy: $m(\nu(e)) < m(\nu(\mu)) < m(\nu(\tau))$, and hence the decays $\nu(\mu) \rightarrow \nu(e) + X$ or $\nu(\tau) \rightarrow \nu(\mu) + X$ or $\nu(\tau) \rightarrow \nu(e) + X$—with X an unobserved very weakly interacting Majoron—are much more likely.

The intrinsic lifetime $\tau = 8\pi/m \cdot g^{-2}$ for such a decay (m is the mass of the decaying neutrino, which we assume to be much larger than that of the final neutrino, and g is the dimensionless coupling), becomes in the laboratory frame, $\tau(\text{lab}) = 8\pi E/m^2 \cdot g^{-2}$.

The delay in arrival time of the $\nu(e)$ pulse, due to the initial phase of slower propagation as heavy neutrinos, is then

$$\Delta t = \tau(\text{lab}) \cdot m^2/2E^2 = 4\pi/E \cdot g^{-2} = 10^{-21}\ \text{sec}/g^{-2}$$

for $E = 10$ MeV. Thus, if $g = 10^{-10}$, the decay $\nu(e)$ will contribute to the observed $(\nu(e))$ pulse.

The standard supernova model predicts average temperatures of the heavier neutrino species, which are a factor of 2 higher than that of $\nu(e)$. Thus the decay neutrinos will have energies comparable to the initial $\nu(e)$ and enhance the detected $\nu(e)$ and $\bar{\nu}(e)$ signals by factors of 2 to 3.

A very specific range of couplings, $g = 10^{-11}$, would cause the decay neutrinos to come with a few seconds delay with respect to the primary electron neutrinos.

(3) In the triplet Majoron model interactions of the neutrinos with the ~3-degree Majoron background may disperse the signal and degrade its energy.[19] The cross section for scattering a supernova neutrino of energy E on a background Majoron of

energy = temperature = T is

$$\sigma = g^4/ET \cdot 8\pi = g^4/10 \text{ MeV} \cdot 10^{-3} \text{ eV} \cdot 8\pi = 1.6/10^{-17} \cdot g^4 \text{ cm}^2,$$

the number of collisions suffered on transit from the supernova is $n\sigma l = 0.5 \cdot 10^{12} \cdot g^4$. The requirements of no signal dispersal due to such collisions implies $g^2 = 2 \cdot 10^{-5}$, which is inferior to the lower bounds from the double beta decay experiments.

A modified Majoron model with a massive neutrino background was recently suggested by Raffelt and Silk.[20] The interactions of the neutrinos prevent free streaming, making them much better dark matter candidates. In this case the local density of the massive neutrinos in the halos of our and the Lesser Magellanic Cloud (LMC) galaxies is 10^6 times larger than the cosmological density $10^3/\text{cm}^3$, used in the previous estimate for the Majoron background. The requirement of no signal dispersal in this case will allow only 10^{-6} times lower cross sections. This implies $g_e g_H < 10^{-8}$, where g_e, g_H are the Majorons couplings to ν_e and ν_H, respectively.

(4) As pointed out by Dar,[21] $\nu(L) \leftrightarrow \nu(R)$ flips due to an anomalous magnetic moment, suggested to explain the solar neutrino suppression, and in particular, its possible correlation with the sun-spot activity, are operative also in supernovas. The latter effect involves the spin-flip scattering $\nu(L) + e$ to $\nu(R) + e$ in the dense core, followed by the escape of $\nu(R)$ from the core and subsequent return flip $\nu(R)$ to $\nu(L)$ in the strong magnetic field outside. This is a fascinating possibility that allows quick transport of high-energy neutrinos from the core.

For this mechanism to operate we need $10^{-12} < \mu(\nu(e)) < 10^{-10}$, for the electron neutrino magnetic moment in units of electron Bohr magnetons, so that the incoherent $\nu_{eL} \rightarrow \nu_{eR}$ will occur during the trapping in the core, but will not be reversed while ν_{eR} is escaping from the core. We also need to assume that the ν_e magnetic moment exceeds that of ν_τ and ν_μ and also any transition moments (that arise in many models). Such transition moments could induce $\nu_{eL} \rightarrow \nu_{\mu R}$ (say), and in the magnetic field outside the core we will then have $\nu_{\mu R} \rightarrow \nu_{\mu L}$, but not the desired ν_{eL}.

Significant $L \rightarrow R$ precessions can occur in the near vicinity of the core for a nominal neutron star magnetic field of 10^{12} G, even if the magnetic moment in the 10^{-14} to 10^{-12} range causes a factor of 2 suppression of the original signal. Matter effects—that is, different interactions of ν_e^L and ν_e^K will "quench" the $\nu_L \leftrightarrow \nu_R$ precission if $B < G_F N_e/\mu$. No quenching will happen according to delayed explosion scenarios where the density right outside the neutrino sphere is $\rho \leq 10^5$–10^4 g/cm^3. This range is quite acceptable from the point of view of many (left–right symmetric and/or supersymmetric) extensions of the Standard Model.

Such $\nu(L) \rightarrow \nu(R)$ flips can be generated by the galactic and intergalactic fields ($\sim 3.10^{-6}$ and 10^{-8} G, respectively). Since all the neutrinos traverse exactly the same path and the B-induced precession is momentum-independent, all the neutrinos suffer a common precession that a priori is incalculable and could make the number of left-hand neutrinos arriving at the detectors completely arbitrary. However, the previously mentioned precession in the strong fields in the immediate vicinity of the core is different for the different neutrinos emerging from different parts of the core. The neutrinos emitted from the star already form an equal mixture of $\nu(L)$ and $\nu(R)$, and further precessions do not effect this 1:1 ratio

(5) The radiative decays: $\nu(\text{heavy}) \rightarrow \nu(\text{light})$ + electron + positron + photon or

$\nu_H \rightarrow \nu_i + \gamma$ translate, for $\Phi(E)$, the flux of the heavy neutrino $= 10^{10}/\text{cm}^2$ into a gamma-ray flux of $10^{10}/\tau(\text{lab})$ (or a value 10^3 smaller in the first case due to the $\alpha/2\pi$ factor), where $\tau(\text{lab})$ is the lifetime for the decay in the lab frame. Evidently even inefficient γ-ray detectors could exclude a very large range of radiative decay lifetimes.

(6) The energy of the $\bar{\nu}_e$ component has been estimated by the Kamiokande collaboration, obtaining $W(\bar{\nu}_e) = 8 \cdot 10^{52}$ ergs. A thermal neutrino scenario is strongly indicated if the few second dispersal of the neutrino events is viewed as actual trapping time in the dense core.

In such a scenario the total energy in all ν species, $W_{\nu e} + W_{\bar{\nu} e} + \cdots + W_{\bar{\nu}\tau}$ is 8.3 $W(\bar{\nu}_e)$ giving then the first estimate W_ν

$$W_\nu = 6.68 \cdot 10^{53} \text{ ergs} = 0.37\ M_\odot.$$

If the two events consistent with $\theta = 0$ are identified with neutrino electron scatterings slightly[19] higher W_ν values are found.

The preceding W_ν values are too high, or at best marginal for neutron star binding energies (BE), which, as previously indicated, typically agree with one-half that value.[22]

These large apparent values of W_ν suggest very strong conclusions. In particular, effects that would decrease the ν signal are unlikely, whereas those new physics aspects that enhance the signal are favored. Thus:

(a) The type I supernova is ruled out.
(b) A remnant neutron star would favor the hard equation of state for nuclear matter, which allows for a higher total binding energy.
(c) The very massive neutron star required is inconsistent with an ordinary supernova with a standard core. This may be related to the optical magnitude 3.5, which is weaker than expected.
(d) We have to avoid strong $L \rightarrow R$ precession and a typical ensuing factor of 2 depletion. If the star is formed with an initial magnetic field $B = \epsilon \cdot 10^{12}$ G, this would exclude ν_e anomalous magnetic moments in the excess of $\chi \approx 2 \cdot 10^{-15}/\epsilon$.
(e) Additional ν generation tends to dilute the ν_e ($\bar{\nu}_e$) signals. This is yet another astrophysical argument limiting the number of neutrino species to less than or equal to 5 or 6.
(f) Decays of ν_τ, ν_μ to ν_e could enhance the ν_e ($\bar{\nu}_e$) signals by factors of 2 to 3, thus alleviating the energy difficulty.
(g) A simple way out of all the preceding constraints is that a black hole is formed rather than a neutron star.

ACKNOWLEDGMENT

We have benefited from many discussions with Y. Aharonov, G. Alexander, S. Bludman, A. Dar, P. Langacker, I. Goldman, B. Kozlovsky, and Y. Rephaeli.

REFERENCES

1. See CLINE, D. B. 1987. Missing transverse energy events and a search for additional neutrino families and a fourth-generation lepton. This volume; DESHPANDE, N. G. 1987. Neutrino counting from $\bar{p}p$ and e^+e^- machines. This volume; BARGER, V. 1987. Some new aspects of fourth-generation physics. This volume; SCHRAMM, D. N. 1987. Cosmological limit on the number of families. This volume.
2. BERKVIST, K. E. 1985. *In* Proc. 1985 Int. Symp. on Leptons and Photons.
3. SHIN, M. 1987. Fourth-family quark masses and the Kobayashi–Maskawa phenomenology from the regularities in the structure of quark mass matrices. This volume; LANGACKER *et al.* 1986. Cern Th. 4421/86; ACHIMAN, Y. 1986. Dortmunt preprint; HARARI, H. & Y. NIR. 1987. SLAC Pub. 4224.
4. BAHCALL, J. N. *et al.* 1985. Astrophys. J. Lett. **292:** L79.
5. BILENKY, S. M. & B. PONTECORVO. 1978. Phys. Rev. C.
6. MIKHEYEV, S. P. & A. YU. SMIRNOV. 1986. Nuov. Cimento **9C:** 17.
7. BAHCALL, J. N. *et al.* 1986. Phys. Lett. **181B:** 369.
8. NUSSINOV, S. 1976. Phys. Lett **63B:** 201.
9. BRYMAN, D. A. *et al.* 1983. Phys. Rev. Lett. **50:** 1546.
10. BERGSMA, F. *et al.* 1984. Phys. Lett. **147B:** 481.
11. DICUS, D. A., E. W. KOLB & V. L. TEPLITZ. 1977. Phys. Rev. Lett **39:** 168.
12. SARKAR, S. & A. M. COOPER. 1984. Phys. Lett. B **148:** 347; KRAUSS, L. M. 1984. Phys. Rev. Lett. **53:** 1976.
13. DAR, A., J. GOODMAN & S. NUSSINOV. 1986. IAS preprint and references therein.
14. SHAPIRO, S. L. & S. A. TEUKOLSKI. 1983. Black Holes, White Dwarfs and Neutron Stars. Wiley, N.Y.
15. See, for example, WOOSLEY, S. E. & T. A. WEAVER. 1986. Ann. Rev. Astron. Astrophys. **24:** 205.
16. HIRATA, K. *et al.* 1987. Preprint LC3712.
17. BIONTA, R. M. *et al.* 1987. UCI Neutrino 87-10. Preprint.
18. BAHCALL, J. N. & S. L. GLASHOW. 1987. IAS preprint.
19. NUSSINOV, S. & M. RONCADELLI. 1983. Phys. Lett. **B122:** 157.
20. RAFFELT, G. & J. SILK. 1987. Berkeley preprint.
21. DAR, A. 1987. IAS preprint.
22. A more detailed version of this section appears in GOLDMAN, I., Y. AHARONOV, G. ALEXANDER & S. NUSSINOV. 1987. Taup 1543-87.

Invariant Functions of Mass Matrices: A New Approach to Family Problems[a]

C. JARLSKOG

Department of Physics
University of Stockholm
Stockholm, Sweden

INTRODUCTION

In the Standard Electroweak Model, the quark masses and mixings are dictated by the quark mass matrices. For n families, let m and m' denote, respectively, the n-by-n mass matrices for the charge $2/3$ and $-1/3$ quarks. In this paper, we take a closer look at these mass matrices. There are good reasons for doing so, as follows:

(1) The mass matrices are of fundamental importance and therefore merit to be understood better.
(2) There are a large number of models for m and m', and many theorists spend a great deal of effort in studying the associated phenomenology.
(3) Some attempts have been made in the past to restrict the form of the mass matrices by imposing further conditions, such as discrete symmetries and the so-called maximal *CP*-violation.

Before bringing m and m' more into focus, let us make the following two observations.

OBSERVATION 1

The elements of the mass matrices are not all measurables. Thus we may consider the mass matrices as partially "measurable quantities." This follows from the fact that, with n quark families, m and m' represent $4n^2$ real parameters. However, we know that the measurables, in the Standard Model, are the $2n$ masses and the $(n - 1)^2$ parameters of the quark mixing matrix. Thus

$$4n^2 - (2n + (n - 1)^2) = 3n^2 - 1$$

parameters in m and m' are not accessible quantites. In fact in the Standard Model what is relevant is not m and m' but their "Hermitian squares,"

$$S = mm^\dagger, \qquad S' = m'm'^\dagger. \tag{1}$$

For simplicity of notation, in the following we assume that m and m' are Hermitian. This assumption is perfectly justified due to the absence of right-handed charged

[a]This work was supported by the Swedish National Research Council (NFR).

currents in the Standard Model. However, an awkward feature of using m and m' is that they are not positive (semidefinite) matrices because their eigenvalues, that is, the quark masses, are not required to be positive, in the Lagrangian. Although S and S' are manifestly positive matrices, we use m and m' because the generalization of the obtained results is trivial.

OBSERVATION 2

The mass matrices m and m', again in the Standard Model, are not unique. In other words, there are not only two "good" mass matrices that we have to find in order to be happy, because then we will understand the mass and mixing problem, but there is an infinite number of such good mass matrices. The point is that the transformation

$$m \rightarrow XmX^\dagger, \qquad m' \rightarrow Xm'X^\dagger, \tag{2}$$

where X is an arbitrary unitary matrix, leaves the eigenvalues (quark masses) as well as the quark mixing matrix invariant. In other words, the pair (m, m') and the rotated pair $(XmX^\dagger, Xm'X^\dagger)$ lead exactly to the same physics. Thus we may conclude that all measurables must be invariant functions of m and m' where a function $f(m, m')$ is defined to be an invariant function of mass matrices if it satisfies the relation[1,2]

$$f(m, m') = f(XmX^\dagger, Xm'X^\dagger). \tag{3}$$

The unitary rotation in this equation may be thought of as causing a change of the "coordinate system" in the flavor space. The invariant functions are thus independent of the choice of the coordinate frame in the flavor space. There is an immediate analogy with the usual space–time variables. For example, energy and momentum are not invariant concepts. They depend on the choice of the Lorentz frame, but the Møller–Mandelstam variables s, t, u, and functions of them are useful because they do not depend on the specific choice of the frame.

EXAMPLES OF INVARIANT FUNCTIONS

Quantities such as

$$\mathrm{tr}\,(m^k),\ \mathrm{tr}\,(m'^k),\ \mathrm{tr}\,(m^k m'^l),\ \mathrm{tr}\,(m^k m'^l m^r m'^s),$$

where k, l, r, and s are constants, are manifestly invariant functions. We now give a few examples.

Example 1

Let us show how the concept of invariant functions may help avoid pitfalls. First, examine the following definition[3] of the so-called maximal CP-violation, in the Standard Model with $n = 3$.

Let m and m' be Hermitian and go to the frame where the mass matrix for the charge $\frac{2}{3}$ quarks is diagonal: $m = (m_u, m_c, m_t)$. We say that CP is maximally violated if in this frame m' has purely imaginary nondiagonal elements, that is

$$m' = \begin{pmatrix} a & ir & is \\ -ir & b & it \\ -is & -it & c \end{pmatrix}.$$

Here m' is the mass matrix for the down-type quarks, and a, b, c, r, s, and t are assumed to be real numbers. It is obvious that the preceding definition, as it stands, cannot make sense, because we may rotate both mass matrices with, for example, the unitary matrix

$$X = \text{diag}\,(i, 1, 1).$$

The term m is invariant under this rotation, but the new m' has no longer purely imaginary nondiagonal elements. Thus, the property of having purely imaginary nondiagonal elements is not an invariant concept. After all, we require an important concept, such as maximal CP-violation, to be at least independent of the choice of the coordinate frame. In previous publications,[1,4] we examined a bit more in detail whether one could generalize the preceding definition by asking the product of the off-diagonal elements to have maximal phase. There were several phases and there was no way to know which one should be maximized. In our opinion, the only sensible definition of maximal CP-violation is in terms of the determinant of the commutator found in reference 5. We give a definition in a forthcoming paper.[6]

Measurable Quantities as Invariant Functions

We expect the measurable quantities to be invariant functions of mass matrices. The question is then what kind of invariant functions do we get? It turns out that by far the most difficult task is to express the masses (eigenvalues) in terms of invariants, as is well known from quantum mechanics, for example. In principle the solution to this problem has been known for a very long time. The eigenvalues are the solutions of the characteristic equation

$$r^n - r^{n-1}\,\text{tr}\,(m) - r^{n-1}(\text{tr}\,(m^2) - \text{tr}^2\,(m)) + \cdots + (-1)^n \det\,(m) = 0.$$

Of course, the determinant is also a function of the traces. In principle, this equation gives the masses, but in practice, if there are more than five families, we cannot express the solutions in terms of elementary functions (Abel's theorem). We assume that the eigenvalues have been determined (perhaps by perturbative methods) and so focus on other measurables. As an example consider the measurables of the quark mixing matrix. These may all be expressed as invariant functions.[1,2] For example, the measurable $|V_{\alpha a}|^2$, which enters in the rate of $W \rightarrow \alpha + \bar{a}$, is given by[6]

$$|V_{\alpha a}|^2 = \frac{\text{tr}\,[\text{v}_\alpha\,(m)\,v_a'(m')]}{(vv')}. \tag{4}$$

Here the charge $\frac{2}{3}$ $(-\frac{1}{3})$ quarks have been denoted by Greek (Latin) indices.

Furthermore, the v and v' are Vandermonde-type determinants.[7] For n families

$$v = v(m_1, \ldots, m_n),$$

where the m_α, $\alpha = 1 - n$ are the quark masses ($\alpha = 1$ corresponds to the up-quark, and so on). The quantity v is given by

$$v = \begin{pmatrix} 1 & 1 & 1 & \cdots & 1 \\ m_1 & m_2 & m_3 & \cdots & m_n \\ m_1^2 & m_2^2 & m_3^2 & \cdots & m_n^2 \\ & & & & \vdots \\ m_1^{n-1} & m_2^{n-1} & m_3^{n-1} & \cdots & m_n^{n-1} \end{pmatrix}$$

This quantitiy is well known to be given by

$$v = v(m_1, \ldots, m_n) = \Pi\ (m_\beta - m_\alpha),$$

where α, $\beta = 1 - n$, with the restriction $\beta > \alpha$. Thus $v(m_1, m_2, m_3) = (m_3 - m_1)(m_3 - m_2)(m_2 - m_1)$, etc. The quantity v_α is obtained from v by replacing m_α by the mass matrix m. All other eigenvalues are just simply multiplied by the unit n-by-n matrix. Thus this quantity is an n-by-n matrix. For example, for $n = 3$

$$v_1 = (m_3 - m_2)(m_3 - m)(m_2 - m).$$

The primed quantities refer to the charge $-1/3$ quarks and are obtained from the previous expressions by simple substitutions, $m \rightarrow m'$, etc. For example,

$$v' = v(m_1', m_2', \ldots, m_n').$$

The generalization to S and S' is indeed trivial. All we have to do is to replace $m(m')$ by $S(S')$ and the masses by their squares throughout. The point of the preceding considerations is that once the masses are computed (perturbatively, or otherwise) the determination of the measurables of the quark mixing matrix is trivial, in the sense that it only involves elementary operations.

Example 2

As a second example consider the quantities such as

$$V_{\alpha j}V_{\beta k}V^*_{\alpha k}V^*_{\beta j}, \qquad V_{\alpha j}V_{\beta k}V_{\gamma l}V^*_{\alpha l}V_{\beta j}V^*_{\gamma k}, \text{ etc.},$$

where the indices are not summed over. Note that such quantities are independent of the choice of phases of the quark fields, a fact that has been noted by many authors.[8] Since these are also measurables, they may be expressed in terms of invariant functions.[1,2,6] For example,

$$V_{\alpha j}V_{\beta k}V^*_{\alpha k}V^*_{\beta j} = \frac{\mathrm{tr}\ [v_\alpha v'_k v_\beta v'_j]}{(v^2 v'^2)}. \tag{5}$$

The Commutator for n = 3

As a third example consider the commutators of functions of S and S'. These were found to have very remarkable *CP*-properties. Indeed, for three families, one has[5]

$$\det\,[S, S'] \neq 0 \tag{6}$$

iff *CP* is violated. Of course, this determinant is an example of the invariant functions we have been discussing. The computation of this determinant is easy. We computed it by taking the $\mathrm{tr}(C^3)$, where C is the commutator (see references 5 and 9 for details). One finds

$$\det C = -2vv'J,$$

where v and v' are the Vandermonde determinants discussed before, and J is an invariant of the Kobayashi–Maskawa (KM) matrix, which emerges in *CP* computations. It can be found in many papers.[8] We found the following nice formula

$$\mathrm{Im}\,(V_{\alpha j}V_{\beta k}V^*_{\alpha k}V^*_{\beta j}) = J\Sigma\,\epsilon_{\alpha\beta\gamma}\epsilon_{jkl}.$$

In order to get *CP*-violation, 14 conditions have to be satisfied. These are based on the following principles:

(1) No two quarks with the same charge are allowed to be degenerate in mass. This gives six conditions, $m_u \neq m_c$, $m_d \neq m_s$, etc.
(2) None of the angles in the quark mixing matrix is permitted to assume its maximally or minimally allowed value (e.g., $\theta_i \neq 0, \pi/2, i = 1$–$3$ and $\delta \neq 0, \pi$, in the KM parametrization).

These 14 conditions are unified in the relation stating that the determinant of the commutator must be nonzero (equation 6). This relalation is the if and only if condition for having *CP*-violation with three families. In fact, the case $n = 3$ does seem very special. Perhaps there is a deep reason for it.

It is interesting to note that the preceding commutator is the simplest among an infinite number of commutators that all have vanishing determinants if *CP* is conserved. Taking any Hermitian functions $f(S)$ and $g(S')$, one can show that the determinant of the commutator of f and g vanishes if *CP* is conserved.[5]

FEYNMAN AMPLITUDES AS ALMOST INVARIANT FUNCTIONS

A Feynman diagram is, by itself, not neceassarily a measurable quantity. We usually add several diagrams and square to get a measurable gauge invariant. Nevertheless, one may write[1,9] the flavor amplitudes as "almost invariant functions of mass matrices." The results are pretty. We haven't the space to discuss them here, but recommend that those who are interested see references 1 and 8. The major point is to write the diagrams, as far as possible, in terms of the *mass matrices*. The advantage of the formalism is that the symmetry properties become more transparent. Thus one may, for example, quickly check the *CP* properties (Hermiticity) of the amplitudes.

The expressions one gets are concise and more transparent than in the usual formalism.

ADVANTAGES OF THE INVARIANT FORMALISM

Following is a list of advantages of the invariant formalism:

(a) For any number of families, one can express the measurables of the quark mixing matrix as invariant functions of the underlying mass matrices. The expressions are quite concise and easy to program. One does not need to find explicitly the matrices that diagonalize m and m' or S and S', multiply them together to find the quark mixing matrix, worry about which phases are physical and which ones are removeable, etc.

(b) By writing the Feynman diagrams in terms of the mass matrices one can easily examine their Hermitic properties that are very essential when doing *CP* calculations.

(c) The formalism is good for avoiding pitfalls, as shown in an example in this paper.

(d) For $n = 3$ there are further very powerful applications, for example, given a specific model for mass matrices, one may use the commutator formalism to study the *CP* properties of the model, without having to first diagonalize the mass matrices (for an application, see reference 10). One may also, for example, check whether or not the model proposed gives *CP*-violation as certain parameters in the model approach zero.

(e) Det C appears in physics[9] in the renormalization of the θ-parameter of quantum chromodynamics (QCD). We know that QCD does not automatically respect *CP*, due to the presence of the so-called θ-term in its Lagrangian. Electroweak interactions renormalize the θ-parameter, $\theta \longrightarrow \theta + \delta\theta$, where the last term is expected due to *CP*-violation in electroweak interactions. The general form of $\delta\theta$ has been known for several years. In fact, by simple manipulations, one may easily show that[9,1]

$$\delta\theta \sim \det(C).$$

Finally, the commutator formalism has recently been used[11] to compute the renormalization of the observed low-energy *CP*-violation, when one goes to the energy scale of the grand unifing theories.

SUMMARY

We have reviewed some of the work done, in the past two years, on the quark mass matrices and *CP*-violation. The main topic discussed was the invariant functions approach, whereby the measurables of the quark mixing matrix are expressed as manifestly invariant functions of the underlying mass matrices. Several examples were given. The Feynman amplitudes may also be written in "almost invariant forms," in terms of mass matrices.

REFERENCES

1. JARLSKOG, C. 1987. Matrix representation of symmetries in flavor space, invariant functions of mass matrices and applications. Phys. Rev. **D35:** 1685–1692.
2. JARLSKOG, C. & A. KLEPPE. 1987. Measurables of the quark mixing matrix as invariant functions of mass matrices. Nucl. Phys. **B286:** 245–252.
3. GRONAU, M. 1986. Lessons to be learned from Kobayashi Maskawa matrix. *In* Proc. Int. Symp. on Production and Decay of Heavy Hadrons. Heidelberg, West Germany; SCHUBERT K. R. & R. WALDI. 1986. DESY Rep. pp. 391–394.
4. JARLSKOG, C. 1986. Mysteries in the Standard Model. loc. cit. pp. 331–343.
5. JARLSKOG, C. 1985. A basis independent formulation of the connection between quark mass matrices, CP violation and experiment. Z. Phys. C—Particles and Fields **29:** 491–497, and references therein.
6. JARLSKOG, C. 1987. Flavor projection operators and applications to CP-violation. Stockholm Univ., Stockholm, Sweden, preprint. To be published.
7. See, for example, GANTMACHER, F. R. 1959. Matrizenrechnung II. Deutscher Verlag, Berlin.
8. See, for example, CHAU, L.-L. & W.-Y. KEUNG. 1984. Comments on the parametrization of the Kobayashi-Maskawa matrix. Phys. Rev. Lett. **53:** 1802–1805.
9. JARLSKOG, C. 1986. A new theoretical approach to CP-violation. *In* Proc. 21st Rencontre de Moriond. J. Tran Thanh Van Ed.: 389–398. Editions Frontieres.
10. WOLFENSTEIN, L. 1986. CP-violating Fritzsch mass matrices. Phys. Rev. **D34:** 897–898.
11. ATHANASIU, G. G., S. DIMOPOULOS & F. J. GILMAN. 1986. CP nonconservation at the unification scale. Phys. Rev. Lett. **57:** 1982–1985.

Experimental Status of *B* Decays

NAHMIN HORWITZ

Syracuse University
Syracuse, New York 13244-1130

INTRODUCTION

One can seek evidence for physics beyond three generations by studying various aspects of *B*-meson decays. This paper reviews the status of a number of current *B*-meson decay topics. One conclusion is that there is no evidence yet for departures from the Standard Model. Topics reviewed are: *B* reconstruction, penguins and rare decay modes, is there a "charm deficit"?, V_{bu}/V_{bc}, new limit on FCNC. Most of the results presented are from the CLEO collaboration[a] at CESR.

B RECONSTRUCTION

FIGURE 1 shows a recent CLEO mass distribution[1] containing some 40 events in which the *B*-meson decays via a 2- or 3-body mode, usually including a *D* or D^*. Fits to appropriate explicit final states yield the masses and branching ratios (BRs) given in TABLES 1 and 2. Here and in what follows, (1) any mode implies also the charge conjugate mode, (2) *D* branching ratios when used are the MK III 1986 values.[2]

Similar work has been done by ARGUS.[3] They have only reported D^{*+} modes, but have included modes with one π^0. CLEO and ARGUS results are in reasonable agreement.

If *p*-wave phase space determines the ratio $R \equiv (\Upsilon(4S) \rightarrow B^0\overline{B}^0)/(\Upsilon(4S) \rightarrow B^+B^-)$, then the CLEO values in TABLE 1 yield $R = 43/57$. The CLEO values in the model of Ono[4] give $R \simeq 50/50$.

PENGUIN-ENHANCED DECAYS

Various authors have urged measurement of rare decays whose BR in the Standard Model is expected to be $\leq 10^{-4}$ and for which an observation of a much larger value would be a signal of new physics. Typically these are mediated by penguin diagrams (FIGURE 2). A few examples of theoretical predictions are given in TABLE 3.

CLEO[5] has searched for rare decays $K^-\pi^+$, $K^-e^+e^-$, etc., by selecting all $-+$, $-+-$, etc., track combinations, applying cuts to reject continuum (non-$B\overline{B}$) events, calculating $M = (E^2_{\text{beam}} - (\Sigma \mathbf{P}_i)^2)^{1/2}$, and looking for a bump in the M distribution at $M = M_B$. Using the indicated beam energy constraint results in mass resolutions σ_M that are typically of order 5 MeV.

[a] Carnegie-Mellon, Cornell, Florida, Harvard, Ithaca College, Ohio State, Purdue, Rochester, SUNY Albany, Syracuse, and Vanderbilt

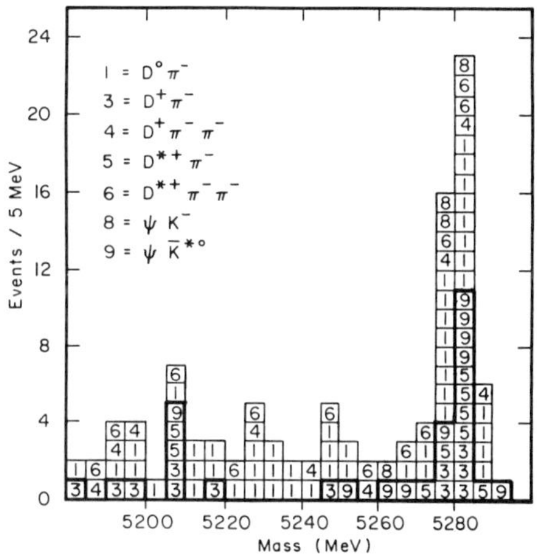

FIGURE 1. CLEO mass distribution of reconstructed B^--decay candidates for seven low multiplicity modes.

FIGURE 3a and b shows the distribution of ΔM ($\equiv M - E_{\text{beam}}$) obtained by CLEO for the mode $K^-\pi^+$ when running on the $\Upsilon(4S)$, and the continuum just below the $\Upsilon(4S)$, respectively. FIGURE 3c, the continuum subtracted distribution, shows no evidence for a bump in the shaded region corresponding to $M = M_B \pm \sigma_M$. A 90% C.L. upper limit is calculated from

$$\text{BR}\ (B^0 \longrightarrow K^-\pi^+) = \frac{N^{0BS} + 1.3\ \sigma_N^{0BS}}{\epsilon N_B}, \tag{1}$$

where N^{0BS} is the entire number of events with $M = M_b \pm \sigma_M$ [i.e., *no* "smooth" background subtraction (see FIGURE 3c)], ϵ is efficiency for detecting the decay $B^0 \longrightarrow K^-\pi^+$, and N_B is the number of B^0 mesons in the sample.

TABLE 4 gives the resultant 90% C.L. limits reported by CLEO for 17 penguin-enhanced and also for various $B \longrightarrow l^+l^-$ modes. In no case is there evidence for a signal. The 90% C.L. upper limit BRs are typically between 10^{-3} and 10^{-4}.

Conclusion: The penguin 2-body modes are small ($<10^{-3}$) compared to D or D^* 2-body modes ($\simeq 10^{-2}$).

FIGURE 4 shows the BR for $B \longrightarrow Kl^+l^-$ inclusive as a function of various fourth-generation parameters as calculated by Hou *et al.*[6] The shaded region is excluded by the CLEO limit given in TABLE 4 if one assumes the Kl^+l^-x inclusive BR is twice the Kl^+l^- exclusive one (as was predicted by Eilam *et al.*[7] for the case of three generations).

TABLE 1. *B*-Meson Mass (MeV)

	CLEO	ARGUS
M_0	5280.6 ± 0.8 ± 2.0	5279.6 ± 1.0 ± 3.0
$M_0 - M_+$	2.0 ± 1.1 ± 0.3	2.4 ± 1.6
$M_{\Upsilon(4S)} - 2M_0$	18.8 ± 1.7 ± 2.0	

TABLE 2. B-Meson BRs (%)

	CLEO	ARGUS
$D^{\circ}\pi^-$	$0.36^{+0.12}_{-0.10}{}^{+0.08}_{-0.05}$	
$D^+\pi^-\pi^-$	$0.20^{+0.33}_{-0.18}{}^{+0.19}_{-0.06}$	
$D^{*+}\pi^-\pi^-$	$0.16^{+0.11}_{-0.10}{}^{+0.06}_{-0.03}$	$0.4 \pm 0.2 \pm 0.2$
$D^{*+}\pi^-\pi^-\pi^{\circ}$	$0.48^{+0.26}_{-0.23}{}^{+0.12}_{-0.10}$	$3.5 \pm 1.1 \pm 2.1$
$D^+\pi^-$		
$D^{*+}\pi^-$	$0.25^{+0.13}_{-0.10}{}^{+0.08}_{-0.05}$	$0.25 \pm 0.15 \pm 0.15$
$D^{*+}\pi^-\pi^{\circ}$		$1.1 \pm 0.6 \pm 0.6$
$D^{*+}\pi^-\pi^-\pi^+$		$2.4 \pm 0.7 \pm 1.1$
ψK^-	$0.09 \pm 0.06 \pm 0.02$	<0.20
$\psi K^{*\circ}$	$0.41 \pm 0.19 \pm 0.03$	0.44 ± 0.27

IS THERE A CHARM DEFICIT IN B DECAY?

Evidence that the ratio of Kobayashi–Maskawa (KM) matrix elements $|V_{bu}/V_{bc}| \ll 1$ comes from the momentum spectrum of leptons emitted in semileptonic (SL) B decay. This is discussed in the following section. Accepting that result, one then expects $b \rightarrow c$ in essentially 100% of B-meson decays; additionally the production of a $\bar{c}s$ pair from the W vertex in about 15% of decays (FIGURE 5). Thus we expect $\simeq 1.15$ c-quarks per B decay. Do we see them?

TABLE 5 displays CLEO and ARGUS results. The value $c\bar{c} = 0.04 \pm 0.01$ is from the measured BR $(B \rightarrow \psi x)$ from CLEO and ARGUS, BR $(B \rightarrow \psi' x)$ from ARGUS, a theoretical estimate of $\eta_c{:}\psi{:}\chi$ to get the η_c and χ contribution, and finally a factor of 2 for $c\bar{c}$. $\Lambda_c = 0.1 \pm 0.05$ was a guess based on p and Λ^0 production. CLEO now has evidence for BR $(B \rightarrow \Lambda_c x) = 0.074 \pm 0.029$ (this is discussed in the following subsection).

The essential conclusion of TABLE 5 is that we observe 0.86 ± 0.09 rather than 1.15 c/B, that is,

$$\text{A deficit of } (\simeq 0.3 \pm 0.1)\ c/B.$$

If one accepts the deficit as statistically significant, the following are possible explanations, some of which are interesting:

(1) The assumption $|V_{bu}/V_{bc}| \ll 1$ is wrong.
(2) CLEO and ARGUS have made large measurement errors.

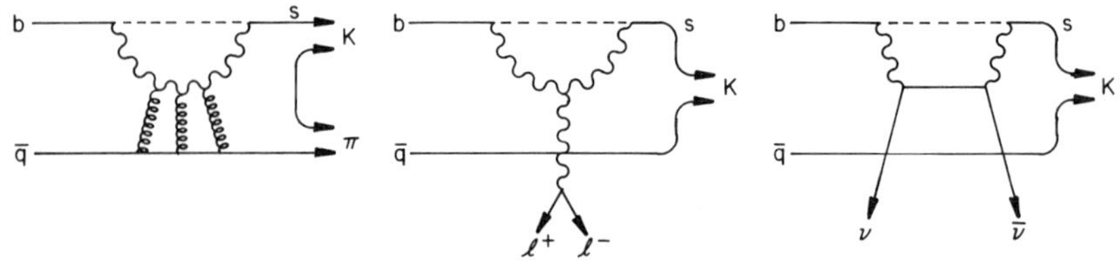

FIGURE 2. Some diagrams for penguin-enhanced B-meson decay.

TABLE 3.

Mode	BR	Author
Sum of all penguin decays	$\simeq 1.5\%$	Whisnan *et al.*
Kl^+l^- exclusive if 3 generations	$1\text{–}1.6 \cdot 10^{-6}$	Eilam *et al.*
Kl^+l^-x if 4 generations	Up to 10^{-3}	Hou *et al.*
$K\nu\bar{\nu}x$ if 4 generations	Up to 10^{-3}	Hou *et al.*

(3) Unexpectedly large charmed baryon production.
(4) $\Upsilon(4S)$ does not decay to $B\bar{B}$ 30% of the time.
(5) BR for penguin-enhanced decays is of order 30%.
(6) The MK III *D* decay BRs that were used by CLEO and ARGUS when measuring D^0/B and D^+/B are 50% too big.

The basis for eliminating explanation (1) is given in the following section. Are CLEO and ARGUS measurements in error? FIGURE 6 shows the CLEO D^0 and D^+ mass distributions from which the numbers in TABLE 5 were extracted. The signals are clean and clear. TABLE 6 shows CLEO and ARGUS measured values for the product $\sigma \cdot B$ for continuum production of D^0 and D^+. The values agree with each other and with extrapolations of Spear and HRS values taken at lower and higher $\sqrt{s}$. *Conclusion:* No larger error in CLEO and ARGUS measurements.

$BR\,(B \rightarrow \Lambda_c x) < 11.2\%$

Is large charmed baryon production the explanation? FIGURE 7 shows some of the diagrams one expects to be important for baryon production in *B* decay. CLEO has just completed a new study.[8] New measurements of Λ and p branching fractions are

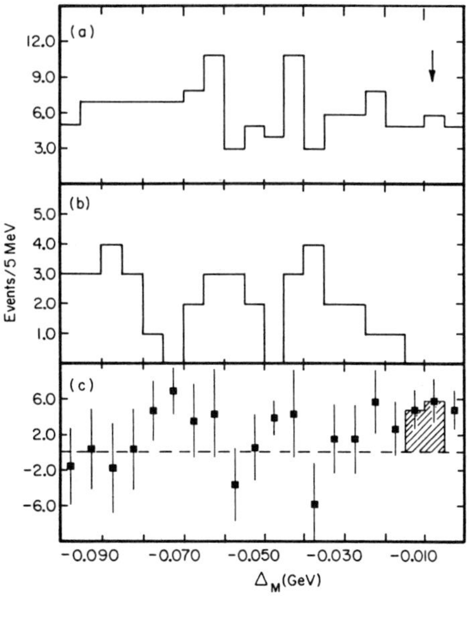

FIGURE 3. The distribution in ΔM ($= M_{k\pi} - E_{\text{beam}}$) for $K^-\pi^+$ candidates. (a) At the $\Upsilon(4S)$. (b) At the continuum below the $\Upsilon(4S)$. (c) The net distribution from $B\bar{B}$ events. The shaded region is $\pm\,\sigma_M$ centered on where a B^0 signal should appear.

TABLE 4. A Summary of Our Search for the Indicated Exclusive Decay Modes of *B*-Mesons

Mode	Number of Events 90% C.L. Limits	Detection Efficiency	Branching Fraction 90% C.L. Limits
$K^+\pi^-$	15.3	0.46	3.2×10^{-4}
$K^\circ\pi^+$	5.3	0.05	6.8×10^{-4}
$K^{*\circ}\pi^+$	6.8	0.17	2.6×10^{-4}
$K^{*+}\pi^-$	2.3	0.03	7.0×10^{-4}
$\rho^\circ K^+$	10.1	0.25	2.6×10^{-4}
$\rho^\circ K^\circ$	3.4	0.04	8.0×10^{-4}
ϕK^+	3.9	0.12	2.1×10^{-4}
ϕK°	3.9	0.03	13.0×10^{-4}
$\phi K^{*\circ}$	3.9	0.08	4.7×10^{-4}
$\rho^\circ K^{*\circ}$	19.1	0.16	11.5×10^{-4}
$K^{*\circ}\gamma$	22.0	0.11	20.5×10^{-4}
$K^{*+}\gamma$	3.2	0.014	18.4×10^{-4}
$K^+\mu^+\mu^-$	11.8	0.24	3.2×10^{-4}
$K^+e^+e^-$	5.8	0.18	2.1×10^{-4}
$K^\circ\mu^+\mu^-$	2.3	0.05	4.5×10^{-4}
$K^\circ e^+e^-$	5.3	0.03	6.5×10^{-4}
e^+e^-	2.3	0.29	0.8×10^{-4}
$\mu^+\mu^-$	3.9	0.40	0.9×10^{-4}
$\mu^\pm e^\mp$	2.3	0.26	0.9×10^{-4}

Note: The 90% confidence level upper limit on the number of events is the sum of the mode indicated and its charge conjugate. The 90% confidence level upper limit branching fraction applies to the mode given or its charge conjugate but not the sum.

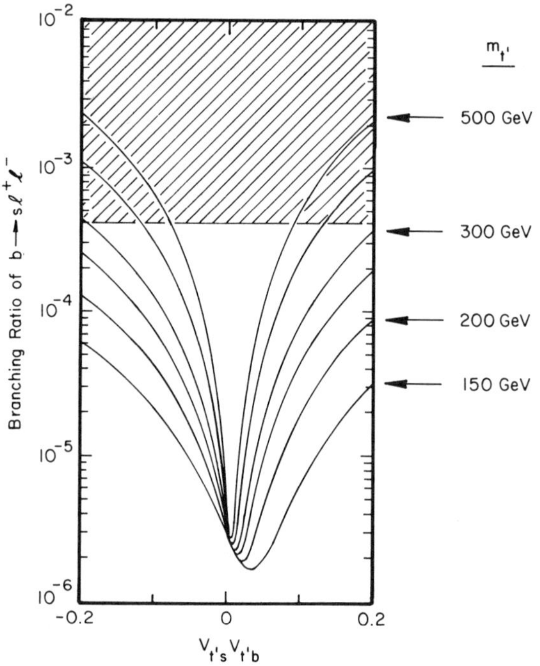

FIGURE 4. Fourth-generation $B \to Kl^+l^-x$ prediction of Hou *et al.*, assuming $M_t = M_l^{(4)} = 50$ GeV. Shaded region is excluded by CLEO limit if $Kl^+l^-/Kl^+l^-x > 0.5$.

FIGURE 5. If $|V_{bu}/V_{bc}| \ll 1$, then expect $\simeq 1.15$ c/b.

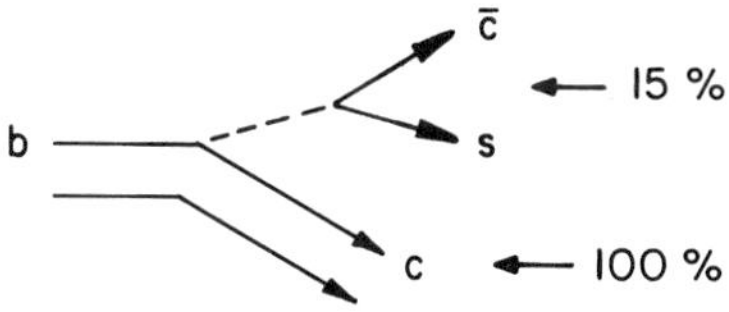

reported. If baryons are produced at the decay vertex of the W (FIGURE 7b, c), one might expect to observe $\Lambda\bar{\Lambda}$ pairs. The number of $B \rightarrow \Lambda\bar{\Lambda}x$ candidates observed was 5 ± 13. Thus null result leads to the limit:

$$\frac{\text{BR }(B \rightarrow \Lambda\bar{\Lambda}x)}{\text{BR }(B \rightarrow \Lambda X)} < 0.12 \qquad 90\% \text{ C.L.}$$

The absence of $\Lambda\bar{\Lambda}$ pairs suggests that Λ's produced in B decay come from Λ_c's, and that the Λ_c's are produced via $b \rightarrow c \rightarrow$ hadronization (rather than at the decay vertex of the virtual W). This hypothesis can be tested further because it implies that if Λ's are produced along with high momentum (>1.4 GeV$/c$) leptons, then $\bar{\Lambda}^0$'s will occur with l^- (and Λ^0's with l^+). This correlation comes about as follows: The restriction to high momentum ensures that the l^- came from the semileptonic decay of a $\bar{B}$ and not a $\bar{D}$. The high momentum also means the c from that $\bar{B}$ will hadronize to a D meson rather than a Λ_c baryon because of limited available energy. Thus if a Λ is observed and if all Λ's come from Λ_c's, then the "Λ" must come from the other meson, that is, the B, and it must be a $\bar{\Lambda}$ and not a Λ. So expect only $l^+\Lambda$'s and $l^-\bar{\Lambda}$'s. This is illustrated in FIGURE 8.

The CLEO observation is:

$$27 \pm 8 \qquad l^+\Lambda,\ l^-\bar{\Lambda}$$

$$3 \pm 5 \qquad l^-\Lambda,\ l^+\bar{\Lambda}.$$

Thus, the observed Λl correlation supports the conclusion: All Λ^0 in B decay come from Λ_c, and the Λ_c contain the c from $b \rightarrow c$.

To proceed, CLEO further assumes:

(a) BR $(B \rightarrow \Lambda_c\bar{p}x)$ = BR $(B \rightarrow \Lambda_c\bar{n}x)$
(b) BR $(\Lambda_c \rightarrow px)$ = BR $(\Lambda_c \rightarrow nx)$
(c) No baryons are produced in the decay of the vitual W.

TABLE 5. Measured Charm Production in B-Meson Decay

	CLEO		ARGUS
D°	$0.39 \pm 0.05 \pm 0.04$		$0.50 \pm 0.07 \pm 0.08$
D^+	$0.17 \pm 0.04 \pm 0.04$		$0.23 \pm 0.07 \pm 0.05$
D_S	0.11 ± 0.06		0.07 ± 0.04
$c\bar{c}$	0.04 ± 0.01		0.04 ± 0.01
Λ_C	0.1 ± 0.05		0.1 ± 0.05
Sum	0.81 ± 0.12		0.94 ± 0.15
Average		0.86 ± 0.09	
Expected		1.15	

Source: M. G. D. Gilchriese, reference 3.

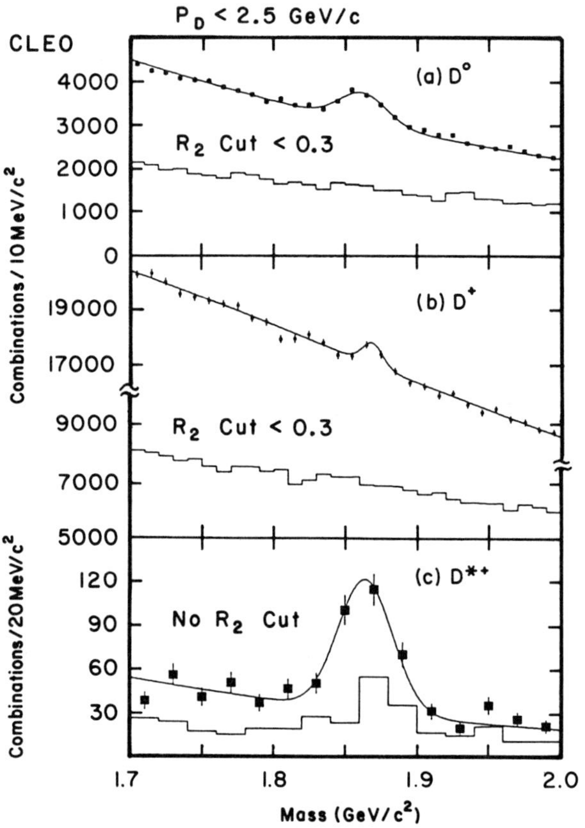

FIGURE 6. Mass distributions from which CLEO obtained values for D^0/B, D^+/B, and D^{*+}/B.

These assumptions, along with measured values for BR ($B \rightarrow \Lambda x$) and BR ($B \rightarrow px$) enable CLEO to calculate

$$\text{BR}\ (B \rightarrow \Lambda_c x) = (7.4 \pm 2.9)\%.$$

This conclusion is not sensitive to assumptions (a) and (b). If assumption (c) is wrong, it will lower the BR. Therefore, we get a model independent limit

$$\text{BR}\ (B \rightarrow \Lambda_c x) < 11.2\% \qquad 90\%\ \text{C.L.}$$

TABLE 6. $\sigma \cdot B$ for Continuum Charm Production

	CLEO	ARGUS
$\sigma(D°) \cdot B(D° \rightarrow K^-\pi^+)$	0.046 ± 0.002 ± 0.004	0.042 ± 0.006 ± 0.005
$\sigma(D^+) \cdot B(D^+ \rightarrow K^-\pi^+\pi^+)$	0.047 ± 0.006 ± 0.004	0.050 ± 0.010 ± 0.005

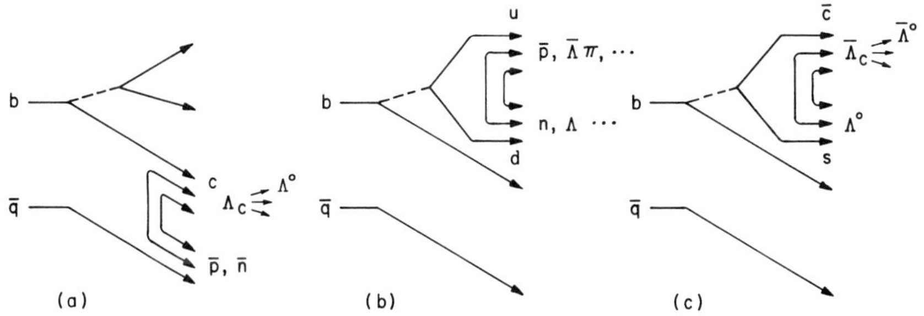

FIGURE 7. Some diagrams for baryon production in *B*-meson decay.

Thus, this new study provides the first estimate for BR ($B \rightarrow \Lambda_c x$) and rules out large Λ_c production as the explanation of a charm deficit.

BR ($\Upsilon(4S) \rightarrow B\bar{B}$)

Could $\Upsilon(4S)$ decay to hadrons without making real $B\bar{B}$ pairs? Lipkin[9] suggested substantial violation of OZI may occur just above $B\bar{B}$ threshold via the following two OZI allowed steps:

(i) $\Upsilon(4S) \rightarrow B\bar{B}$
(ii) $B\bar{B}$ rescatter inelastically to bottomless hadrons (see FIGURE 9).

CLEO[1,10] has explored this interesting possibility in the following two ways:

(1) Comparison of dilepton and single-lepton yields at the $\Upsilon(4S)$. Let
$f \equiv 1 - \text{BR}(\Upsilon(4S) \rightarrow B\bar{B})$;
$\bar{b} \equiv$ mean BR for semileptonic *B* decay (averaged over B^0, B^+);
$\overline{b^2} \equiv$ mean squared BR for semileptonic *B* decay.
Then the observed yields of single lepton and dileptons are given by

$$N_l = 2\bar{b}\epsilon(1 - f)N_{BB} \tag{2}$$

$$N_{ll} = \overline{b^2}\epsilon^2(1 - f)N_{BB}. \tag{3}$$

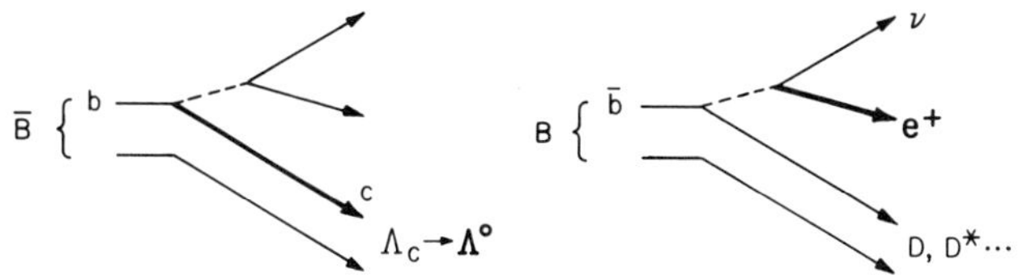

FIGURE 8. If Λ_c are produced via hadronization of the *c* (in $b \rightarrow c$), then expect Λl^+, $\bar{\Lambda} l^-$ correlation.

FIGURE 9. $\Upsilon(4S)$ decay to bottomless hadrons.

Equations 2 and 3 imply

$$f = 1 - \frac{\overline{b^2}/(\bar{b})^2}{[4N_{BB}N_{ll}/N_l^{\,2}]} \tag{4}$$

therefore

$$f < 1 - \frac{1}{[4N_{BB}N_{ll}/N_l^2]}\,.$$

CLEO measures the quantity in brackets to be 1.02 ± 0.1, which yields

$$f < 16\% \qquad 90\%\ \text{C.L.}$$

(2) Search for high x particles ($x \equiv p/E_{\text{beam}}$). For daughters of B decay $x \leq 0.5$, but if $\Upsilon(4S)$ can decay to hadrons directly, then $0 < x < 1$. FIGURE 10 shows the observed x distribution for both continuum and $4S$ after continuum subtraction events. The latter distribution does indeed cut off near $x = 0.5$. There are 103 ± 407 tracks with $x > 0.5$ out of 125,000 $\Upsilon(4S)$ decays. To convert this observation into a limit of f one needs to know the x distribution hadrons would

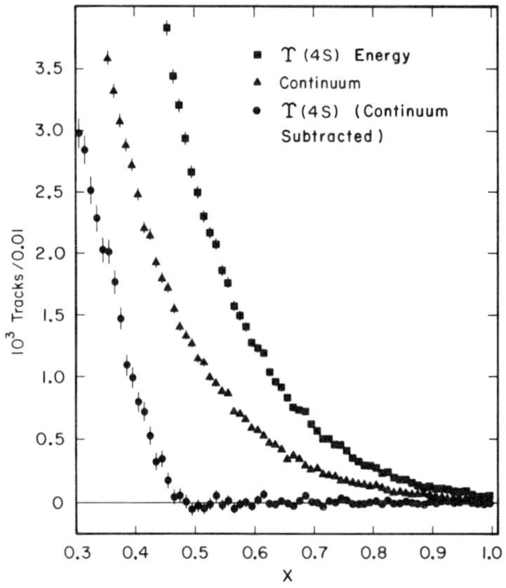

FIGURE 10. The x distribution of hadrons from $\Upsilon(4S)$ events as observed by CLEO.

have if $\Upsilon(4S) \rightarrow$ hadrons directly. CLEO has calculated the limit assuming this x distribution is the same as for hadrons produced in

(a) continuum,
(b) three gluon decays of the $\Upsilon(1S)$.

The corresponding 90% C.L. limits on f are:

(a) $f < 3.8\%$,
(b) $f < 13\%$.

These low limits on f make it unlikely $\Upsilon(4S) \rightarrow B\bar{B}$ is the explanation of the observed charm deficit.

Penguins

As stated in the second section there is no evidence for penguins in *B* decay. On the other hand, the limits quoted apply only to exclusive final states. The measured limits sum to $\simeq 1\%$. There is as yet no theoretical justification for assuming that the inclusive penguin rate is not an order of magnitude or more greater than the exclusive rates for the individual modes studied.[11]

TABLE 7. Comparison of $D°/c$ and D^{*+}/c for SL and All *B*-Decays

	SL Decays	all *B*-Decays
BR ($B \rightarrow D°x$)	0.35 ± 0.09	$0.39 \pm 0.05 \pm 0.04$
BR ($B \rightarrow D^+x$)	?	$0.17 \pm 0.04 \pm 0.04$
BR ($B \rightarrow D^{*+}x$)	0.2 ± 0.1	$0.22 \pm 0.04 \pm 0.06$

Thus the experimental limits on rare decay modes and/or the theoretical limit on the inclusive/exclusive ratio must be reduced before this approach can rule out penguins as the explanation of the observed deficit.

Is There a Charm Deficit in Semileptonic B *Decays?*

The question asked by the subsection title is important because if the answer is yes, one cannot blame a charm deficit on either penguins or $\Upsilon(4S) \rightarrow B\bar{B}$. The current status of CLEO measurements[12] is shown in TABLE 7, which compares D^0/c and D^{*+}/c for SL *B* decays with all *B* decays. The yields are similar. The D^+ yield has not yet been measured. Since D^{*+} decays more often to D^0 than D^+, and D^{*0} decays only to D^0, the total D^+ yield will be less than that for D^0 if the direct D^+ and D^0 yields are comparable. Therefore, the data in TABLE 7 imply

$$\frac{D^0 + D^+}{c} < 0.7 \pm 0.18.$$

In semileptonic decays we expect $c/B = 1.0$ and expect very little D_s or ψ production.

Conclusion: Looks like there exists a similar charm deficit in SL B decays; however, not yet statistically compelling. It is important to measure the D^+ yield.

Next we turn to the question of the value of $|V_{bu}/V_{bc}|$.

CURRENT LIMIT ON $|V_{bu}/V_{bc}|$

Determining the values of the KM matrix elements V_{bu} and V_{bc} is of great importance quite independent of the "charm deficit" in B decay. Extreme values could signal new physics. For example, on the basis of the Standard Model one expects[13]

$$|\epsilon| = 3B\,|V_{bu}|\,|V_{bc}|\,\sin\phi\left[-0.7 + 0.4\ln x^2 + 0.6x^2\,|V_{bc}|^2\left(1 - \frac{V_{bu}}{V_{su}V_{bc}}\cos\phi\right)\right], \quad (5)$$

where

$$\epsilon = \frac{1}{3}(2\eta_{+-} - \eta_{00});$$

$$B \propto \langle K^0\,|\Delta S = 2|\,\bar{K}^0\rangle;$$

$$x = \frac{M_t}{M_c};$$

$$\phi = \text{CP invariant phase.}$$

Since the left-hand side of equation 5 is nonzero, V_{bu} cannot be zero. If one uses $\epsilon = 2.27\ 10^{-3}$, $B = 1$, $M_c = 1.35$ GeV, $\phi = \pi/2$, and $V_{bc} = 0.045$, then equation 5 yields the $|V_{bu}/V_{bc}|$ vs. M_t curve shown in FIGURE 11. It is clearly of interest to check this relationship experimentally.

Some History

The best limits on $|V_{bu}/V_{bc}|$ have come from the momentum spectrum of the lepton emitted in semileptonic B decay. The idea is that for $b \rightarrow u$ one may expect the

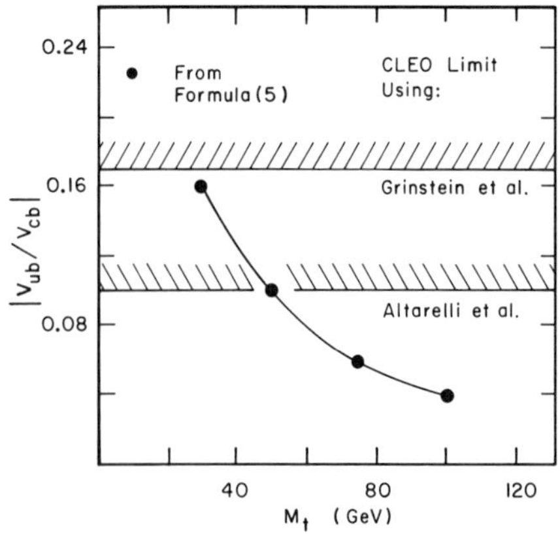

FIGURE 11. $|V_{bu}/V_{bc}|$ vs. m_t from equation 5 and the CLEO limits on $|V_{bu}/V_{bc}|$ described in the fifth section of the paper.

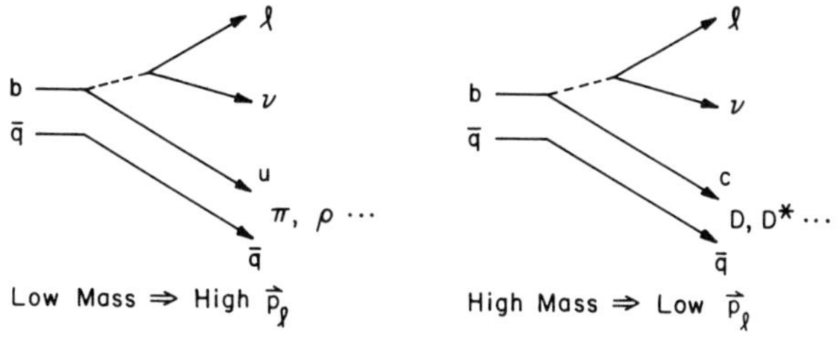

FIGURE 12. $(b \rightarrow u)/(b \rightarrow c)$ via the shape of the lepton spectrum.

hadronic system x to consist of a low mass π or ρ whereas for $b \rightarrow c$, M_x must be at least the mass of a D. Consequently the lepton spectrum should be harder for $b \rightarrow u$ and softer for $b \rightarrow c$ (see FIGURE 12).

In 1984 both CLEO and CUSB reported $\Gamma(b \rightarrow u)/\Gamma(b \rightarrow c) < 0.04$ by fitting the entire observed lepton spectrum to shapes obtained using the unmodified model of Altarelli *et al.* At Kyoto in 1985, CLEO reported (1) that with the addition of new data, the model could no longer fit the observed spectrum, and (2) the $(b \rightarrow u)/(b \rightarrow c)$ result was sensitive to the Fermi momentum parameter used in that model. At the Berkeley conference in 1986 Gilchriese summarized the situation by saying he didn't feel justified in quoting a 90% C.L. limit on $|V_{bu}/V_{bc}|$ lower than 0.30.

New CLEO Results Based on High P *Leptons*

The maximum lepton momentum (in the B rest frame) for semileptonic decay via $b \rightarrow c$ is 2.35 GeV/c. Thus the number of leptons with $p > 2.4$ GeV/c is independent of $b \rightarrow c$ processes. The total number of $b \rightarrow c$ decays is expected to be close to that predicted by the free quark model. To that extent we can predict the fraction of leptons with $p > 2.4$ in terms of $|V_{bu}/V_{bc}|$ independent of $b \rightarrow c$ models. The price paid for eliminating $b \rightarrow c$ model dependence is increased statistical uncertainty due to severely limiting the size of the measured spectrum. Using this technique CLEO recently reported[14] two model-dependent 90% C.L. limits:

$$\left|\frac{V_{bu}}{V_{bc}}\right| < 0.10, 0.17.$$

The following gives some of the relevant details. If Δp is a lepton momentum interval above the maximum p via $b \rightarrow c$, then

$$\frac{\Gamma_{\text{SL}}(b \rightarrow u)}{\Gamma_{\text{SL}}(b \rightarrow c)} = \left[\frac{\sigma_{\text{SL}}(\Delta p)}{\sigma_{\text{SL}}(\text{all})}\right]\left\{\frac{1}{f_u(\Delta p) - [\sigma_{\text{SL}}(\Delta p)]/\sigma_{\text{SL}}(\text{all})]}\right\}, \tag{6}$$

where the ratio $\sigma_{\text{SL}}(\Delta p)/\sigma_{\text{SL}}(\text{all})$ is the observed fraction of leptons with p within Δp, and $f_u(\Delta p)$ is the fraction of $b \rightarrow u$ decays with p within Δp. One momentum range used by CLEO is Δp equals 2.4 to 2.6 GeV/c. For simplicity we focus on this one. A model is needed to get f_u. CLEO used those of Altarelli *et al.*[15] and Grinstein *et al.*[16] The term $\sigma_{\text{SL}}(\text{all})$ is measured to be (235 ± 14)pb for electrons and (237 ± 14)pb for muons. The term $\sigma(\Delta p)$ is measured to be (0.06 ± 0.17)pb. The measured $\sigma_{\text{SL}}(\Delta p)$

[and therefore $\Gamma(b \rightarrow u)/\Gamma_{SL}(b \rightarrow c)$] is consistant with zero. The problem then becomes one of quoting an appropriate upper limit. The 90% C.L. limits on $|V_{bu}/V_{bc}|$ obtained from the data and the models of Altarelli and of Grinstein *et al.* are 0.10 and 0.17, respectively. For $\Gamma_{SL}(b \rightarrow u)/\Gamma_{SL}(b \rightarrow c)$, the Altarelli model leads to a 90% C.L. limit of about 0.02. Grinstein *et al.* do not calculate $\Gamma_{SL}(b \rightarrow u)$, but guess it is somewhat smaller than the free quark rate. The corresponding 90% C.L. limit on Γ_{SL} $(b \rightarrow u)/\Gamma_{SL}(b \rightarrow c)$ using CLEO data, and the Grinstein model is then somewhat smaller than 0.06. Perhaps the most useful way to report the current status of the CLEO $\Gamma_{SL}(b \rightarrow u)/\Gamma_{SL}(b \rightarrow c)$ limit is as follows:

$$\frac{\Gamma_{SL}(b \rightarrow u)}{\Gamma_{SL}(b \rightarrow c)} = (0.00 \pm 0.013)\left[\frac{0.058}{f_u}\right], \tag{7}$$

where f_u is the fraction of expected leptons with $2.4 < p < 2.6$ GeV/c from $b \rightarrow u$ processes, as determined by your favorite model.

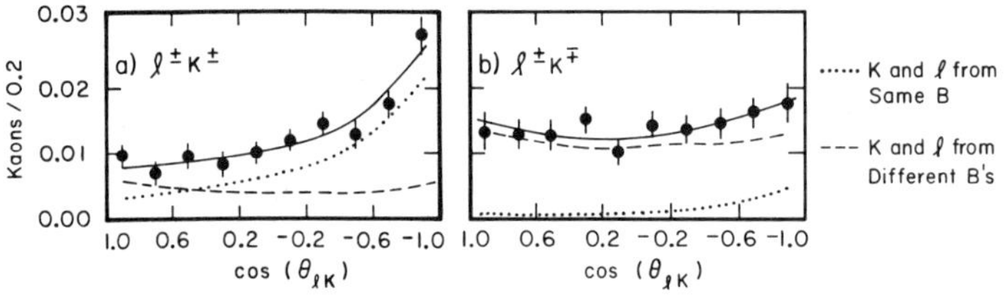

FIGURE 13. Angular distributions of $K^{\pm}$ with respect to lepton direction in $\Upsilon(4S)$ events containing a >1.4 GeV/c lepton.

A Different Approach to (b → u)/(b → c) – K-*Lepton Events*

Since $c \rightarrow s$, it follows that $b \rightarrow c$ should result in a higher yield of K-mesons than $b \rightarrow u$. However, past attempts to use the observed K yield to extract a quantitative value for $(b \rightarrow u)/(b \rightarrow c)$ floundered on uncertainties in the amount of $s\bar{s}$ popping and $W \rightarrow \bar{c}s$. On the other hand, to a very good approximation BR $(\bar{B} \rightarrow s)$ – BR $(\bar{B} \rightarrow \bar{s})$ is independent of amount of $s\bar{s}$ popping and $W \rightarrow \bar{c}s$. Therefore:

$$\text{BR}(\bar{B} \rightarrow s) - \text{BR}(\bar{B} \rightarrow \bar{s}) = \text{BR}(\bar{B} \rightarrow c). \tag{8}$$

To estimate the left-hand side of equation 8, CLEO has measured $K^+/\bar{B}$ and $K^-/\bar{B}$. The trick is to use events with a high momentum lepton. The lepton charge tags whether the lepton came from a B or $\bar{B}$. One then plots the angular distribution of all identified $K^{\pm}$ with respect to $\hat{p}_l$. The K's from the same B as the lepton will be peaked at 180°; those from the other B will be isotropic. The exact shapes of these two distributions are determined by Monte Carlo (MC) simulation. The observed angular distribution of K^- from K^-e^- events is fit to a sum of these two MC shapes and the amount of $K^-/\bar{B}$ (peak distribution) is determined for SL decays as well as the amount of K^-/B (isotropic) for all decays. Similarly the observed K^-l^+ events yield the $K^-/\bar{B}$ for all decays and K^-/B for SL decays. FIGURE 13a shows the two MC distributions for like sign Kl events as well as the continuum subtracted data. The amount of each distribution shown is the amount needed to fit the data. The solid line is the sum of the two distributions. FIGURE 13b shows similar quantities for unlike sign Kl events. The resultant fits give:

$$\frac{K^-}{\bar{B}} = 0.66 \pm 0.05 \pm 0.07$$

$$\frac{K^+}{\bar{B}} = 0.19 \pm 0.05 \pm 0.02$$

and by subtraction:

$$\frac{K^- - K^+}{\bar{B}} = 0.47 \pm 0.07.$$

An MC simulation yields:

$$\frac{K^- - K^+}{\bar{B}} = 0.48 \qquad \text{for } 100\%\ b \to c$$

$$= 0 \qquad \text{for } 100\%\ b \to u.$$

This implies:

(i) $\dfrac{b \to c}{\text{all}} = 0.98 \pm 0.16 \pm 0.12$

(ii) $\dfrac{b \to c}{\text{all}} > 0.68 \qquad 90\%$ C.L.

This method for getting $(b \to c)/\text{all}$ is interesting because it is different and independent from the lepton spectrum method. However, it is not yet accurate enough to compete.

(b → u)/(b → c) *via the Explicit Final State* $\rho^0 l\nu$*—Very Preliminary*

Yet another approach is to look for explicit final states that arise from $b \to u$. An important example is $B \to \rho^0 l\nu$. Grinstein *et al.*[16] and Wirbel *et al.*[17] calculate both the shape of the lepton momentum spectrum in $B \to \rho^0 l\nu$ and also the fraction of $\rho^0 l\nu$ in terms of $|V_{bu}/V_{bc}|^2$.

CLEO has looked at 260,000 B-decays and found -10 ± 51 $\rho^0 l\nu$ candidates using cuts with an efficiency $\epsilon \simeq 0.2$. This yields the 90% C.L. limit

$$\text{BR}\,(B \to \rho^0 l\nu) < 2.3 \cdot 10^{-4}\frac{1}{f_u}.$$

Both calculations agree $f_u \simeq 12\%$, so BR $< 0.21\%$. The Grinstein calculation plus BR $(B \to l\nu x) = 0.11$ then yields $|V_{bu}/V_{bc}| < 0.26$, while that of Wirbel plus $\tau_B = 1.3 \cdot 10^{-12}s$ yields $|V_{bu}/V_{bc}| < 0.15$. These results are very preliminary. They indicate that the method may be competative with using the high-momentum lepton spectrum. However, the method is subject to similar model uncertainties.

Summary Concerning the "Charm Deficit" Problem

Most of the possibilities considered here can be ruled out as individually explaining all of the apparent 30% charm deficit in the decay products of B-mesons. The situation is summarized as follows:

Error in measurement?	No	
Large Λ_c production?	No	<11%
Penguins?	?	
$\Gamma(b \rightarrow u)/\text{all} \simeq 30\%$?	No	< 6%
BR $(\Upsilon(4S) \rightarrow BB) \simeq 30\%$?	No	<16%
MK III BRs too high?	?[b]	

NEW LIMIT ON FCNC IN B-DECAY

The existence of FCNC in B decay at the tree level would be inconsistant with the Standard Model. The expected value for BR $(B \rightarrow l^+l^-x)$ via penguin and box diagrams is of order 10^{-5}. CLEO (in 1984) reported a 95% C.L. limit of 0.3%. JADE and MARK-J have reported 95% C.L. limits of 0.7%. Using 4S data obtained in 1985 CLEO now reports[18]

$$\text{BR}\ (B \rightarrow l^+l^-x) < 0.17\% \qquad 95\%\ \text{C.L.}$$

REFERENCES

1. BEBEK, C. *et al.* 1987. Phys. Rev. D **36:** 1289.
2. BALTRUSAITIS, R. M. *et al.* 1986. Phys. Rev. Lett. **56:** 2140.
3. GILCHRIESE, M. G. D. 1986. *In* Proc. 23rd Int. Conf. on High Energy Physics. Berkeley, Calif.
4. ONO, S. 1984. Acta Phys. Pol. **B15:** 201.
5. AVERY, P. *et al.* 1987. Phys. Lett. **B183:** 429.
6. HOU, W., R. S. WILLEY & A. SONI. 1986. Pitt-86-08 and UCLA/86/TEP/43.
7. EILAM, G., A. SONI, G. L. KANE & N. G. DESHPANDE. 1986. UCLA/TEP/86/7.
8. ALAM, M. S. *et al.* 1987. Phys. Rev. Lett. **59:** 22.
9. LIPKIN, H. J. 1986. Argonne Preprint ANL-HEP-PR-86-43.
10. BEHRENDS, S. *et al.* 1986. CLEO contributed paper. *In* Proc. 23rd Int. Conf. on High Energy Physics. Berkeley, Calif.
11. SONI, A. Private communication.
12. BORTOLETTO, D. *et al.* 1987. Phys. Rev. **D35:** 19.
13. This form for the expression for $|\epsilon|$ appears, for example, in GRONAU, M. & J. SCHECHTER. 1985. Phys. Rev. Lett. **54:** 385.
14. BEHRENDS, S. *et al.* 1987. Phys. Rev. Lett. **59:** 407.
15. ALTARELLI, G. *et al.* 1982. Nucl. Phys. **B208:** 365.
16. GRINSTEIN, B., M. WISE & N. ISGUR. 1986. Phys. Rev. Lett. **56:** 298; GRINSTEIN, B. *et al.* 1968. CALT-68-1311.
17. WIRBEL, M. 1986. Dortmund Preprint DU-TH 86/11.
18. BEAN, A. *et al.* 1987. Phys. Rev. **D35:** 3533.

[b] Subsequent to the First International Symposium for the Fourth Family of Quarks and Leptons, MK III reported at Moriond values for D BRs 20 to 25% lower than those of reference 2. This reduces the charm deficit to $0.16 \pm 0.1\ c/B$.

Heavy Flavor Production and the Evidence for $B - \bar{B}$ Mixing in the UA1 Experiment

KARSTEN EGGERT AND HANS-GÜNTHER MOSER

Physics Institute
Aachen, West Germany

INTRODUCTION

Inelastic proton–antiproton interactions at the CERN collider represent a prolific source of heavy flavors, which are produced by various quantum chromodynamic (QCD) processes. Due to its large center-of-mass energy (630 GeV) the wide-band beam of quarks and gluons is still effective in beauty pair creation down to values of $x = 2p_{\text{l}}/\sqrt{s} \approx 0.02$. FIGURE 1 shows the increase of the cross section for heavy flavor production with the center-of-mass energy. The calculation was done with Isajet,[1] using only lowest order QCD contibutions. It is sensitive to the gluon content in the proton at low x.

QCD calculations become reliable at large transverse momenta, where nonperturbative effects can be neglected. As an example, the large transverse momentum cross section for jets[2] is well described by QCD over 7 orders of magnitude (FIGURE 2). We demonstrate that QCD also reproduces the beauty production, since the mass of the beauty quark of 5 GeV/c^2 introduces a natural cutoff. This contrasts the situation at the ISR, where nonperturbative effects make a QCD calculation for charm production less certain.

Heavy flavors are tagged via their semileptonic decays into muons, which offer the experimental advantage that they can be identified at low transverse momenta ($p_t > 3$ GeV/c) and even in the presence of hadronic activity. First-generation decays of pair produced b- and c-quarks give unlike-sign muon pairs where the muons are, in general, accompanied by hadronic activity from the quark fragmentation and decay.

Muons can also be due to second-generation b-decays resulting in same-sign dimuons. However, the large amount of like-sign dimuons in the UA1 data can only be understood in terms of $B - \bar{B}$ oscillations, analogous to the well-known phenomena in the K-complex.[3]

It is plausible that the dominant source of dimuons with large p_t is $b\bar{b}$ production rather than $c\bar{c}$. At transverse momenta larger then the quark masses involved, that is, ignoring threshold effects, b- and c-quarks are produced with an identical p_t spectrum. The harder fragmentation of the beauty quark compared to charm gives, on average, a larger fraction of the original quark momentum to the B-meson. Hence, the harder p_t-spectrum of muons from beauty quarks compared with charm quarks favors $b\bar{b}$-events over $c\bar{c}$, once a p_t-cut is applied to the muons.

This effect is demonstrated in FIGURE 3, where the p_t-spectra of the quarks, the mesons, and the resulting muons are plotted. Muon pairs can also arise from the decay of hidden heavy flavors, like J/ψ and Y, and from virtual photons (Drell–Yan). These unlike-sign events can be separated from open heavy flavors with the help of the

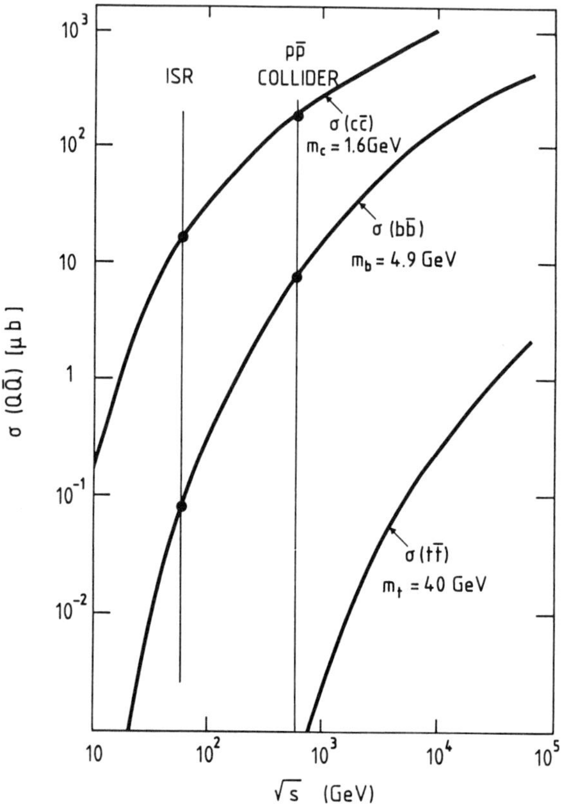

FIGURE 1. Cross section for heavy flavor production. (Isajet, lowest order.)

isolation of the muons. In this case no hadronic activity in addition to the underlying event is expected in a cone around the muon.

DETECTOR, EVENT SELECTION, AND CLASSIFICATION

The UA1 detector has been described in detail elsewhere.[4] In brief, a muon from the $p\bar{p}$ interaction traverses the central detector, where the momentum is determined by the deflection in the central magnetic dipol field of 0.7 T. After penetrating the electromagnetic and hadronic calorimeters and the additional iron absorber (a total of more than nine interaction lengths) the muon is finally detected in the muon chambers, which form the outer shell of the UA1 detector and cover 70% of the pseudorapidity range $|\eta| < 2.0$.

A fast hardware muon trigger, based on the hit information in the drift tubes of the muon chambers, selects muon candidates by requiring a track pointing to the interaction region within a cone of aperture ± 150 mrad. Data were recorded using an inclusive muon trigger with a pseudorapidity acceptance of $|\eta| < 1.5$ and a dimuon trigger within $|\eta| < 2.0$.

The data presented in this paper were collected during three collider periods from 1983 to 1985.[5] The total integrated luminosity amounts to 659 nb^{-1}, 108 nb^{-1} at $\sqrt{s}$ = 546 GeV (1983), and 551 nb^{-1} at $\bar{s}$ = 630 GeV (1984/85). Muon events were selected with the following cuts:

(1) single muons: $p_t > 6$ GeV/c;
(2) muon pairs: $p_t > 3$ GeV/c for each muon.

The background to the muon sample has been studied extensively and is discussed in detail elswhere.[6] The probability that a hadron simulates a muon either by

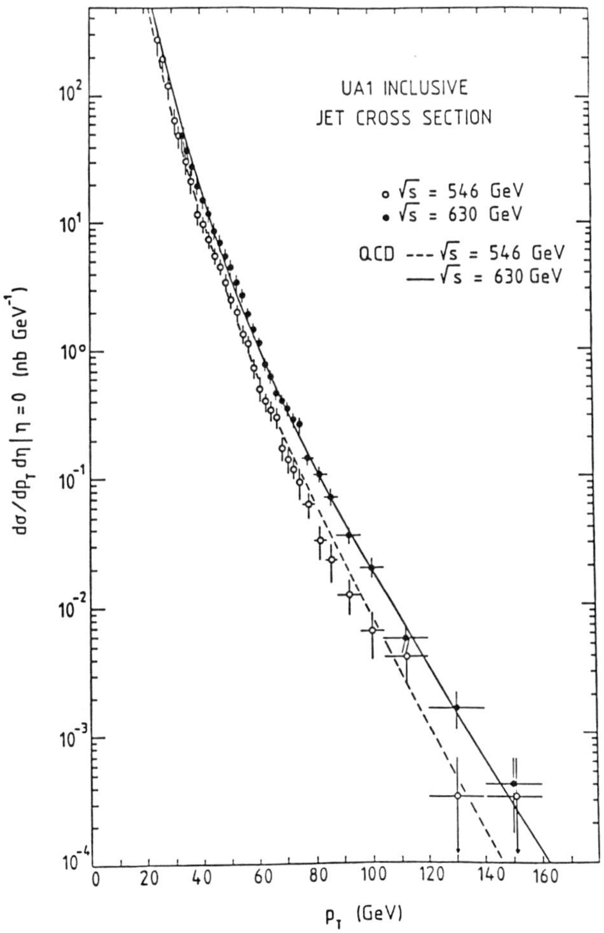

FIGURE 2. $d\sigma/dp_t$ for jets at 630 (546) GeV. (UA1 data.)

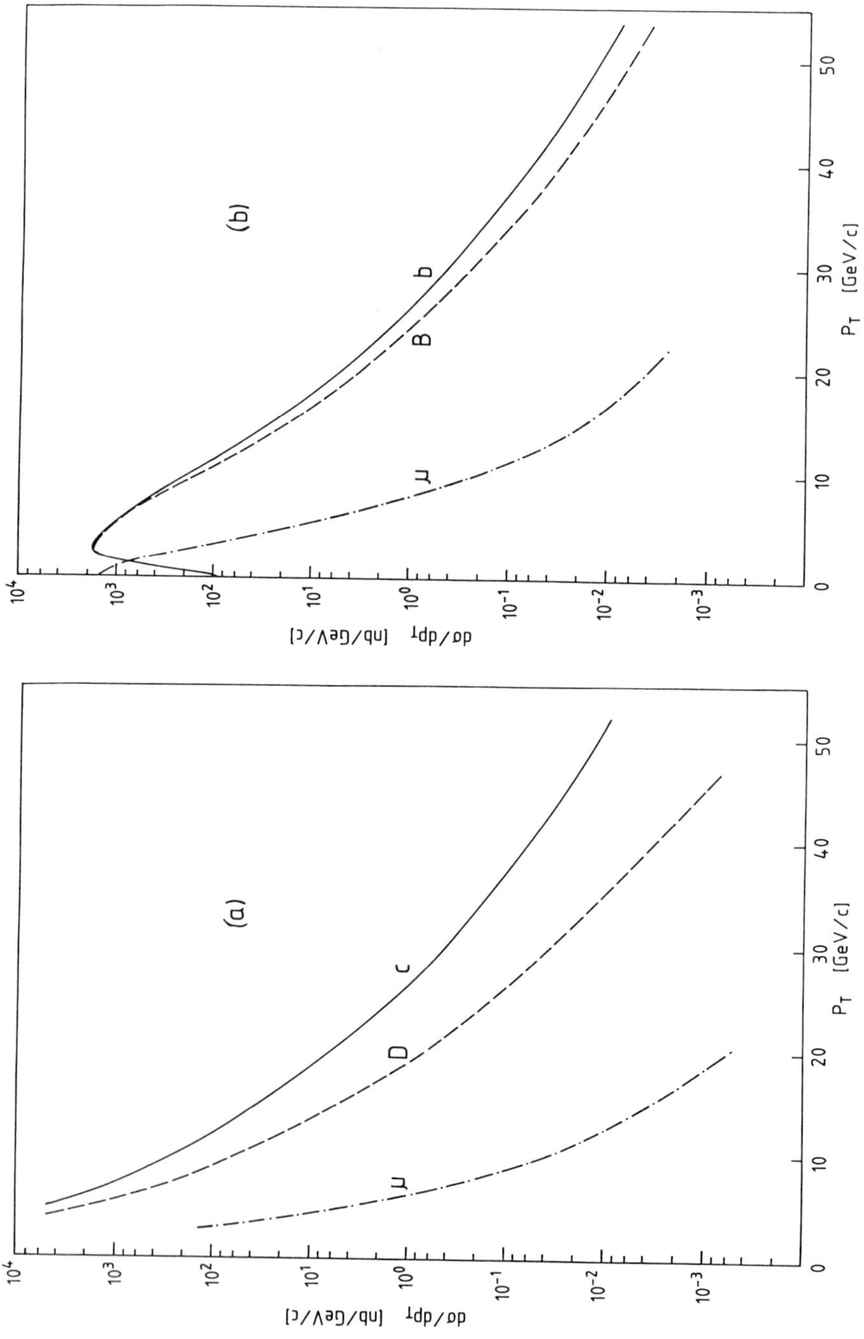

FIGURE 3. p_t spectra of quarks, mesons, and muons. (a) $c\bar{c}$. (b) $b\bar{b}$. (Isajet, lowest order.)

TABLE 1. Inclusive Muon and Dimuon Samples and Estimated Background

		Data	Background (%)
Single μ	$p_t > 6$ GeV/c	22,000	65
	$p_t > 10$ GeV/c	1,733	38
Dimuons	$p_t > 3$ GeV/c	847	22
	$M\mu\mu < 6$ GeV/c^2	335	17
	$M\mu\mu > 6$ GeV/c^2	512	26

penetrating the absorber without interaction or by leakage of the induced hadron shower was found to be small ($\ll 10^{-4}$). Severe background arises from pions (kaons) that decay in flight in the central detector with a probability of $0.02/p_t$ ($0.11/p_t$). The number of events for the different data samples and their estimated background is listed in TABLE 1.

The muon pairs are classified according to their charge (neutral or double charged) and the degree of muon isolation. Since the muon momentum is rather low, the particle sign is well measured. The isolation is characterized by the transverse energy around the muon, that is, the sum ΣE_t of the transverse energy of all calorimeter cells within a cone of $\Delta R < 0.7$ about the muon in the pseudorapidity η – azimuth ϕ – space (ϕ about the muon axis). The expected energy deposit of the muon itself is subtracted.

$$\Sigma E_t \text{ in a cone of } \Delta R = \sqrt{\Delta\phi^2 + \Delta\eta^2} < 0.7$$

$$S = (\Sigma E_t^{(1)})^2 + (\Sigma e_t^{(2)})^2.$$

The isolation quantities are added in quadrature for both muons. They are considered as isolated, if $S < 9$ GeV^2. Eighty-two percent of the muons from W- and

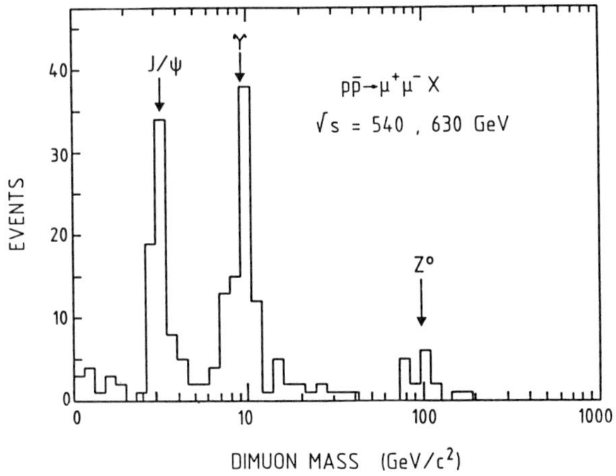

FIGURE 4. Dimuon mass distribution for unlike-sign isolated events. (Background not subtracted, not corrected for acceptance.)

TABLE 2. Classification of the Dimuon Events ($M\mu\mu > 6$ GeV/c^2) in Terms of Sign and Isolation

	Unlike Sign	Like Sign
Isolated	99	17
Nonisolated	256	140

Z-decays, which should be isolated, survive this cut. For the low mass events the isolation cut has been modified. If the two muons are so close together that the respective cones overlap, a 3-GeV cut on the transverse energy in a single cone around the common dimuon axis is applied instead.

We observe 256 isolated unlike-sign events. Their mass distribution—not corrected for acceptance—is plotted in FIGURE 4. The mass peaks due to the vector particles decaying into muon pairs are clearly visible. The J/ψ particles can also occur nonisolated in jets, and furthermore their acceptance is difficult to calculate, since they cluster near to our p_t cut. Therefore a mass cut of 6 GeV/c^2 was applied in the following analysis. The classification of these events according to the dimuon charge and the isolation is given in TABLE 2.

DRELL-YAN AND UPSILON

The 99 isolated unlike-sign events should be mainly caused by the Drell–Yan process and by upsilon decays with some contamination from open heavy flavor decays and from background. Their mass distribution is dominated by the upsilon peak above the Drell–Yan continuum and the heavy flavor decays (FIGURE 5). After subtracting the background and taking into account the different shapes of the mass and isolation distributions we obtain the following cross sections for Drell–Yan and upsilon [we

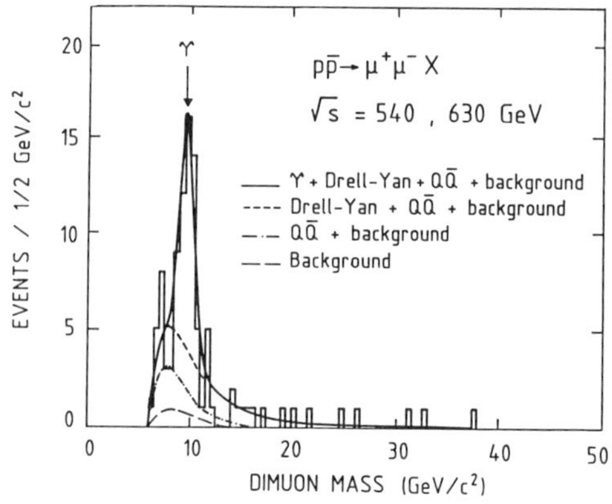

FIGURE 5. $\partial\sigma/\partial m^{\mu\mu}$ of the isolated unlike-sign events. The fitted contributions from Drell–Yan, upsilon, heavy flavor, and background are indicated.

assumed that $N(Y){:}N(Y'){:}N(Y'') = 1{:}0.30{:}0.15]^7$:

$$\sigma(DY, m^{\mu\mu} > 11\ \text{GeV}/c^2) = 0.26 \pm 0.08_{\text{stat}} \pm 0.05_{\text{sys}}\ \text{nb}$$

$$\frac{\partial^2\sigma}{\partial m \partial y}(y = 0, m^{\mu\mu} = 10\ \text{GeV}/c^2) = 15 \pm 5_{\text{stat}} \pm 3_{\text{sys}}\ \text{pb}/\text{GeV}/c^2$$

$$(\text{BR})\sigma(p\bar{p} \rightarrow Y, Y', Y'') = 1.0 \pm 0.2_{\text{stat}} \pm 0.2_{\text{sys}}\ \text{nb}$$

$$\partial\sigma/\partial y(p\bar{p} \rightarrow Y, Y, Y'', y = 0) = 0.20 \pm 0.04_{\text{stat}} \pm 0.04_{\text{sys}}\ \text{nb}.$$

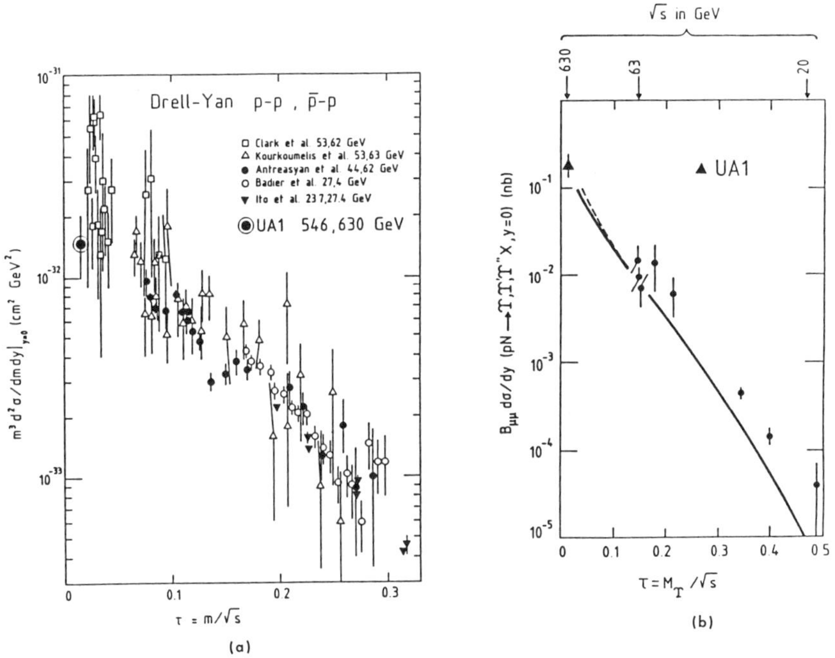

FIGURE 6. (a) UA1 cross section for Drell–Yan compared with data at lower energies (reference 8). (b) UA1 cross section for upsilon compared with data at lower energies (reference 7). The solid line represents a theoretical calculation from reference 9.

The results are compared with extrapolations from low-energy data in FIGURE 6 and show approximate scaling behavior, that is, independence of $\sqrt{s}$, if plotted versus the scaling variable $\sqrt{\tau} = m/\sqrt{s}$.

PRODUCTION OF HEAVY FLAVORS

The nonisolated dimuon events and all like-sign events are expected to come from open heavy flavor decays. Various QCD processes contribute to the associated production of heavy flavors. In lowest order heavy quark pairs are mainly created by gluon–gluon fusion and less often by quark–antiquark annihilation:

$$\text{Flavor creation:} \quad gg \rightarrow Q\bar{Q}, \qquad q\bar{q} \rightarrow Q\bar{Q}.$$

Higher order processes ($\approx\alpha_s^3$) have not yet been completely calculated. However, it was shown, that some of the 2 → 3 processes contribute a sizable fraction to the heavy flavor cross section and can be dominant for selected topologies. Besides initial- and final-state gluon radiadion, the following 2 → 3 processes were considered:

$$\text{Flavor excitation:} \quad Qg \rightarrow Qg \quad \text{and} \quad Qq \rightarrow Qq$$

$$\text{Gluon splittimg:} \quad g \rightarrow Q\bar{Q}.$$

In the case of flavor excitation the heavy quark already exists in the sea. Heavy quark pairs can also result from the QCD evolution of a large p_t gluon jet via gluon splitting into quarks. FIGURE 7 shows the calculated[10] fraction of gluons splitting into heavy quark pairs. Our measured point was determined by the number of D^* inside jets. Splitting into beauty pairs is mass suppressed, whereas charm pairs can still be created in cascade processes with gluons splitting into other gluons in the first step.

In order to confront the muon data with QCD calculations, elaborate Monte Carlo programs—Isajet[1] and Eurojet[11]—had to be used. These calculations can be factorized into the structure functions, obtained from lepton nucleon scattering data,[12] the hard QCD process, and the parton fragmentation and particle decay, which were both carefully adjusted to the wealth of the heavy flavor data from e^+e^- colliders.[13] In FIGURE 8 the inclusive muon distribution is compared with QCD calculations for single muons in the pseudorapidity intervall $/\eta/ < 1.5$, and also for muons from the nonisolated dimuon events. Both QCD calculations agree with the single- and dimuon data in shape and absolute cross section to better than a factor of 2, which is the uncertainty of these approximate calculations. The dashed curves correspond to lowest order (flavor creation) only; the solid lines include the previously mentioned higher order processes. These processes are definitely needed to explain the single-muon yield, but add less to the dimuons. In these events the higher order processes, like flavor

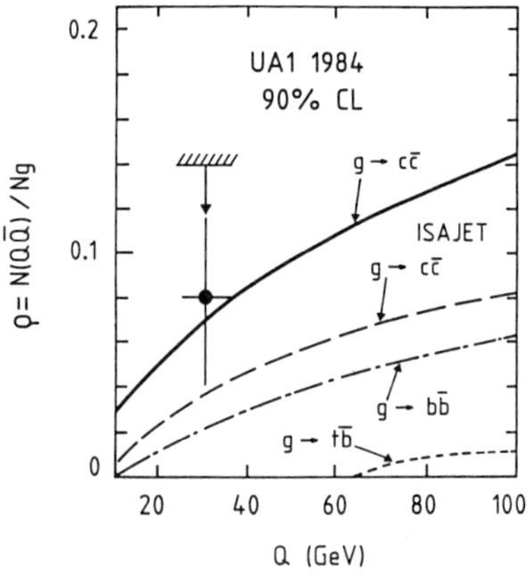

FIGURE 7. Fraction of gluon jets splitting into heavy quark pairs. (Q is the mass of the virtual gluon.)

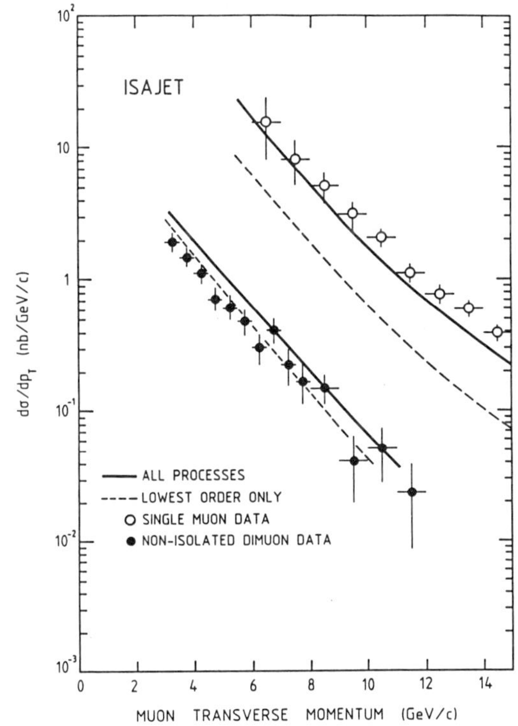

FIGURE 8. $d\sigma/dp_t$ of inclusive muons. Single muons with 6 GeV/c < p_t < 15 GeV/c. For comparison, the inclusive p_t distribution of the dimuons with p_t > GeV/c is shown, too. Lowest order Isajet calculations are shown as dashed lines, calculations including higher order processes as solid lines.

excitation and gluon splitting, are suppressed by the dimuon mass cut of 6 GeV/c^2, which forces muons with $p_t \approx 3$ GeV/c to be predominantly back-to-back in the transverse plane. However, a 25% fraction of higher order processes is also present in the dimuon data. This can be best demonstrated with the $\Delta\phi$-distribution of the two muons (ϕ is the azimuthal angle), which is sensitive to higher orders. Lowest order processes alone cannot explain the events with $\Delta\phi < 100°$ (FIGURE 9). TABLE 3 summarizes the contributions of the different processes, as calculated in Isajet, to the single and dimuon data and also the beauty and charm content. Due to the harder fragmentation of the beauty quark the dimuons rise to 90% and the single muons to 75% from beauty quarks.

What are the direct experimental evidences for the large beauty content in the dimuon data? First of all, the large amount of like-sign dimuons can only come from beauty quarks, either by cascade decays or even more important by beauty oscillations. Furthermore, the transverse muon momentum p_t^{rel} with respect to the jet axis reflects the mass of the fragmenting quark. A fit to the p_t^{rel}-distribution (FIGURE 10), seperate for like and unlike dimuons, results in a charm content in the dimuon data of 8 ± 6%,

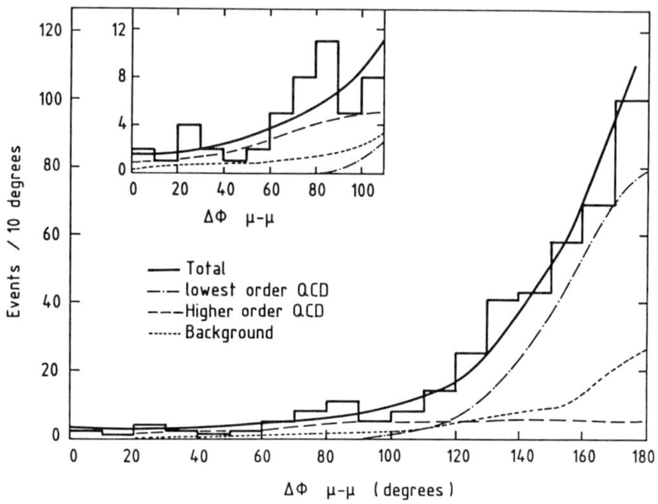

FIGURE 9. $\Delta\phi$ between both muons (nonisolated events). Superimposed are the contributions from Isajet, separately for lowest order and higher order contributions and the expected background.

consistent with the Monte Carlo calculations. We extract from the data the beauty pair production cross section in lowest order:

$$\sigma(p\bar{p} \to b\bar{b}) = 1.1 \pm 0.1_{\text{stat}} \pm 0.4_{\text{sys}}\ \mu\text{b}$$

for

$$p_t > 5\ \text{GeV}/c \text{ (both } b\text{-quarks)}$$

$$|\eta| < 2 \text{ (both } b\text{-quarks).}$$

The Isajet cross section is 1.7 μb for the lowest order beauty production. The good description of the beauty production by QCD gives us confidence in the calculation of the top cross section. The heavy top mass makes a perturbative QCD estimate even more reliable. If the mass is less than $\approx$80 GeV/c^2, W-decays are an additional source for top. FIGURE 11 shows the expected number of semileptonic top decays versus the top mass for an integrated luminosity of 690 nb^{-1}. The needed p_t and topology cuts to

TABLE 3. Contributions to the Single Muon ($p_t > 10$ GeV/c) and Dimuon ($m_{\mu\mu} > 6$ GeV/c^2) Samples from Different Processes

	Single Muons	Dimuons
$b\bar{b}$ lowest order	33	71
$b\bar{b}$ higher order	42	19
$c\bar{c}$ lowest order	6	8
$c\bar{c}$ higher order	19	2

Note: Higher order includes "flavor excitation" and "gluon splitting" (Isajet).

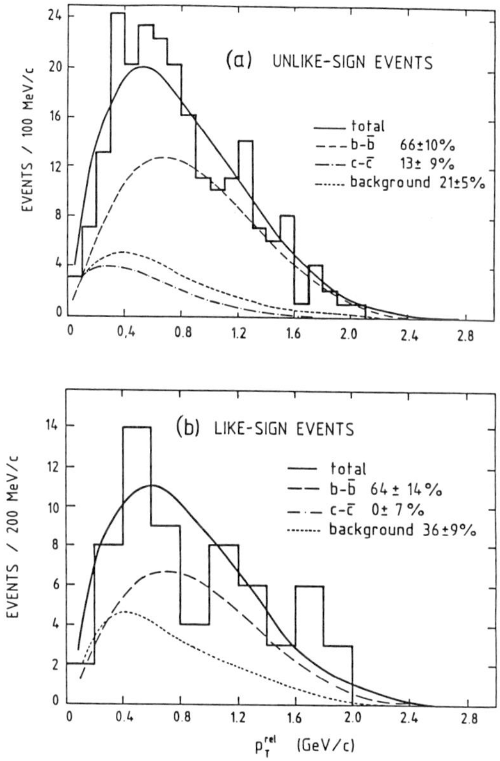

FIGURE 10. p_t of the muon with respect to the jet axis. The curves are the result of a fit of Monte Carlo generated p_t^{rel} distributions for $c\bar{c}$, $b\bar{b}$, and background. (a) Unlike-sign events. (b) Like-sign events.

FIGURE 11. Number of muon events at 690 nb^{-1}.

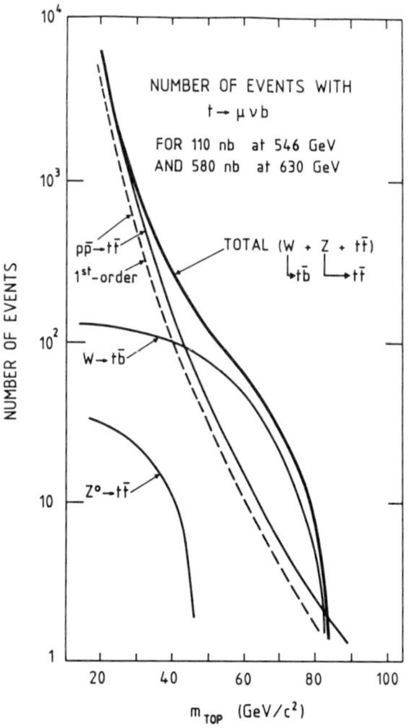

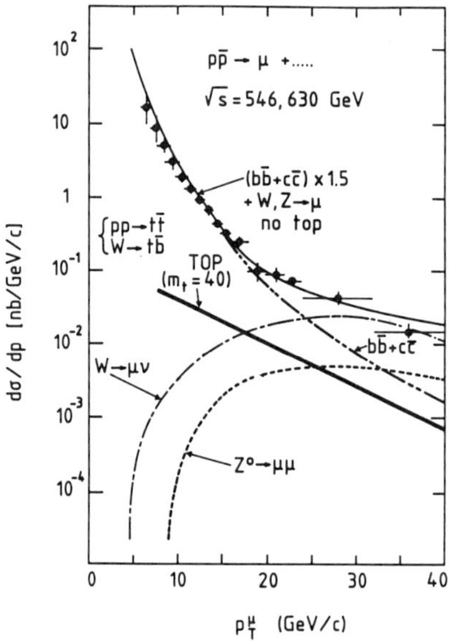

FIGURE 12. $d\sigma/dp_t(\mu)$ for various processes. (Also shown are the inclusive muon data.)

separate top from bottom are not yet applied. In the mass range between 50 and 80 GeV/c^2 W-decays are the dominant source. They give rise to spectacular events with two large p_t like-sign muons and jet activity. The calculated muon transverse momentum spectra from the different sources is plotted in FIGURE 12 together with the single-muon data. For a top mass of 40 GeV/c^2 the signal-to-background ratio is only 0.1 in the best case at a muon p_t of about 15 GeV/c. Additional criteria, like the muon isolation, jet activity, topology, and missing energy, have to be applied to single out the top events, but they drastically reduce the number of top events. UA1 has performed a careful top analysis and estimates at a 95% confidence limit a lower top mass limit of about 30 GeV/c^2 from the dimuon data and about 50 GeV/c^2 from the combined single-muon and electron data.[14] Similar limits hold for a fourth-generation bottom quark. In the future, with ten times increased statistics due to ACOL, top will be found, if the mass is below 100 GeV/c^2. At such a high mass the top decay into a W offers a spectacular signature.

The heavy flavor production and all kinematic distributions are in good agreement with QCD. There is only one exception: The large fraction of like-sign dimuons, which can only be explained in terms of substantial $B° - \overline{B}°$ oscillations.[15]

PHENOMENOLOGY OF BEAUTY OSCILLATIONS

As in the $K° - \overline{K}°$ system, transitions between a $B°$- and a $\overline{B}°$-meson are allowed by second-order weak interaction. Hence $B°$ and $\overline{B}°$ are a linear combination of the mass eigenstates B_H and B_L (assuming no CP violation):

$$B° = \frac{(B_H + B_L)}{\sqrt{2}} \qquad \overline{B}° = \frac{(B_H - B_L)}{\sqrt{2}}$$

These states may have different masses and decay widths, m_H, Γ_H, and m_L, Γ_L respectively, that result in a time-dependent phase difference between the B_H and B_L wave functions. With the present experimental resolution it is not possible to measure this time dependence directly. However, it is possible to measure time-integrated quantities, like the probability that a $B°$-meson decays into a "wrong"-sign muon ($\Delta m = m_H - m_L$, assuming $\Gamma_H = \Gamma_L = \Gamma$):

$$\chi = \frac{P(B° \rightarrow \bar{B}° \rightarrow \mu^-)}{P(B° \rightarrow \mu^{\pm})} = \frac{1}{2} \frac{(\Delta m/\Gamma)^2}{[1 + (\Delta m/\Gamma)^2]} .$$

In the case of maximal mixing, that is if the decay time $\tau_{\text{decay}} = 1/\Gamma$ is much longer than the oscillation time $\tau_{\text{osc}} = 1/\Delta m$, χ approaches 0.5.

The mass difference Δm can be calculated from the box diagram[16] (FIGURE 13):

$$\Delta m = \frac{G_f^2}{6\pi^2} B f_B^2 m_B m_w^2 F(m_c, m_t),$$

where G_f is the Fermi coupling constant, m_B the B mass, m_W the W mass, B is the bag parameter ($0.5 < B < 1.5$, $B = 1$ is the vacuum saturation approximation), and f_B is the B-decay constant ($f_B \approx 110$ MeV for the B_d^0 and 150 MeV for the B_s^0).[17]

The term $F(m_c, m_t)$ describes the dependence on the mass of the exchanged quarks. A good approximation is

$$B_d^0\text{:} \quad F \approx \eta_{cc}(V_{cb}^* V_{cd}) \frac{m_c^2}{m_W^2} + \eta_{tt}(V_{tb}^* V_{td}) \frac{m_t^2}{m_W^2}$$

$$+ 2\eta_{ct}(V_{cb}^* V_{cd} V_{tb}^* V_{td}) \frac{m_c m_t}{m_W^2}$$

$$B_s^0\text{:} \quad F \approx \eta_{cc}(V_{cb}^* V_{cs}) \frac{m_c^2}{m_W^2} + \eta_{tt}(V_{tb}^* V_{ts}) \frac{m_t^2}{m_W^2}$$

$$+ 2\eta_{ct}(V_{cb}^* V_{cs} V_{tb}^* V_{ts}) \frac{m_c m_t}{m_W^2},$$

where V_{ij} are the respective Kobayashi–Maskawa matrix elements, η_{ij} are QCD corrections (which are close to 1),[18] and m_i the masses of the involved heavy quarks. The dominant term comes from the top exchange, since the transitions are proportional to the mass squared of the exchanged particle.

As many of these parameters are rather unknown, precise predictions of the strength of $B° - \bar{B}°$ oscillations are not possible. Theoretical estimates of χ are in a range of 0.1 to 0.48 for B_s^0 and less than 0.2 for B_d^0.

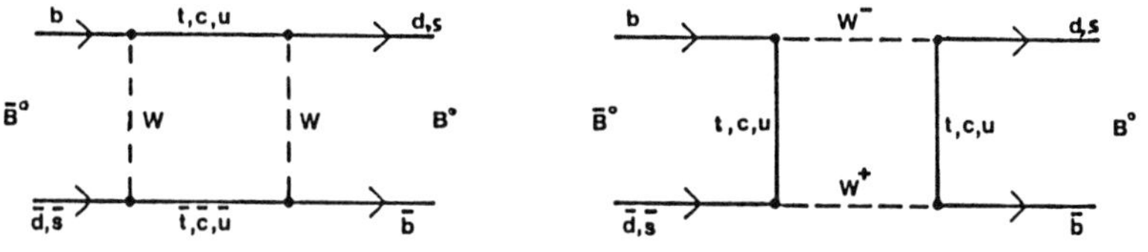

FIGURE 13. Box diagram for $\Delta B = 2$ weak transitions.

The main difference between B_d° and B_s° is due to the Kobayashi–Maskawa matrix elements V_{td} and V_{ts}, which favors mixing in the B_s° channel. In the Standard Model with three quark families these matrix elements are already constrained by the unitarity of the Kobayashi–Maskawa matrix.[19]

$$0.002 < |V_{td}| < 0.017 \qquad 0.037 < |V_{ts}| < 0.048.$$

This can be used to give an upper limit on χ_d as a function of χ_s. Assuming $(Bf_B\tau)_{Bs} = (Bf_B\tau)_{Bd}$, we obtain:

$$\frac{(\Delta m/\Gamma)_{Bd}}{(\Delta m/\Gamma)_{Bs}} = \frac{|V_{td}|^2}{|V_{ts}|^2} < 0.21.$$

This transforms into:

$$\chi_d < \frac{0.21^2\chi_s}{1 - 2\chi_s(1 - 0.21^2)}.$$

The advantage of this approach is that most of the theoretical uncertainties cancel, especially the dependence on the top mass.

EVIDENCE FOR BEAUTY OSCILLATIONS

The experimental signature of $B^\circ - B^\circ$ oscillations are like-sign dimuons in excess of the expected small number of like-sign events from a beauty–charm cascade:

$$b \rightarrow c \rightarrow \mu^+.$$

The cascade decays result in a softer p_t spectrum of the second-generation muons. Therefore these events can be suppressed by a cut on the muon p_t. First-generation decays of the pair-produced beauty quarks result in unlike-sign events. However, if one b hadronizes into a $\bar{B}_d^\circ$ or a $\bar{B}_s^\circ$, which than oscillates and decays into a μ^+, like-sign events can also arise from first-generation decays.

We observe 256 unlike-sign nonisolated dimuons and 140 like-sign nonisolated events. After subtracting the background of 58.5 events in either sample, the following ratio is obtained:

$$R = \frac{N(++,--)}{N(+-)} = \frac{140 - 58.5}{256 - 58.5} = 0.41 \pm 0.07_{\text{stat}} \pm 0.03_{\text{sys}},$$

TABLE 4. Predictions for R without $B^\circ - \bar{B}^\circ$ Oscillations ($\chi = 0$) and for Three Values of the Mixing Parameter χ

Reference	$\chi = 0$	$\chi = 0.05$	$\chi = 0.10$	$\chi = 0.15$	Charm Content
Barger *et al.*	0.25	0.31	0.36	0.42	0.23
Halzen *et al.*	0.25	0.33	0.41	0.48	0.11
Isajet (reference 1)	0.26	0.34	0.42	0.50	0.10
Eurojet (reference 11)	0.21	0.28	0.36	0.43	0.15

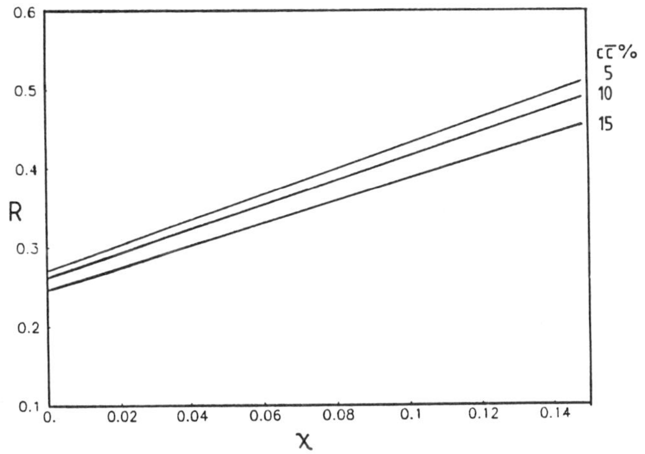

FIGURE 14. Dependence of R on the mixing parameter χ. (Isajet calculation for different $c\bar{c}$ contributions: 5, 10, and 15%.)

where the systemic error reflects the uncertainty of the background subtraction. This number ignores the contribution from Drell–Yan and Upsilon. Because of the limited efficiency of the isolation cut we expect 15 ± 7 events from Drell–Yan and Upsilon in the nonisolated unlike-sign sample. Taking this into account, we get:

$$R = 0.45 \pm 0.08 \pm 0.04.$$

These numbers have to be compared with the expectations from heavy flavor production in the case of no mixing:

$$R = \frac{N_s}{(N_f + N_c)},$$

where N_f is the number of dimuon events with both muons from first-generation decays, N_s is the number of events with one muon from first- and one muon from second-generation decay, and N_c is the contribution from $c\bar{c}$ production. Using Isajet to estimate N_f, N_s, and N_c we expect:

$$R_{\text{Isajet}} = 0.26 \pm 0.03.$$

The error is due to uncertainties from the b and c fragmentation functions and the semileptonic branching ratios. The term R_{Isajet} agrees well with independent calculations (TABLE 4).

If we consider $B° - \bar{B}°$ oscillations, the probability of a $b(\bar{b})$ to decay into a "wrong"-sign muon becomes (if all B-hadrons have the same semileptonic branching ratio):

$$\chi = \frac{P(b \rightarrow \bar{b} \rightarrow \mu^+)}{P(b \rightarrow \mu^\pm)} = f_{s\bar{s}}\,\chi_s + f_{d\bar{d}}\chi_d,$$

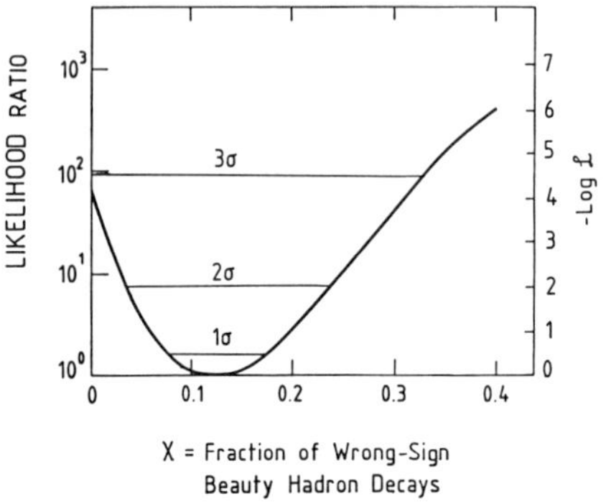

FIGURE 15. Likelihood function of the mixing parameter χ.

where $f_{s\bar{s}}(f_{d\bar{d}})$ is the probability that the b picks up an $\bar{s}$-quark ($\bar{d}$-quark) and forms a $\bar{B}^\circ_s(\bar{B}^\circ_d)$ meson. Based on the ratio K^+/π^+ at high p_t, which was measured at the ISR,[20] we use (allowing for 10% of B-baryons):

$$f_{s\bar{s}} = 0.18 \qquad f_{d\bar{d}} = 0.36.$$

The ratio of like-sign to unlike-sign events is related to χ by:

$$R = \frac{2\chi(1-\chi)N_f + [(1-\chi^2) + \chi^2]N_s}{[(1-\chi^2) + \chi^2]N_f + 2\chi(1-\chi)N_s + N_c}.$$

The dependence of R on χ is shown in FIGURE 14 based on Isajet calculations. TABLE 4 contains the results from several independent calculations. According to these calculations, the preferred value of χ is between 0.10 and 0.15.

Muons from second-generation decays have a softer p_t spectrum than primary muons. This difference can be used to determine the mixing parameter χ more precisely. The two-dimensional p_t distributions of first- and second-generation beauty decays, charm decays, and background were fitted to the data. The maximum likelihood fit accounted for errors of the Monte Carlo calculations and of the background determination. The fit results in (FIGURE 15):

$$\chi = 0.121 \pm 0.047$$

$$\chi > 0.065 \text{ at } 90\% \text{ C.L.}$$

The alternative of no mixing ($\chi = 0$) is disfavored by 2.9 standard deviations.

TABLE 5. Results on $B^\circ - \bar{B}^\circ$ Oscillations from e^+e^- Experiments

	$b\bar{b}$ source	$\sqrt{s}$/GeV	Result
CLEO (reference 23)	$Y(4s) \rightarrow B^\circ_d \bar{B}^\circ_d$	10.6	$\chi_d < 0.19$ at 90% C.L.
ARGUS (reference 24)	$Y(4s) \rightarrow B^\circ_d \bar{B}^\circ_d$	10.6	$\chi_d = 0.19 \pm 0.07$
MARK II (reference 25)	$e^+e^- \rightarrow b\bar{b}$	29	$\chi < 0.12$ at 90% C.L.
JADE (reference 26)	$e^+e^- \rightarrow b\bar{b}$	34	$\chi < 0.13$ at 90% C.L.

Subtracting the contribution form Drell–Yan and Upsilon we obtain:

$$\chi = 0.158 \pm 0.059$$

$$\chi > 0.091 \text{ at } 90\% \text{ C.L.,}$$

which is 3.3 standard deviations away from $\chi = 0$.

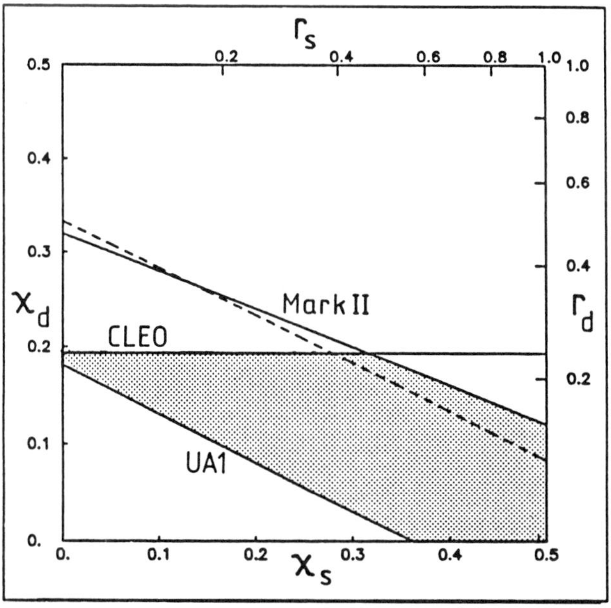

FIGURE 16. Limits on χ_d and χ_s MARK II, CLEO, and UA1. The solid lines are the 90% C.L. limits. The dashed line indicates the UA1 central value of $\chi = 0.12$. The UA1 data were evaluated with $f_{s\bar{s}} = 0.18$, while $f_{s\bar{s}} = 0.15$ was used for MARK II data. This is justified by the lower Q^2 of the e^+e^- data.

COMPARISON WITH OTHER EXPERIMENTS

We can compare our result with measurements from e^+e^- experiments. Searches for $B^\circ - \bar{B}^\circ$ oscillations were made by ARGUS,[24] CLEO,[23] JADE,[26] and MARK II.[25] Their results are summarized in TABLE 5. CLEO and ARGUS use the $Y(4s)$ resonance, which decays into $B^\pm$ and B_d°; thus they are sensitive only to B_d° mixing. MARK II and JADE use continuum data well above the B_s° threshold, and should therefore be sensitive to B_d° and B_s°. CLEO, MARK II, and JADE report upper limits. These limits are shown together with the UA1 result in FIGURE 16. ARGUS has recently reported a positive signal for B_d° oscillations. The ± 1 standard deviation bands of the UA1 and ARGUS measurements are shown together in FIGURE 17. The two measurements are consistent within a large area in the $\chi_s - \chi_d$ plane. Thus ARGUS is in best agreement with UA1 and confirms our result.

FIGURE 18 shows the allowed region in the $\chi_s - \chi_d$ plane for a fit that combines the likelihood function of all measurements. We used the following data:

- UA1: Gaussian distribution: $\chi = 0.121 \pm 0.047$;
- CLEO: Gaussian distribution: $\chi_d = 0.096 \pm 0.096$;
- ARGUS: Gaussian distribution: $\chi_d = 0.187 \pm 0.069$;
- MARK II: Gaussian distribution: $\chi = 0 \pm 0.075$.

All experimental results are consistent and allow a large region in the $\chi_s - \chi_d$ plane. While χ_d is restricted between 0.08 and 0.20 (mainly due to ARGUS), there is no restriction on χ_s from experimental data: All values between 0 and 0.5 are possible. We also show the upper limit on χ_d as a function of χ_s, which can be derived from existing limits on the Kobayashi–Maskawa matrix elements V_{ts} and V_{td} (i.e., $|V_{td}|^2/V_{ts}|^2 < 0.21$). If this constraint is combined with the experimental limits, only a small allowed region in the $\chi_s - \chi_d$ plane remains, indicating almost maximal mixing in the B_s^o channel. Thus a lower limit on $\chi_s > 0.40$ (90% C.L.) can be derived ($\chi_s > 0.30$ from UA1 alone).

At present the data are consistent with the expectations of the Standard Model. Future experiments with a more precise determination of χ_s and χ_d are required to test the Standard Model seriously. With ACOL UA1 is expected to collect about 20,000

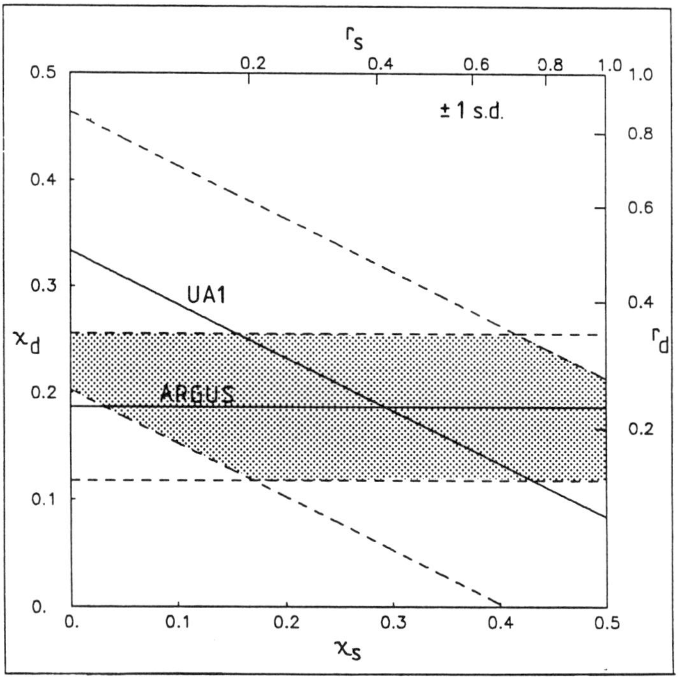

FIGURE 17. ±1 standard deviation bands from UA1 and ARGUS. (The same values for $f_{s\bar{s}}$ as in FIGURE 16.)

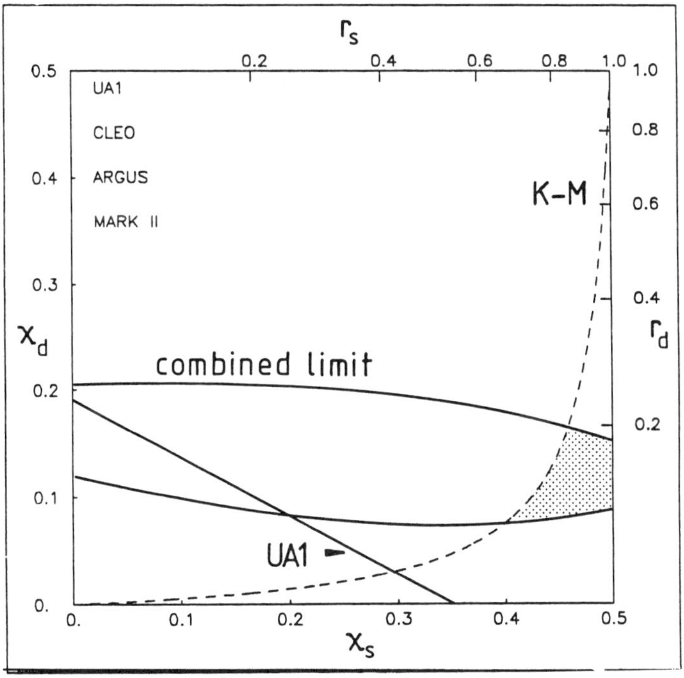

FIGURE 18. Combined limit on χ_d and χ_s from UA1, MARK II, CLEO, and ARGUS. (Solid lines, 90% C.L.) The dashed line indicates the limits on χ_s and χ_d from the Kobayashi–Maskawa matrix. The dotted region is allowed by both constraints. (UA1 > indicates the UA1 lower limit $\chi > 0.065$, 90% C.L.)

dimuon events. This will allow us to determine χ with an accuracy of 10%, which will improve the present accuracy by a factor of 4.

SUMMARY

We have discussed the heavy flavor production at the CERN proton–antiproton collider ($\sqrt{s} = 630$ GeV). The heavy quarks were tagged in UA1 via their semileptonic decays into muons. The production cross section for beauty particles and the topology of the events were in good agreement with QCD calculations, once higher processes, like flavor excitation and gluon splitting, were taken into account. The large amount of like-sign dimuons was interpreted as evidence for beauty oscillations. These data were discussed in detail and compared with similar measurements at e^+e^- colliders.

REFERENCES

1. Paige, F. E. & S. D. Protopopescu. 1986. Brookhaven National Laboratory, BNL-38034.

2. ARNISON, G. *et al.* 1986. Phys. Lett. **B162:** 461.
3. LEE, T. D. & C. S. WU. 1966. Annu. Rev. Nucl. Sci. **16:** 511.
4. BARRANCO LUQUE, M. *et al.* 1980. Nucl. Instrum. Methods **176:** 175; CALVETTI, M. *et al.* 1986. Nucl. Instrum. Methods **A234:** 45; CORDEN, M. J. *et al.* 1985. Nucl. Instrum. Methods **A238:** 273; EGGERT, K. *et al.* 1985. Nucl. Instrum. Methods **176:** 273; BAUER, G. *et al.* 1987. Nucl. Instrum. Methods **A253**.
5. ALBAJAR, C. *et al.* 1987. Phys. Lett. **B186:** 237; ALBAJAR, C. *et al.* 1987. Phys. Lett. **B186:** 247.
6. MOSER, H.-G. 1987. Ph.D. thesis. Univ. of Aachen, Aachen, West Germany.
7. ANGELIS, A. L. S. *et al.* 1979. Phys. Lett. **B87:** 398; KOURKOUMELIS, C. *et al.* 1980. Phys. Lett **B91:** 481; ANTREASYN, D. *et al.* 1979. *In* Proc. EPS Int. Conf. on High Energy Physics. Geneva, Switzerland, p. 779; YOH, J. S. *et al.* 1978. Phys. Rev. Lett. **41:** 481.
8. CLARK, A. G. *et al.* 1978. Nucl. Phys. **B142:** 29; KOURKUMELIS, C. *et al.* 1980. Phys. Lett. **B91:** 475; ANTREASYN, D. *et al.* 1982. Phys. Rev. Lett. **48:** 302; BADIER, J. *et al.* 1985. Z. Phys. **C26:** 481; ITO, A. S. *et al.* 1981. Phys. Rev. **D23:** 604.
9. BARGER, V. & A. D. MARTIN. 1985. Phys. Rev. **D31:** 1051.
10. MUELLER, A. H. & P. NASON. 1985. Phys. Lett. **B157:** 226.
11. VAN EIJK, B. 1985. *In* Proc. 5th Topical Workshop on Proton-Antiproton Collider Physics. St. Vincent, Aosta Valley, Italy, p. 165.
12. EICHTEN, E. *et al.* 1984. Rev. Mod. Phys. **56:** 579.
13. SAXON, D. H. 1985. *In* Proc. EPS Meeting on High Energy Physics. Bari, Italy, p. 889.
14. REVOL, J. P. 1987. Presented at the UA1 seminar at CERN. Paper by UA1 in preparation.
15. EGGERT, K. 1986. *In* Proc. New Particles 1985, Wisconsin. V. Barger, D. Cline & F. Halzen, Eds.: 207. World Scientific, Singapore; EGGERT, K. *In* Proc. 21th Recontre de Moriond. Les Arcs, France. Edition Frontieres, 1986. Progress in electroweak interaction. p. 369; MOSER, H.-G. *In* Proc. 21th Recontre de Moriond. Les Arcs, France. Edition Frontieres, 1986. Strong interactions and gauge theories. p. 331.
16. BURAS, A. J., W. SLOMINSKI & H. STEGER. 1984. Nucl. Phys. **B245:** 3691.
17. SHIFMAN, M. A. & M. V. VOLOSHIN. 1986. ITEP Rep. 54: 86.
18. HAGELIN, J. S. 1981. Nucl. Phys. **B193:** 123.
19. KLEINKNECHT, K. & B. RENK. 1987. Z. Phys. **C34:** 209.
20. BREAKSTONE, A. *et al.* 1984. Phys. Lett. **B135:** 510; A recent review can be found in GEIST, W. M. 1987. Experimental comments on the suppression of strangeness and diquarks in parton fragmentation. Presented at the 22nd Recontre de Moriond.
21. BARGER, V. & R. J. N. PHILLIPS, MAD/PH/155: 239, 266.
22. HALZEN, F. & A. MARTIN. 1984. DTP/84/14.
23. BEAN, A. *et al.* 1987. Phys. Rev. Lett. **58:** 183.
24. SCHROEDER, H. 1987. Presented at the DESY Seminar. Hamburg, Germany.
25. SCHAAD, T. *et al.* 1985. Phys. Lett. **B160:** 188.
26. BARTEL, W. *et al.* 1984. Phys. Lett. **B146:** 437.

Possible Signatures of the Fourth Generation at Present Energies[a]

EMMANUEL A. PASCHOS

Institut für Physik
Universität Dortmund
Dortmund 50, West Germany

INTRODUCTION

It is evident from the papers in this volume that a fourth generation of quarks and leptons is a very attractive and quite reasonable extension of the Standard Model. At this time, although none of the observations demands their existence, several measurements are at the borderline allowed by the Standard Model with three generations. Thus improved accuracy in the measurements could lead to contradictions. Furthermore, there is a renewed interest in heavier quarks and leptons motivated partly by the ability of the next generation of accelerators and experiments to settle several questions, such as the number of generations (by measuring the number of neutrinos), the production of charged heavy leptons, and the observation of indirect effects through the contribution of heavy quarks in intermediate states. Before we discuss the limitations and evidences for heavy quarks and leptons, we briefly review some of their properties as we understand them today. Articles concerning their properties began as early as 1980 and we refer to them for more details.[1]

Masses

Theoretically the renormalization group equations give upper bounds for the quark masses[2–5]

$$m_Q \lesssim 200\ \text{GeV}/c^2. \tag{1}$$

Charged leptons, on the other hand, are constrained by the $SU(5)$ assumption, which states that the heavy lepton and the corresponding down-quark masses start equal and gives the prediction[3]

$$m_L \approx \tfrac{1}{3}\, m_Q \lesssim 50\text{–}70\ \text{GeV}. \tag{2}$$

Experimentally,

$$m_L > 23\ \text{GeV}/c^2$$

from DESY[6] and

$$\geq 41\ \text{GeV}/c \tag{3}$$

[a]This paper is dedicated to Professor U. Bonse in honor of his 60th birthday.

from the results[7] of UA1. Finally, the ratio of single leptons to opposite-sign dileptons observed in $\bar{p}$-p reactions correlates the masses[8] of the top quark to the masses of E and b', the heavy lepton and heavy bottom quark of the fourth generation, respectively.

Mixing Angles

There are thorough investigations of the matrix elements of the Kobayashi–Maskawa (KM) matrix, which give restrictive ranges summarized in V_3. The ranges that we use are somewhat larger than in other papers, since we consider more conservative values[9,10] for the experimental and theoretical errors:

$$V_3 = \begin{bmatrix} 0.972\ldots0.976 & 0.218\ldots0.222 & 0.000\ldots0.012 \\ 0.217\ldots0.222 & 0.973\ldots0.975 & 0.048\ldots0.057 \\ 0.000\ldots0.024 & 0.044\ldots0.058 & 0.998\ldots0.999 \end{bmatrix}. \qquad (4)$$

A good summary for the matrix is given in the Maiani notation

$$V_3 = \begin{bmatrix} c_\theta & s_\theta & \beta \\ -s_\theta - \beta\gamma e^{i\delta} & c_\theta & \gamma e^{i\delta} \\ -\beta + \gamma\theta e^{-i\delta} & -\gamma e^{-i\delta} - \beta\theta & 1 \end{bmatrix} \qquad (5)$$

with

$$\sin\theta = 0.221 \pm 0.006$$

$$0.049 \leq \gamma \leq 0.057$$

$$\beta \leq 0.012.$$

Within the errors unitarity is comfortably satisfied, but there is a restriction for the new element

$$|V_{uB}| \lesssim 0.08. \qquad (6)$$

Several of the elements are determined directly from the data, without any references to the number of generations, and hold for the four generations as well. The extension to 4 × 4 matrix restricts the new elements as follows:[11]

$$V_4 = \begin{bmatrix} 0.972\ldots0.976 & 0.218\ldots0.222 & 0.0\ldots0.012 & 0.0\ldots0.08 \\ 0.18\ldots0.23 & 0.84\ldots0.98 & 0.04\ldots0.06 & 0.0\ldots0.50 \\ 0.0\ldots0.13 & 0.0\ldots0.47 & 0.0\ldots0.99 & 0.0\ldots0.99 \\ 0.0\ldots0.14 & 0.0\ldots0.49 & 0.0\ldots0.99 & 0.0\ldots1.0 \end{bmatrix}. \qquad (7)$$

We note that several of the new elements can be very large, yielding substantial couplings of heavy-to-light quarks. As a consequence heavy quarks can make substantial contributions as intermediate states of several processes.

Evidence for a fourth generation can be *direct,* through the production of new states, or *indirect,* through their effects in intermediate states. The state most likely to be produced directly is the heavy charged lepton, as was discussed in the talk by Cline.[12] In the absence of any direct evidence at this time, we analyze processes where indirect effects are likely to show up. They include

(1) $B_d - \bar{B}_d$ mixing;
(2) *CP*-violating parameters of the *K*-meson;
(3) the rare decay $K^+ \rightarrow \pi^+ \nu\bar{\nu}$.

In each case we first give the predictions of the three generations and then the enlarged ranges brought in by the heavy intermediate states.

MATTER–ANTIMATTER OSCILLATIONS IN THE B_d-SYSTEM

The big news at the conference was the observation by the Argus collaboration[13] of $B_d - \bar{B}_d$ mixing at the 20% level. In general, the mixing is described by the parameter[14]

$$r = \left| \frac{1 - \epsilon_B}{1 + \epsilon_B} \right|^2 \frac{(\Delta M)^2 + (\Delta\Gamma)^2}{2\Gamma^2 + (\Delta M)^2 - [(1/2)\Delta\Gamma]^2}, \tag{8}$$

which for the B_d system reduces to

$$r_d \approx \frac{(\Delta M/\Gamma)^2}{2 + (\Delta M/T)^2} \tag{9}$$

because $\Delta\Gamma \ll \Delta M$ and ϵ_B is practically purely imaginary. The experimental result gives $\Delta M/\Gamma \approx 0.70$, which is somewhat large. The analysis in the Standard Model can be taken from the published articles, where most of them expected much smaller mixing,[15] but a few suggested[16] as early as 1983 "that $B_d - \bar{B}_d$ mixing at the 10–20% level is reasonable."

The $\Delta B = 2$ transition amplitudes are dominated by box diagrams. All three quarks appear in intermediate states. The *u*-quark intermediate state will be neglected because of its small mass. An estimate of the relative contributions from *c* and *t* states, including the mixing angles, shows that the *t*-quark dominates by a factor of $(m_t/m_c)^2$. The matrix element keeping only the top intermediate state is

$$M_{12} = \frac{G^2}{16\pi^2} X_B \tilde{\eta} |V_{tb} V_{td}^*|^2 M_W^2 E(x_t) \tag{10}$$

with

$$X_B = \langle B_0 | [\bar{d}\gamma_\mu(1 - \gamma_5)b]^2 | \bar{B}_0 \rangle, \tag{11}$$

where $x_t = m_t^2/M_W^2$, $\tilde{\eta}$ is the QCD correction that slightly suppresses the amplitude, and

$$E(x_t) = \frac{3}{2} \frac{x_t^2}{(1 - x_t)^3} \ln x_t - x_t \left[\frac{1}{4} + \frac{9}{4} \frac{1}{1 - x_t} - \frac{3}{2} \frac{1}{(1 - x_t)^2} \right] \tag{12}$$

is the structure function resulting from the four-dimensional integral.[17] This formula holds for all values of the top-quark mass. A typical value is $\tilde{\eta} = 0.87$ for $m_t = 40$ GeV and $\Lambda_{QCD} = 0.1$ GeV.

$$\Delta M_B = M_H - M_L = 2 \cdot |M_{12}| \qquad (13)$$

with the indices H, L denoting heavy and light, respectively. The mass difference reduces to this form because the absorptive part γ_{12} is smaller (reference 16, first two papers) than the dispersive part m_{12} by a factor 5 to 10. The same property gives an ϵ_B that is largely imaginary.

The mass difference is computed from equations 10 and 13, and carries uncertainties associated with the range of the angular function $|V_{tb}V_{td}^*|$, the mass of the top quark, and the hadronic matrix element X_B. The mass of the top quark will be treated as a parameter and is allowed to vary between 30 and 70 GeV. For each value of m_t, standard analyses of the CP-violating parameter ϵ_K give a range for the angular function $V_{tb}V_{td}^*$, which is included in the results. The major uncertainty lies in the matrix element X_B, which is very difficult to estimate reliably. A popular parametrization is

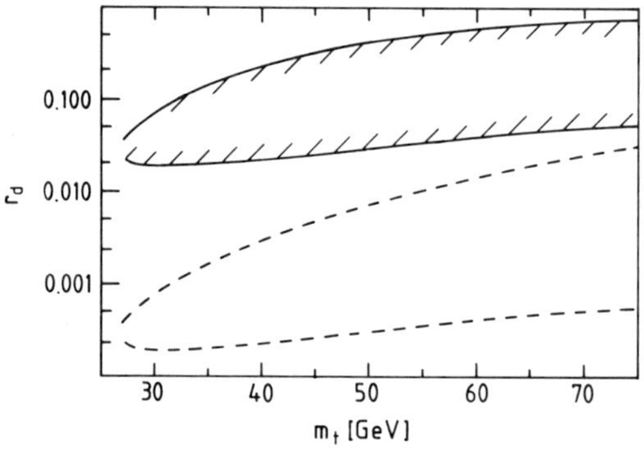

FIGURE 1. Upper and lower bounds for the mixing parameter r_d of the $B_d - \overline{B}_d$ system as a function of the mass of the top quark. Solid line is for $B_B = 1.0$; dashed line is for $B_d = 0.1$.

$$X_B = {}^4\!/_3 f_B^2 M_B B_B, \qquad (14)$$

with the term ${}^4\!/_3(f_B^2 M_B)$ representing the vacuum insertion and the rescaling factor B_B summarizing the effects of additional contributions. Since many intermediate states are possible, the factor B_B could be very different from unity. Furthermore, even the decay coupling constant f_B has not been measured yet, and one must rely on theoretical estimates that yield $f_B = 100$–300 MeV. The preceding remarks explain the wide range of values for X_B used by the many authors and point out that larger values of $f_B^2 B_B$ cannot be excluded. We consider it quite reasonable to present in FIGURE 1 the Standard Model predictions for r_d as a function of m_t for $f_B = 300$ MeV and two values of $B_B = 1.0$ and 0.10. Evidently for $m_t \geq 40$ GeV the mixing of the $B_d - \overline{B}_d$ states at the 10–20% level is possible.

With the B_d mixing observed at the 20% level we can be more confident for the mixing of the B_s states. The ratio of the mass differences is

$$\frac{\Delta M_{Bs}}{\Delta M_{Bd}} = \frac{X_{Bs}\,|V_{ts}|^2}{X_{Bd}\,|V_{td}|^2}. \tag{15}$$

We expect the hadronic matrix elements to be similar, with X_{B_s} slightly bigger than X_{Bd}. The range allowed for the angular functions restricts

$$\Delta M_{Bs} \gtrsim 4\Delta M_{Bd}. \tag{16}$$

This in turn makes the B_s mixing very large, that is, larger than 80% and very likely maximal. This is an old prediction[18] that appeared before the measurement of the long B lifetime, and it still holds after the improved estimates for the mixing angles.

The only other positive experimental result on mixing comes from the UA1 collaboration.[19] In p-$\bar{p}$ collisions, they observe an excess of same-sign dimuons that they attribute to $B_0 - \bar{B}_0$ oscillations. Their result is consistent with the Argus report and also with maximal $B_s - \bar{B}_s$ mixing.

With four generations, it is easier to account for $B_d - \bar{B}_d$ mixing, because now both the new heavy quark and the top quark contribute.[20] In particular the dispersive element in the mass matrix is

$$M_{12} = -\frac{G^2}{16\pi^2} X_B M_W^2 \{\tilde{\eta}\xi_t^2 E(x_t) + 2\xi_t\xi_T E(x_t, x_T) + \xi_T^2 E(x_T)\} \tag{17}$$

with

$$\xi_i = \begin{cases} V_{ib}^* V_{id}, & \text{for the } B_d\text{-meson} \\ V_{ib}^* V_{is}, & \text{for the } B_s\text{-meson,} \end{cases}$$

and $\tilde{\eta}$ is again the QCD correction. The other factors $E(x_t)$, $E(x_t, x_T)$ are functions of $x_i = m_i^2/M_W^2$ resulting from the computation of the box diagrams. Values for them occur in TABLE 1 for $m_c = 1.5$ GeV and the arbitrary values $m_t = 40$ GeV and $m_T = 200$ GeV. The table shows that the structure functions increase fast with increasing masses and that they are substantial for very heavy quarks. Thus even for small mixing angles heavy quarks can make sizable contributions.

The range of the new angles and phases is still very large, so that an exhaustive study is not very informative. Instead we choose to present in TABLE 2 three solutions[20] that are consistent with ϵ_K and demonstrate that a wide range for the mixing parameters r_d is allowed. This indicates that smaller values of $f_B^2 B_B$ can also be accommodated in the enlarged model. It is interesting to note that in solution 3 the element $\beta = V_{ub}$ vanishes.

MODIFICATION OF *CP*-VIOLATING EFFECTS

In three generations all *CP*-violating effects vanish for $\beta = V_{ub} = 0$, because in this case the *CP*-violating phase of the KM matrix can be transformed away.

With four generations an ϵ_K is generated from the heavier quarks through the new

TABLE 1. The Box Diagram Functions $E(x_i, x_i)$ and $E(x_i, x_j)$, Relevant for $K - \overline{K}$ and $B - \overline{B}$ Mixing

$-E(x_i, x_j)$	x_c	x_t	x_T
x_c	3.51×10^{-4}	2.26×10^{-3}	3.07×10^{-3}
x_t	2.26×10^{-3}	2.23×10^{-1}	5.99×10^{-1}
x_T	3.07×10^{-3}	5.99×10^{-1}	3.18

Note: The functions are evaluated for $m_c = 1.5$ GeV, $m_t = 40$ GeV, and $m_T = 200$ GeV.

phases. There is still a large space of solutions, some with very small or vanishing V_{ub}, as illustrated by the solutions in TABLE 2.

The ϵ'_K parameter is more complicated, so we discuss it in greater detail. Parameter ϵ'_K is related to the relative phase of the $I = 2$ to the $I = 0$ amplitudes of $K_L \rightarrow 2\pi$ decays. More specifically, one defines

$$\eta_{+-} = \frac{A(K_L^0 \rightarrow \pi^+\pi^-)}{A(K_s^0 \rightarrow \pi^+\pi^-)} = \epsilon + \frac{\epsilon'}{(1 + \omega/\sqrt{2})} \quad \textbf{(18)}$$

$$\eta_{00} = \frac{A(K_L^0 \rightarrow \pi^0\pi^0)}{A(K_s^0 \rightarrow \pi^0\pi^0)} = \epsilon - \frac{\epsilon'}{(1 - \sqrt{2}\omega)} \quad \textbf{(19)}$$

with

$$\omega = \frac{\mathrm{Re}\, A_2}{\mathrm{Re}\, A_0} = 0.045 \quad \textbf{(20)}$$

and

$$\epsilon'_K = \frac{1}{\sqrt{2}} \frac{\mathrm{Im}\, A_2}{A_0} e^{i(\delta_2 - \delta_0)}. \quad \textbf{(21)}$$

The phases δ_0 and δ_2 arise from final state interactions of the two pions with isospin zero and two, respectively. A phase from the weak interactions is generated from the penguin diagrams with charm and top-quark intermediate states. In fact, there is a

TABLE 2. Three Solutions with Four Generations

Parameter \ Solution	1	2	3
β	0.003	0.0031	0.00
ρ	0.39	0.008	1.43
σ	0.011	0.011	0.094
τ	0.020	0.038	0.019
δ_1	1.50	0.65	2.45
δ_2	2.09	1.21	0.10
δ_3	1.45	2.52	2.17
r_d	0.096	0.024	0.95
r_s	0.98	0.95	1.00

Note: $\theta = 0.23$, $\gamma = 0.053$, and the other parameters as in the table.

cancellation between the two diagrams, so that the integral involves higher momenta. A short distance expansion is better justified for the imaginary part of the amplitude. In the Wu-Yang phase convention

$$A_2 = |A_2|\, e^{-i\xi} \tag{22}$$

and

$$\frac{\epsilon'}{\epsilon} = -15.5\xi. \tag{23}$$

The phase ξ involves ratios of hadronic matrix elements and can be bounded provided that we have approximate estimates of the hadronic ratios. They are in turn estimated either by current algebra[21] or other methods.[22] A crucial assumption requires that the penguin amplitude interferes constructively with the other amplitudes in order to produce an enhanced $\Delta I = 1/2$ amplitude. The last assumption also gives a positive sign for ϵ'/ϵ. The results of such calculations[9,23] are shown in FIGURE 2 for two values of B_K. The allowed domains are the shaded regions between the curves.

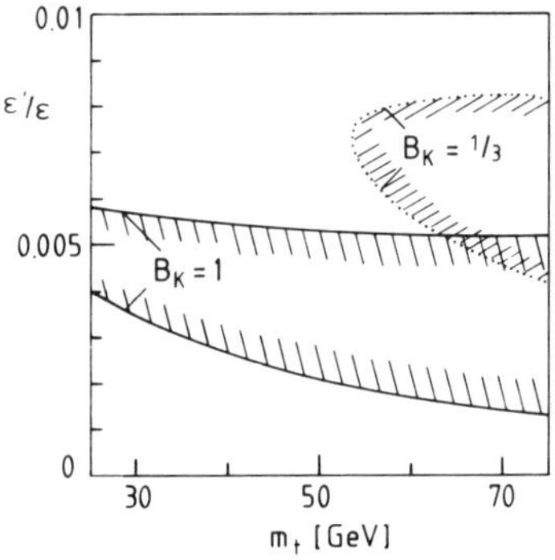

FIGURE 2. Upper and lower bounds for the ratio ϵ'/ϵ versus the mass of the top quark. Solid (dotted) lines are for $B_K = 1/3$ (1).

The preceding results can be modified by the photopenguin diagrams and diagrams with $(\eta^0 - \pi^0)$ mixing. Photopenguins are obtained from the normal penguins by replacing the gluon with a photon. For a while it was argued[24] that both of the new contributions subtract from the standard analysis and reduce the ϵ'/ϵ by a factor of 2 to 3. Subsequent work by Buras and Gérard[25] points out that the two new terms come with opposite signs and the photopenguin operators are substantially smaller. The end effect is to reduce the prediction of FIGURE 2 by 27%, which in view of the uncertainties occurring in the entire calculation is a minor reduction. The conclusion is that with three generations the bounds from FIGURE 2 hold and they are consistent with the measurements

$$\left(\frac{\epsilon'}{\epsilon}\right)_K = (3.5 \pm 3.0 \pm 2.0) \times 10^{-3} \qquad \text{Fermilab}^{26} \tag{24}$$

and

$$\left(\frac{\epsilon'}{\epsilon}\right)_K = (1.7 \pm 8.4) \times 10^{-3} \qquad \text{Yale-BNL.}^{27} \tag{25}$$

Thus improvements of the experiments will either confirm the Standard Model prediction or find a value outside the allowed region.

Should the latter case prevail, then the fourth generation will be very helpful. Repeating the preceding analysis we obtain[20]

$$\left(\frac{\epsilon'}{\epsilon}\right)_K = (46.7\,\beta\gamma \sin\delta_1 - 52.0\,\sigma\tau \sin\delta_2)\,\frac{\langle I = 0\,|Q_6|\,K\rangle}{1.4\ \text{GeV}^3}, \tag{26}$$

where σ and τ are the new angles of the fourth generation, δ_1 is the old phase, and δ_2 is one of the new phases. The hadronic matrix element $\langle 2\pi, I = 0\,|Q_6|\,K\rangle$ is equal to 1.4 GeV^3 in bag model estimates. It is now possible for $(\epsilon'/\epsilon)_K$ to have a range of values positive, zero, or negative, and thus be consistent if the experiments find a small or negative value.

RARE DECAYS

The rare decay $K^+ \longrightarrow \pi^+\nu\bar{\nu}$ has additional intermediate states[28] and the upper bound for the branching ratio is relaxed.

$$\begin{aligned} Br\,(K^+ \longrightarrow \pi^+\nu\bar{\nu} &\le 10^{-10} && \text{(Standard Model)} \\ &\le 8 \times 10^{-9} && \text{(four generations}^{28}\text{)} \\ &\le 1.4 \times 10^{-7} && \text{(experiment).}^{29} \end{aligned}$$

An experiment at the Brookhaven Laboratory is searching for this decay mode.

SUMMARY

Several measurements are at the borderline allowed by the Standard Model with three generations. Among them are

(1) the small value of $V_{ub} \lesssim 0.012$;
(2) the small value of $(\epsilon'/\epsilon)_K$;
(3) the relatively large $B_d - \bar{B}_d$ mixing.

We can interpret the results in two very different scenarios.

Scenario 1: We can account for all the effects with three generations by readjusting the parameters of the Standard Model and/or increasing the hadronic matrix elements. For instance, the $B_d - \bar{B}_d$ mixing can be accounted for with $m_t \ge 40$ GeV and $f_B\,\sqrt{B_B} = (\text{few}) \times 100$ MeV. In this case, improved measurements should find V_{ub} and $(\epsilon'/\epsilon)_K$ close to the values discussed in this article.

Scenario 2: One could also argue that we have already seen the effects of a fourth generation with a large part of r_d coming from it. This scenario becomes more convincing if

(1) V_{ub} turns out to be very small and ϵ_K receives a large contribution from the fourth generation;
(2) if $(\epsilon'/\epsilon)_K$ turns out very small (less than 1.5×10^{-3}) or negative.

It is too early to decide between the two cases and improved measurements, especially in a B-meson factory, as well as future measurements of Z^0-decays will be of great value in deciding the issue.

ACKNOWLEDGMENT

We wish to thank Drs. A. Datta, U. Türke for theoretical discussions, and Prof. D. Wegener for discussions concerning the experimental results of the Argus collaboration.

REFERENCES

1. Bose, S. K. & E. A. Paschos. 1980. Nucl. Phys. B169: 384–402; Barger, V., H. Baer, K. Hagiwara & R. J. M. Phillips. 1984. Phys. Rev. **D30:** 947–960; Gronau, M. & J. Schechter. 1985. Phys. Rev. **D31:** 1668–1675; Anselm, A. A. *et al.* 1985. Phys. Lett. **156B:** 102–108; He, X.-G. & S. Pakvasa. 1985. Phys. Lett. **156B:** 236–242.
2. Cabibbo, N., L. Maiani, G. Parisi & R. Petronzio. 1979. Nucl. Phys. **158B:** 295–305; Pendelton, M. & G. G. Ross. 1981. Phys. Lett. **98B:** 291–294.
3. Hill, C. 1981. Phys. Rev. **D24:** 691–703.
4. Paschos, E. A. 1984. Z. Phys. **C26:** 235–239.
5. Bagger, J., S. Dimopoulos & E. Masso. 1985. Phys. Rev. Lett. **55:** 920–922.
6. Branson, J. G. 1987. Low-thrust hadron events with isolated μ or e from PETRA. This volume; Komamiya, S. 1985. Presented at the Kyoto Conference. Kyoto, Japan.
7. Cline, D. & M. Mohammadi. 1986. Wisconsin preprint 86/269.
8. Datta, A. & R. M. Godbole. 1987. Relative rates of W and Z events at the $p\bar{p}$ collider and nonstandard physics. Phys. Rev. **D37:** 225; Despande, N. G. 1987. Neutrino counting from $\bar{p}p$ and e^+e^- machines. This volume.
9. Paschos, E. A. & U. Türke. 1986. Phys. Lett. **176B:** 185–189.
10. Marciano, W. J. & A. Sirlin. 1986. Phys. Rev. Lett. **56:** 22–25; Marciano, W. J. 1987. Constraints and speculations on fourth-family physics. This volume.
11. Kleinknecht, K. & B. Renk. 1987. Z. Phys. **C34:** 209–213. For most elements we used their results except for those elements where there are better experimental values.
12. Cline, D. B. 1987. Missing transverse energy events and a search for additional neutrino families and a fourth-generation lepton. This volume. Barger, V. *et al.* 1983. Phys. Lett. **133B:** 449.
13. Albrecht, H. *et al.* 1987. Observation of $B^\circ - \bar{B}^\circ$ mixing. Phys. Lett. **192B;** and the talk by Albrecht at this conference.
14. Pais, A. & S. Treiman. 1975. Phys. Rev. **D12:** 2744–2750; Okun, L. B., V. I. Zakharov & B. M. Pontecorvo. 1975. Lett. Nuovo Cimento **13:** 218–220.
15. Buras, A. J., W. Slominski & H. Steger. 1984. Nucl. Phys. **B245:** 369–398; Ali, A. & C. Jarlskog. 1984. Phys. Lett. **144B:** 266–272; Chau, L.-L. & W.-Y. Keung. 1984. Phys. Rev. **D29:** 592–595; Bigi, I. I. & A. I. Sanda. 1984. Phys. Rev. **D29:** 1393–1398; Brown, T. & S. Pakvasa. 1985. Phys. Rev. **D31:** 1661–1667.
16. Paschos, E. A., B. Stech & U. Türke. 1983. Phys. Lett. **128B:** 240–244; Paschos,

E. A. & U. TÜRKE. 1983. Nucl. Phys. **B234:** 29–43; GILMAN, F. J. & J. S. HAGELIN. 1983. Phys. Lett. **133B:** 443–448.
17. INAMI, T. & C. S. LIM. 1981. Prog. Theor. Phys. **65:** 297–314; also Erratum. 1981. Prog. Theor. Phys. **65:** 1772.
18. ALI, A. & Z. Z. AYDIN. 1979. Nucl. Phys. **B148:** 165–188.
19. ALBAJAR, C. *et al.* 1987. Phys. Lett. **186B:** 247; EGGERT, K. & H.-G. MOSER. 1987. Heavy flavor production and the evidence for $B - \bar{B}$ mixing in the UA1 case experiment. This volume; The data are still consistent with $r_s \lesssim 1/2$, a possibility that is ruled out in three generations, but can be realized with a fourth one. DATTA, A., E. A. PASCHOS & U. TÜRKE. 1987. Phys. Lett. **196B:** 376.
20. TÜRKE, U., E. A. PASCHOS, H. USLER & R. DECKER. 1985. Nucl. Phys. **B285:** 313–327; HE, X.-G. & S. PAKVASA. 1986. Nucl. Phys. **B278:** 905–933; HAYASHI, T., M. TANIMOTO & S. WAKAIZUMI. 1986. Prog. Theor. Phys. **75:** 353–365.
21. SHIFFMAN, M. A., A. J. VAINSTEIN & V. I. ZAKHAROV. 1977. JETP Lett. **45:** 670–682; GAVELA, B. *et al.* 1984. Phys. Lett. **148B:** 225–228.
22. BURAS, A. & J.-M. GÉRARD. 1986. Nucl. Phys. **B264:** 371–392, and references therein.
23. GILMAN, F. J. & J. S. HAGELIN. 1983. Phys. Lett. **133B:** 443–448; GILMAN, F. J. & M. WISE. 1979. Phys. Rev. **D20:** 2392–2407; The smaller value for $\langle \pi\pi | Q_6 | K \rangle$ is suggested by WOLFENSTEIN, L. 1985. Comments Nucl. Part. Phys. **14:** 135.
24. DONOGHUE, J. F. *et al.* 1986. Phys. Lett. **179B:** 361; Erratum. 1986. Phys. Lett. **188N:** 511; HOLSTEIN, B. 1987. Presented at the 6th Moriond Workshop on Searches of New and Exotic Phenomena. O. Fackler & J. Tran Thanh Van, Eds.: 79–88. Editions Frontières.
25. BURAS, A. J. & J.-M. GÉRARD. 1987. Isospin breaking contribution to ϵ'/ϵ. Phys. Lett. **192B:** 156–162. 7/87; GÉRARD, J.-M. 1987. Presented at the 6th Moriond Workshop on Searches of New and Exotic Phenomena. O. Fackler & J. Tran Thanh Van, Eds.: 89–96. Editions Frontières.
26. BOCK, G. J. 1987. Status of *CP*-violation experiments at Fermilab. This volume.
27. BLACK, J. K. *et al.* 1985. Phys. Rev. Lett. **54:** 1628–1630.
28. TÜRKE, U. 1985. Phys. Lett. **168B:** 296–298.
29. PARTICLE DATA GROUP. 1986. Phys. Lett. **170B:** 1–343.

Loop-Induced Rare B Decays and the Fourth Generation[a]

GEORGE WEI-SHU HOU[b] AND A. SONI[c,d]

[b]*Department of Physics and Astronomy*
University of Pittsburgh
Pittsburgh, Pennsylvania 15260

[c]*Department of Physics*
University of California
Los Angeles, California 90024

INTRODUCTION

With branching ratios (BRs) reaching down to the 10^{-11}–10^{-12} level within a couple of years, at first sight it may seem that the kaon system is more interesting than the B system in providing tests of loop effects. Observe, however, that (1) $|V_{cb}|^2 \ll |V_{us}|^2$, so relatively the B lifetime is more prolonged; (2) $m_B^2 \gg m_K^2$, making many more channels accessible for the B system, thus allowing for many new interesting possibilities; (3) not only is the B lifetime more prolonged, but because $|V_{ts}V_{tb}/V_{cb}|^2/|V_{td}V_{ts}/V_{us}|^2 \sim 10^5$, loop-induced rare B decays may be enhanced over the corresponding K decay modes by such a large factor, hence making B mesons a better probe of virtual effects. Given that the B-meson has only been available for several years, it may well provide new clues to the unanswered questions of the Standard Model (SM).

The questions we would like to address are: What's the top quark mass? Is there a fourth generation? Is there a real deficit of charm in B decays?[1] We study loop-induced $b \rightarrow s$ transitions, viz., the processes $b \rightarrow sl^+l^-$, $s\nu\bar{\nu}$,[2] $s\gamma$,[3] and s + "g".[4] None of these processes (except $b \rightarrow s\nu\bar{\nu}$) are entirely new, but we have many new observations and insights that reveal the importance of these processes. Furthermore, we explore the impact that the existence of a fourth generation may have on these modes. As a consequence of such a study, we propose a "new" parameterization of the 4×4 mixing matrix that seems to be most suitable for the study of the B system. We do not study the more difficult and uncertain exclusive processes. Also left out are the discussions of $B - \bar{B}$ mixing and the question of CP violation in inclusive processes.[5]

To calculate the matrix elements for these processes, we need the effective

[a]This work was supported by the National Science Foundation and the Department of Energy.
[d]To whom correspondence should be addressed.

vertices:

$$\frac{g^3/c_W}{16\pi^2}\,\nu_i\Gamma_i Z(\xi)\;\bar{s}\gamma_\mu Lb$$

$$\frac{g^3 s_W}{16\pi^2(2M_W^2)}\,\nu_i\,\bar{s}(F_1^i(\xi)(q^2\gamma_\mu - q_\mu \not{q})L + F_2^i\, i\sigma_{\mu\nu}q_\nu m_b R)\,b$$

$$g_s\frac{g^2}{16\pi^2(2M_W^2)}\,\upsilon_i\,\bar{s}\,\frac{\lambda^a}{2}\,(F_1^i(q^2\gamma_\mu - q_\mu\not{q})L + F_2^i\, i\sigma_{\mu\nu}q_\nu m_b R)\,b$$

and the effective box graph

$$\pm\frac{g^4}{16\pi^2(2M_W^2)}\,\upsilon_i\begin{Bmatrix}\bar{C}_i^{\text{box}}(\xi)\\ \bar{D}_i^{\text{box}}(\xi)\end{Bmatrix}(\bar{s}\gamma_u Lb)\begin{Bmatrix}\bar{l}\gamma_\mu Ll\\ \bar{\nu}\gamma_\mu L\nu\end{Bmatrix},$$

where the form factors F_1^i, F_2^i, Γ_i^Z, $\bar{C}_i^{\text{box}}$, $\bar{D}_i^{\text{box}}$, etc., can be taken or adapted from Inami and Lim,[6] and are complicated functions of internal quark masses m_i. In the preceding, $\nu_i \equiv V_{is}^* V_{ib}$, i is summed from 2 to n (number of generations), ξ is the gauge parameter, and note that we have been a bit cavalier in using the same F_1, F_2 for both the $bs\gamma$ and bsg form factors; they are different. Note that F_1 for bsg is gauge independent, while that for $bs\gamma$ is not.

We assume that the inclusive rate for each process is equal to the free quark result, for example, $\Gamma(B \to Kl^+l^- + X) \equiv \Gamma(b \to sl^+l^-)$. Also, to reduce uncertainties in estimating BRs, we always use the usual ansatz of taking ratios, for example,

$$\text{BR}(B \to Kl^+l^- + X) = \frac{\Gamma(b \to sl^+l^-)}{\Gamma(b \to (u,c)l\bar{\nu})} * \text{BR}^{\text{expt.}}(B \to l\bar{\nu} + X). \tag{1}$$

PROCESSES $b \to sl^+l^-$, $s\nu\bar{\nu}$

Previous works[7] on $b \to sl^+l^-$ have not included the important Z^0 contribution. We point out that although the photonic F_1 term contains the leading log that is important for low m_t, the Z^0 contribution cannot be neglected when there are very heavy internal quarks, that is, for $m_Q \gtrsim M_W$.

Using the $bs\gamma$, bsZ vertices and the box diagram contribution, we find the rate

$$\begin{aligned}\Gamma^{b\to sll} &= \chi^2(|\nu_i A_i|^2 + |\nu_i B_i|^2 + 4s_W^2\,\nu_i(A_i + B_i)(\nu_j F_2^j)\\ &\quad + 16 s_W^4\left(\log\frac{m_b}{2m_l} - \frac{2}{3}\right)|\nu_i F_2^i|^2)\Gamma_0,\end{aligned} \tag{2}$$

where $\chi \equiv g^2/16\pi^2$, $A_i \equiv \bar{C}_i - s_W^2(F_1^i + 2\bar{C}_i^Z)$, $B_i \equiv -\,s_W^2(F_1^i + 2\bar{C}_i^Z)$, $\bar{C}_i \equiv \bar{C}_i^Z + \bar{C}_i^{\text{box}}$ (these form-factor combinations are now gauge invariant), $\Gamma_0 = G_F{}^2 m_b^5/(192\pi^3)$ and for each neutrino species, we get

$$\Gamma^{b\to s\nu\nu} = \Gamma^{b\to sll}(s_W \to 0, \bar{C}_i \to \bar{D}_i). \tag{3}$$

For three generations, because V_{ub} is small, it turns out that the BRs are not sensitive to the mild uncertainties in the 3×3 Kobayashi–Maskawa (KM) matrix. The results are plotted in FIGURE 1, where for the $s\nu\bar{\nu}$ mode we have summed over three neutrino species. The qualitative nature can be understood as follows. For the sl^+l^- mode, when m_t is not large, $x_t \equiv (m_t/M_W)^2 < 1$, the process is dominated by charm through the photon contribution, since $F_1^i \sim 2Q \log x_i/3(x_i - 1)$, while the other contributions are suppressed by powers of x_i. But as m_t grows, the Z contribution grows, growing ever more rapidly for $m_t > M_W$, asymptotically approaching a m_t^4 behavior in rates as $\overline{C}_i^Z \sim x_i/4$ for $x_i \gg 1$. Thus, for large m_t, the Z contribution takes over. The $s\nu\bar{\nu}$ mode does not have photonic contributions, and has an even more prominent m_t dependence. Of course, the curves are drawn up to $m_t = 500$ GeV just for the purpose of illustration, since from various grounds we already have a bound on m_t roughly as $m_t \lesssim 200$ GeV.[8]

Why is there such a strong heavy internal quark mass dependence? This apparent violation of the intuitive decoupling theorem[9] comes about because in a spontaneously broken gauge theory with fermion masses generated through Yukawa couplings with the scalar field developing a vacuum expectation value, the fermion mass is a coupling

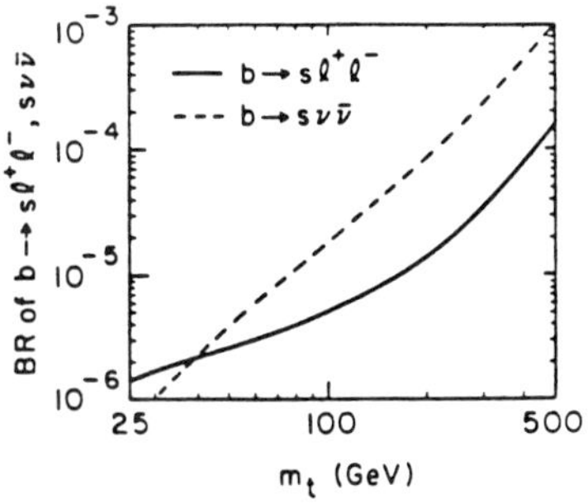

FIGURE 1. BRs for the process $b \to sl^+l^-$, $s\nu\bar{\nu}$ in the three-generation case. For the $b \to s\nu\bar{\nu}$, the 3-neutrino species have been summed over.

constant in disguise. Here it enters into the numerator of the amplitude through the longitudinal component of the W-boson in the virtual loop. Thus, as m_t grows, not only don't we have a power $1/m_t$ suppression (for $m_t > M_W$) from propagator effects, we actually have an amplitude growing as m_t^2. This is analogous to the box diagram contribution to the $K_L - K_S$ mass difference, except now we have a rate $\propto$ (amplitude)^2 $\propto m_t^4$. In comparison, although similar effects do occur in the process $K \to \pi\nu\bar{\nu}$, but in that case the KM angles are constrained such that, to give the biggest rate it actually favors a lower m_t.

Thus, these decay modes are extremely interesting as they test the weak point of the underlying theory (SM), viz., the symmetry breaking *and* the fermion mass generation mechanism. If new heavy quarks get their masses from a different mechanism, their effects would be different. In this respect, it is natural to study the impact of a fourth generation of quarks (if they exist at all) on these decay modes.

But with a fourth generation, not only the new charge 2/3 quark mass $m_{t'}$ is unknown, we also have three more mixing angles and two more phases. Fortunately, the situation is not hopeless, since the following observations greatly reduce the number of essential independent parameters relevant to the problem:

(1) $|\nu_u| \equiv |V_{us}^* V_{ub}| \ll |\nu_c| \equiv |V_{cs}^* V_{cb}| \lesssim 0.05$, which is itself rather small. Thus, from unitarity,

$$\nu_u + \nu_c + \nu_t + \nu_{t'} = 0 \tag{4}$$

and by ignoring ν_u, we can express ν_t as

$$\nu_t = -(\nu_{t'} + \nu_c) \tag{5}$$

and the only crucial mixing parameter is $\nu_{t'}$. We take $\nu_c \approx 0.05$ in subsequent discussions.

(2) Inspection of FIGURE 1 shows that the t contribution is relatively unimportant for say $m_t \lesssim M_W$. If a fourth generation exists, it is plausible that $m_{t'} > M_W > m_t$. Thus, $m_{t'}$ is the single *important* mass. We take $m_t \simeq 50$ GeV in most of the subsequent discussions.

We now plot BR vs. $\nu_{t'}$ for $|\nu_{t'}| \lesssim 0.2$ for $m_{t'} = 150$–500 GeV (FIGURE 2). The current CLEO limit[10] for the exclusive process $B \to Ke^+e^-$ is 2×10^{-4}. Even if one takes $\Gamma(B \to Ke^+e^- + X) \sim 2\,\Gamma(B \to Ke^+e^-)$, it is clear that we are not cutting into the unexplored parameter space of $\nu_{t'}$, $m_{t'}$. But, say this limit gets improved by more than an order of magnitude, we see that it then cuts deeply into the parameter space (if not yielding a positive signal already). Note that for three generations, the BR is less than 1.5×10^{-5}. Thus any observation beyond this level would indicate that a fourth generation may well exist.

As to the neutrino mode, it appears as if the experimental observation would be quite hopeless at present. But recall that kaons have been around for decades, and only now can we explore the $K \to \pi\nu\bar{\nu}$ mode meaningfully. The search for the $b \to s\nu\bar{\nu}$ mode would thus be a challenge to our experimental colleagues. Note that, because of 3 ~ 4 species of neutrinos and the fact that the amplitude for each neutrino species is somewhat larger than the sll mode, the total $b \to s\nu\nu$ BR is more than six times as large as $B \to Kll$. It is clear that for these kinds of modes, a BR that reaches easily above the 10^{-3} level cannot be called "rare" anymore. It would be a pitty if we can't study it.

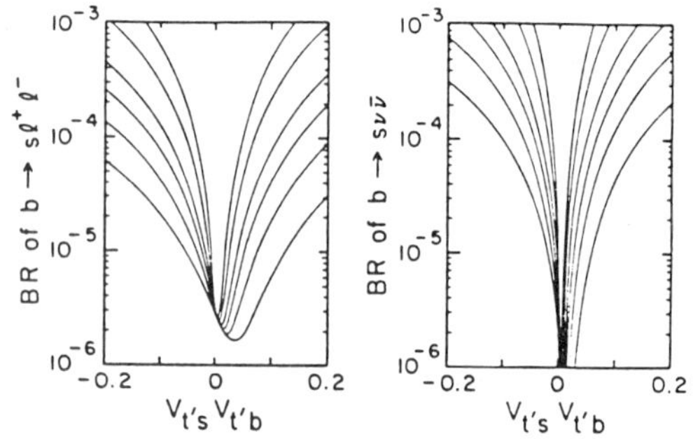

FIGURE 2. The four-generation BRs vs. $\nu_{t'} = V_{t's}^* V_{t'b}$. We have used $V_{cb} = 0.05$, $m_t = m_L = 50$ GeV. Different curves from bottom to top are: $m_{t'} = 150, 200, 250, 300, 400$, and 500 GeV.

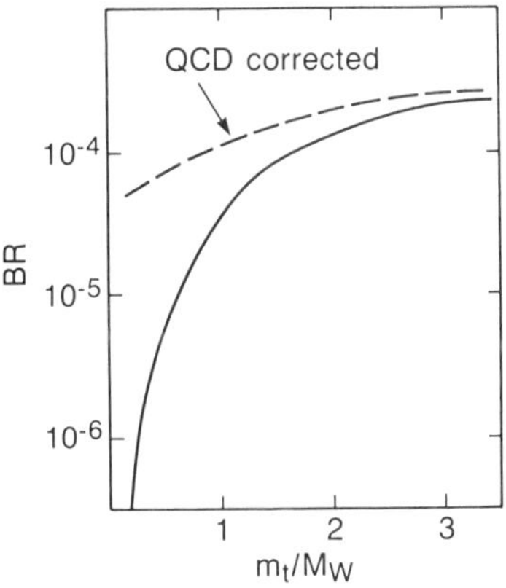

FIGURE 3. BR for $b \to s\gamma$, including QCD corrections within the three-generation SM.

PROCESS $b \to s\gamma$

Clearly, if there is a $b \to sl^+l^-$ mode, there must also be the on-shell photon emission $b \to s\gamma$. This is a magnetic transition and only the F_2 form factor enters. The rate is

$$\Gamma^{b\to s\gamma} = \frac{3\alpha}{2\pi}|v_i F_2^i|^2\,\Gamma_0. \tag{6}$$

The three-generation case has been discussed before in the literature.[11] More recently, it has been pointed out that because the SM weak effect (the F_2 form factor) is suppressed by powers of x_i for small x_i, quantum chromodynamics (QCD) corrections can bring in new log x_i terms at the α_s order that overwhelms the lowest order (in QCD) results.[12] The overall effect gives rise to a much milder m_t dependence for the process $b \to s\gamma$. The BR is roughly 10^{-4}. However, the result is now less precisely known. The current CLEO limit for the exclusive process $B \to K^*\gamma$ is 2×10^{-3}, but significant improvement is expected in several years, and could reach down to the 10^{-5} level. Thus, except for the usual problems of relating the exclusive rates to the inclusive, this mode should be seen soon, and thereby may provide a first test of loop effects (and the large QCD corrections) in the B system.

The large QCD corrections to this process have a significant impact on the four-generation result. From FIGURE 3 it is seen that QCD effects are more prominent for low m_t; therefore, for four generations, the top contribution gets more enhancement compared to the t' contribution. But from equation 5 it is seen that there is a tendency for the t and t' contributions to cancel. Thus, the fact that the t contribution gets more enhanced means that there is a much stronger cancellation between the t and t' contributions compared to the case without QCD corrections. This is illustrated in FIGURE 4. FIGURE 4a shows the result for $m_t = 50$ GeV without QCD corrections. Because of the funnel shape, this would have been a very powerful probe of the parameter $v_{t'} \equiv V_{t's}V_{t'b}$, in that if $b \to s\gamma$ is not observed say below the 10^{-4} level, we could put very stringent bounds on $v_{t'}$, regardless of $m_{t'}$. Unfortunately, there are very large QCD corrections for $m_Q \lesssim 100$ GeV; therefore, for $m_t \lesssim 100$ GeV, the result is similar to that depicted in FIGURE 4b. There is now a much stronger $m_{t'}$ dependence, and although some qualitative features remain, unless one has a precise knowledge of the QCD corrections (and also knowledge of $m_{t'}$), this mode becomes less potent in probing $\nu_{t'}$. Still, note the general feature that the BR may gain roughly by an order of magnitude over the three-generation result, but also could get very suppressed. Of course, if $\nu_{t'}$ vanishes, we recover the three-generation result.

PROCESS $b \rightarrow s$ + "g"

This process is very relevant to the question of charm deficit[1] in B-decays. It demands a much more detailed discussion than can be made here.[4] In the literature,

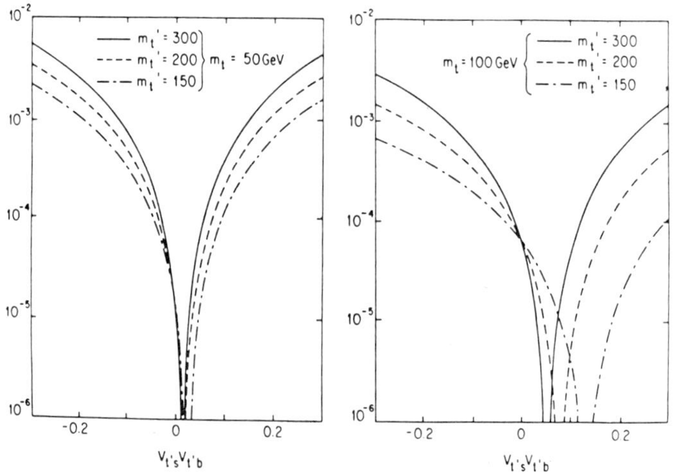

FIGURE 4. BR for $b \rightarrow s\gamma$ for four generations.

QCD-induced $b \rightarrow s$ transitions have been discussed before[13] on many occasions. For the specific problem of an overall inclusive BR, Eilam[e] has discussed the two processes.

However, the first process has been slightly overestimated. For the second process, which was taken from the leading-log, all orders (LLAO) QCD renormalization group analysis of Guberina,[f] only the four quark operators were taken into account, with questionable applicability for $m_t > M_W$. If one wants to explore the full range of heavy quark mass dependences, it is not clear whether the LLAO analysis would be more

[e] Last paper of reference 12.
[f] Third paper of reference 12.

trustworthy than the "exact" 1-loop weak-correction (but lowest order in QCD) calculation. Of greater importance, perhaps, is the fact that the *2 quark–2 gluon operator* has never been taken into account before. We therefore study the parton level processes where for the $Q^2 < 0$ process we use the Guberina result for purposes of comparison; it turns out to make a negligible contribution. Note that the three momentum domains ($q^2 = 0, > 0, < 0$) are distinct at the parton level. Moreover, for $q^2 = 0$ ("on-shell" single gluon emission), only the magnetic form factor F_2 contributes; thus, it is nontrivial to include the $q^2 > 0$ gluon split-up contribution, which does contain the leading F_1 form factor. We thus get the decay rate

$$\begin{aligned}
\Gamma^{b\to s''g''} = {} & \frac{N^2-1}{N}\frac{3\alpha_s}{4\pi}|v_i F_2^i|^2\,\Gamma_0 && (q^2 = 0)\\
& + (n_f + N)\frac{N^2-1}{N}\frac{\alpha_s^2}{32\pi^2}|v_i F_1^i|^2\,\Gamma_0 && (q^2 > 0)\\
& + \left(\frac{f_B}{m_B}\right)^2\frac{(N^2-1)^2}{N^3}6\pi^2|v_i c_5^i|^2\,\Gamma_0 && (q^2 < 0)
\end{aligned} \qquad (7)$$

where f_B is the meson decay constant, since the $q^2 < 0$ penguin is a two-body annihilation process. The α_s dependence for the spacelike term is absorbed into the definition of c_5, which can be found in footnote f. The form factors F_1^i and F_2^i can be derived from the corresponding photonic form factors by dropping the Q (internal quark electric charge) independent piece and then set $Q = 1$; both of them are now gauge invariant. We have maintained some generality in showing the result for the $SU(N)$ color group, and n_f noncharm light fermion flavors; in reality, $N = 3$ and $n_f = 3$ (u, d, s).

In arriving at equation 7, many approximations have been made. First of all, $O(\alpha_s^2)$ QCD corrections for the $q^2 = 0$ case has been ignored at present. When included, it should exhibit similar behavior as in the photonic case,[12] except that the calculation would become even more complicated. However, as shown below, this term is not the dominant piece, therefore we are not making a serious error for not including these corrections. For the $q^2 > 0$ contribution, we have dropped the F_2 term in the amplitude.[g] Finally, we have tacitly assumed that the full hardronic rate is the same as the parton-level decay rate, thereby assuming the partons fragment into hardrons with probability unity. But at the same time, by taking only an incoherent sum of the various processes, interference effects have been ignored. Note that no fragmentation model at present can take interference effects into account. For a more detailed discussion of these subtle questions, see reference 4. We point out that the contribution of $b \to sg^*$, followed by the $g^* \to gg$ split-up, has never been discussed in the literature before. For $SU(3)$, this contribution is as strong as three species of light final state quarks ($b \to sq\bar{q}$).

With these provisions we plot the three-generation result in FIGURE 5. It is clear

[g] A more detailed analysis is given in reference 4, which discusses the importance of interference effects.

that the timelike gluon split-up ($q^2 > 0$) process dominates, even if the $q^2 = 0$ contribution gets slightly enhanced over the one plotted. For the $q^2 < 0$ process, although it has two-body phase space, the $(f_B/m_B)^2$ factor leads to a very strong suppression and can be ignored. Note also that, comparing this LLAO curve to that of the "exact" (but without QCD correction) $q^2 > 0$ curve, the LLAO seems to have too hard an m_t dependence, thus casting some doubt on its validity for $m_t > M_W$.

Why does the $O(\alpha_s^2)$ process ($q^2 > 0$) dominate over the $O(\alpha_s)$ process ($q^2 = 0$)? As indicated earlier, this is because as q^2 vanishes, the F_1 contribution does not enter anymore. But in the $q^2 > 0$ process when q^2 is large, the F_1 term (which contains the leading log) contributes strongly, and overcompensates not only for the extra power of α_s, but also for two-body vs. three-body phase space. This is a nontrivial result.

Thus, for the allowed m_t range, BR ranges from ~ 1% to 2%. This is not a "rare" decay anymore, but we need a way to measure inclusive hardronic rates (B to K

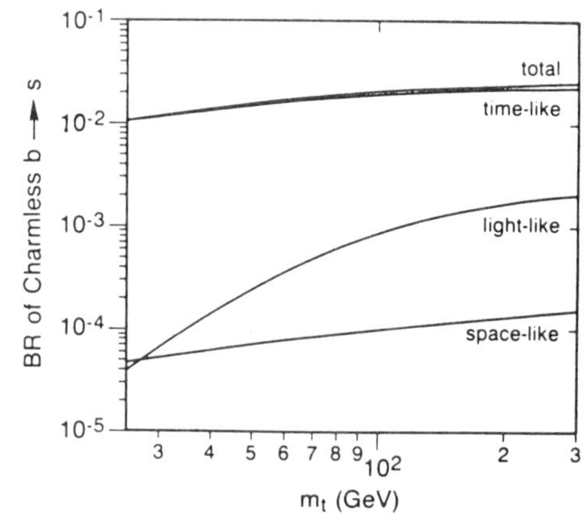

FIGURE 5. BR for $b \rightarrow sg^*$ for the three-generation case.

without charm) directly. For the experimentally more accessible exclusive processes, one not only has smaller rates and the usual difficulties of evaluating hadronic matrix elements that involve two hadronic currents, but it is also difficult to incorporate the important $g^* \rightarrow gg$ split-up process into the calculation (which has not been taken into account in previous discussions).

Note that for large m_t ($\gtrsim$ 100 GeV) the dependence on heavy internal quark masses are very mild. This is because, unlike the photonic case, there is no WWg coupling, so F_1^i approaches a constant for large m_Q. This has implications on the four-generation result, which is plotted in FIGURE 6 for $m_t = 50$ and 80 GeV, respectively.

For 50 GeV top, if KM angles are favorable, one may get a sizable enhancement over the three-generation result, and may well reach the 10% level. For heavier m_t, becuase of the flatness of the m_Q dependence of F_1^Q, there tends to be a stronger cancellation, and the BR range gets reduced. But still, a 5–10% BR for $b \rightarrow s$ without charm would constitute very strong evidence for the existence of a fourth generation. The question is can one realistically search for this decay inclusively? Progress in experimental techniques is needed; we need to veto charm.

It should be pointed out that, because of the extreme flatness of F_1^Q for large m_Q, as exemplified in FIGURE 5, the fourth-generation result will be very sensitive to the terms that were ignored in the preceding discussion. Although this is not essential for the three-generation case, it may considerably enhance the four-generation result (see footnote g).

4 × 4 MIXING MATRIX

It is clear that the B system will turn out a lot of new information. In particular, we have shown that these various modes are sensitive to the presence of a fourth generation; they may, in fact, provide the first indication of new physics.

Within the context of a four-generation SM, what is the most pragmatic parametrization of the 4 × 4 KM matrix? Clearly, this is not a physics question, since physics does not depend on how one parametrizes the KM matrix. Rather, it is a matter of convenience, and we must adapt to the most readily available data in the next few years. Obviously, since V_{ub} and V_{cb} would still be of interest for a while to come, it would be useful to keep it as simple as possible. The element V_{tb} may be studied if t is produced, or through a detailed analysis of $W \rightarrow t\bar{b}$. There is a possibility that the fourth-generation down-quark b' may be lighter than the top. But barring this possibility, the elements $V_{ub'}$, $V_{cb'}$, $V_{tb'}$ may not be accessible for experimental study for a long time. But the B system is readily available, and we have demonstrated that the loop-induced $b \rightarrow s$ transitions may give us information about the new mixing angles (and $m_{t'}$). Thus, there is some reason to keep the elements V_{ub}, V_{cb}, V_{tb}, $V_{t'd}$, $V_{t's}$, $V_{t'b}$, that is, the third column and fourth row of the 4 × 4 mixing matrix, as simple as possible. This can be easily achieved for a unitary matrix, by rearranging the sequence of multiplication of the various rotation matrices that we take from Botella and Chau.[14] Thus

$$V_4 = \left[\begin{array}{ccc|c} & & & 0 \\ & V_3 & & 0 \\ & & & 0 \\ \hline 0 & 0 & 0 & 1 \end{array}\right] \begin{bmatrix} 1 & 0 & 0 & 0 \\ 0 & 1 & 0 & 0 \\ 0 & 0 & c_u & s_u \\ 0 & 0 & -s_u & c_u \end{bmatrix} \begin{bmatrix} 1 & 0 & 0 & 0 \\ 0 & c_v & 0 & s_v e^{-i\phi_2} \\ 0 & 0 & 1 & 0 \\ 0 & -s_v e^{i\phi_2} & 0 & c_v \end{bmatrix}$$

$$\cdot \begin{bmatrix} c_w & 0 & 0 & s_w e^{-i\phi_3} \\ 0 & 1 & 0 & 0 \\ 0 & 0 & 1 & 0 \\ -s_w e^{i\phi_3} & 0 & 0 & c_w \end{bmatrix}$$

$$
= \left[\begin{array}{c|c|c|c}
\begin{array}{c} c_x c_z c_w \\ -s_x c_z s_v s_w e^{i(\phi_3-\phi_2)} \\ -s_z s_u c_v s_w e^{i(\phi_3-\phi_1)} \end{array} &
\begin{array}{c} s_x c_z c_v \\ -s_z s_u s_v e^{i(\phi_2-\phi_1)} \end{array} &
s_z c_u e^{-i\phi_1} &
\begin{array}{c} c_x c_z s_w e^{-i\phi_3} \\ +s_x c_z s_v c_w e^{-i\phi_2} \\ +s_z s_u c_v c_w e^{-i\phi_1} \end{array} \\
\hline
\begin{array}{c} -s_x c_y c_w \\ -c_x s_y s_z c_w e^{i\phi_1} \\ -c_x c_y s_v s_w e^{i(\phi_3-\phi_2)} \\ -s_y c_z s_u c_v s_w e^{i\phi_3} \\ +s_x s_y s_z s_v \\ * s_w e^{i(\phi_3-\phi_2+\phi_1)} \end{array} &
\begin{array}{c} c_x c_y c_v \\ -s_x s_y s_z c_v e^{i\phi_1} \\ -s_y c_z s_u s_v e^{i\phi_2} \end{array} &
s_y c_z c_u &
\begin{array}{c} c_x c_y s_v c_w e^{-i\phi_2} \\ -s_x c_y s_w e^{-i\phi_3} \\ +s_y c_z s_u c_v c_w \\ -c_x s_y s_z s_w e^{-i(\phi_3-\phi_1)} \\ -s_x s_y s_z s_v \\ * c_w e^{-i(\phi_2-\phi_1)} \end{array} \\
\hline
\begin{array}{c} -c_x c_y s_z c_w e^{i\phi_1} \\ +s_x s_y c_w \\ +c_x s_y s_v s_w e^{i(\phi_3-\phi_2)} \\ -c_y c_z s_u c_v s_w e^{i\phi_3} \\ +s_x c_y s_z s_v \\ * s_w e^{i(\phi_3-\phi_2+\phi_1)} \end{array} &
\begin{array}{c} -c_x s_y c_v \\ -s_x c_y s_z c_v e^{i\phi_1} \\ -c_y c_z s_u s_v e^{i\phi_2} \end{array} &
c_y c_z c_u &
\begin{array}{c} c_y c_z s_u c_v c_w \\ -c_x s_y s_v c_w e^{-i\phi_2} \\ -c_x c_y s_z s_w e^{-i(\phi_3-\phi_1)} \\ +s_x s_y s_w e^{-i\phi_3} \\ +s_x c_y s_z s_y \\ * c_w e^{-i(\phi_2-\phi_1)} \end{array} \\
\hline
-c_u c_v s_w e^{i\phi_3} & -c_u s_v e^{i\phi_2} & -s_u & c_u c_v c_w
\end{array}\right]
$$

where

$$
V_3 = \begin{bmatrix} c_x c_z & s_x c_z & s_z e^{-i\phi_1} \\ -s_x c_y - c_x s_y s_z e^{i\phi_1} & c_x c_y - s_x s_y s_z e^{i\phi_1} & s_y c_z \\ s_x s_y - c_x c_y s_z e^{i\phi_1} & -c_x s_y - s_x c_y s_z e^{i\phi_1} & c_y c_z \end{bmatrix}
$$

is the limiting three-generation Maiani–Chau–Keung matrix.[15]

Of course, no parameterization serves every purpose, and here we pay the price in the first three elements of the first and fourth columns. In contrast, the parameterization of Botella and Chau[14] keeps the first row and fourth column simple. Thus, although useful for studying b' decays (in the future), in their parameterization V_{cb} has two terms, V_{tb} and $V_{t'b}$ each are composed of three terms, while $V_{t's}$ and $V_{t'd}$ each have five separate terms. Therefore they are not particularly suitable for studying B physics (especially rare B physics), in the context of four generations. It is clear that, just like

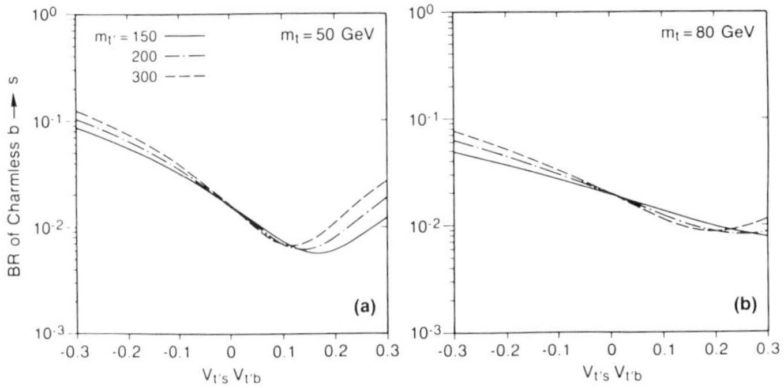

FIGURE 6. BR for $b \rightarrow sg^*$ for the four-generation case.

the case of B production, if b' is produced, b' decays should provide the best way for studying the elements $V_{ub'}$, $V_{cb'}$, and $V_{tb'}$. But before b' is produced, we would have to resort to utilizing loop-induced rare B decays to provide us with information regarding the *new* mixing parameters. This is much like the situation before the advent of B physics, that we had to rely on constraints coming from the kaon system. Let us emphasize once again, though, that for the B system loop effects have larger BRs.

Our parameterization would not be useful for studies of kaon physics, but this is perhaps inevitable if there is a fourth generation, and perhaps one will have to make a choice between having the convenience in the K or the B system. Clearly, if all the new off-diagonal mixing elements are very suppressed (i.e., the fourth generation is effectively decoupled from the low-energy world), all the parameterizations would become similar, reducing to the three-generation result. Only through using and testing could one determine whether our parameterization is useful or not (if a fourth generation indeed exists).

SUMMARY OF RESULTS

We collect our results in TABLE 1; QCD corrections for $b \rightarrow s\gamma$ have been taken into account. The experimental limits quoted are for the exclusive modes $B \rightarrow Kl^+l^-$ or $B \rightarrow K^*\gamma$ (see reference 10), and the reported deficit of charm (reference 1).

TABLE 1.

	Three Family m_t = 25–200 GeV	Four Family $\lvert v_{t'} \rvert < 0.3$, $m_{t'} < 300$ GeV	Current Limit
$b \rightarrow sl^+l^-$	1.5×10^{-6}–1.5×10^{-5}	2×10^{-6}–4×10^{-4}	$\lesssim 2 \times 10^{-4}$
$b \rightarrow s\nu\bar{\nu}$	8×10^{-6}–1×10^{-4}	$<3 \times 10^{-3}$	None
$b \rightarrow s\gamma$†	$\sim 7 \times 10^{-5}$–2×10^{-4}	$<3 \times 10^{-3}$	$\lesssim 2 \times 10^{-3}$
$b \rightarrow s$ "g"	1–2%	$\lesssim$10%	$\lesssim$50% (?)

CONCLUSIONS

We can summarize our conclusions as follows.

- $b \rightarrow sl^+l^-$ can be used as a new way to set bounds on m_t.
- The m_t^4 dependence of the $b \rightarrow sl^+l^-$ rate probes the symmetry breaking and fermion mass generation mechanism. That is, it can be turned around and used as evidence for or against Yukawa couplings.
- $b \rightarrow s\nu\bar{\nu}$ could be very interesting with the turning on of new facilities and detectors.
- $b \rightarrow s\gamma$ may be detected soon at CLEO and Argus. This mode may (?) be the first to show indications of a possible fourth generation, or lead to limits on $|V_{t's}V_{t'b}|$.
- QCD-induced $b \rightarrow s$ transition without charm may not be rare at all. Once the current puzzle of the charm deficit is understood, one should try to measure this rate.
- With the potential activity in studying B physics in mind, we propose a pragmatic parametrization of the 4×4 mixing matrix.

If current experimental limits improve by an order of magnitude, we can probe heavy quark physics, set limits on m_t, and if a fourth generation exists, set limits on $|V_{t's}V_{t'b}|$, $|V_{t'd}V_{t'b}|$, and even m_t, by studying RARE B DECAYS. In fact, QCD-induced charmless B decays may not be rare at all.

SUMMARY

We have discussed loop-induced rare B decays, viz., $b \rightarrow sl^+l^-$, $s\nu\bar{\nu}$, $s\gamma$, and QCD-induced $b \rightarrow s$ transitions ("penguins"). Because of the existence of the Z (and box) diagram contributions, $b \rightarrow sl^+l^-$, $s\nu\bar{\nu}$ decay rates grow rapidly with heavy internal quark masses m_Q. The on-shell process $b \rightarrow s\gamma$ has large QCD corrections, but depends mildly on m_Q. The QCD-induced transition is dominated by a timelike gluon ($q^2 > 0$), which splits into two quark or gluon partons, and has a very mild m_Q dependence. For three generations, with 25 GeV $\lesssim m_t \lesssim$ 200 GeV, the branching ratios of the inclusive processes range roughly as $(1.5 \times 10^{-6} - 1.5 \times 10^{-5})$, $(8 \times 10^{-6} - 1 \times 10^{-4})$, $\sim(7 \times 10^{-5} - 2 \times 10^{-4})$, and $\sim$(1–2%), respectively. If a fourth generation does exist, these rates could grow roughly by one order of magnitude, while the sl^+l^-, $s\nu\bar{\nu}$ modes may gain even more. We have therefore pointed out a pragmatic parametrization of the 4×4 mixing matrix suitable for B physics, and we strongly urge experimental search for these modes.

ACKNOWLEDGMENTS

We wish to thank H. Steger and R. S. Willey for collaboration and discussion.

REFERENCES

1. STONE, S. 1987. Plenary talk; THORNDIKE, E. H. 1987. Presented at the Division of Particles and Fields Meeting. Salt Lake City, Utah.

2. HOU, W. S., R. S. WILLEY & A. SONI. 1986. PITT-86-08 and UCLA/86/TEP/43.
3. HOU, W. S., A. SONI & H. STEGER. 1987. PITT-87-01 and UCLA/87/TEP/1.
4. HOU, W. S., A. SONI & H. STEGER. 1987. PITT-87-02 and UCLA/87/TEP/3.
5. HOU, W. S. & A. SONI. In preparation.
6. INAMI, T. & C. S. LIM. 1981. Prog. Theor. Phys. **65:** 297. (Errata: 1981. Prog. Theor. Phys. **65:** 1772.)
7. DESHPANDE, N. G., G. EILAM, A. SONI & G. L. KANE. 1986. Phys. Rev. Lett. **57:** 1106; O'DONNELL, P. J. 1986. Phys. Lett. **175B:** 369.
8. Aside from ρ parameter bounds, there is the bound coming from a recent detailed analysis of M_W, M_Z, and $\sin^2 \theta_W$; see MARCIANO, W. J. 1987. Constraints and speculations on the fourth-generation physics. This volume.
9. APPELQUIST, T. & H. D. POLITZER. 1975. Phys. Rev. Lett. **34:** 43.
10. AVERY, P. *et al.* 1986. CLNS 86/737.
11. CAMPBELL, B. A. & P. J. O'DONNELL. 1982. Phys. Rev. **D25:** 1989.
12. BERTOLINI, S., F. BORZUMATI & A. MASIERO. 1986. CMU-HEP 86-19; DESHPANDE, N. G., P. LO, J. TRAMPETIC, G. EILAM & P. SINGER. OITS 346.
13. BANDER, M., D. SILVERMAN & A. SONI. 1979. Phys. Rev. Lett. **43:** 242; GUBERINA, B., R. D. PECCEI & R. RÜCKL. 1980. Phys. Lett. **90B:** 169; GUBERINA, B. 1984. Fizika (Zagreb) **16:** 49; EILAM, G. 1982. Phys. Rev. Lett. **49:** 1478.
14. BOTELLA, F. J. & L.-L. CHAU. 1986. Phys. Lett. **168B:** 97.
15. MAIANI, L. 1977. *In* Proceedings of the 1977 Int. Symp. on Lepton and Photon Interactions at High Energies. DESY, Hamburg, West Germany: 867; CHAU, L.-L. & W.-Y. KEUNG. 1984. Phys. Rev. Lett. **53:** 1802.

Radiative B-Decays and Top-Quark Mass

JOSIP TRAMPETIC

Institute of Theoretical Science
University of Oregon
Eugene, Oregon 97403

INTRODUCTION

We investigated the gauge structure of the electroweak theory in one-loop $b \rightarrow s\gamma$ processes. We have found the strong quantum chromodynamics (QCD) enhancement (for $m_t \leq m_w$) that increases the $B \rightarrow K^*\gamma$ and $B \rightarrow K^{**}\gamma$ branching ratios into the experimentally very interesting range of order 10^{-4}.

DISCUSSION

An important question facing both theorists and experimentalists is the location of the top-quark mass. This importance attracted our attention to the neutral-flavor-changing one-loop processes: $B \rightarrow K^*_{892}\gamma$ and $B \rightarrow K^{**}_{1400}\gamma$. Thus, we see that the B-meson has several advantages compared to kaons.

(1) The B-meson is made of a heavy b-quark, thus permitting use of the spectator approximation.
(2) Its lifetime is, relatively speaking, longer than the kaons due to the smallness of the Kobayashy–Maskawa (KM) matrix elements v_{cb} and v_{ub}, allowing $B\bar{B}$ mixing to be studied.
(3) We expect that in the heavy quark systems the long-distance effect ambiguities are going to be absent. This is certainly correct for $B \rightarrow Ke^+e^-$, where, however, the rate is close to the limit of present experiments.
(4) The particular processes, such as $B \rightarrow K^*\gamma$[1] and $B \rightarrow K^{**}\gamma$,[2] have the advantage that the monoenergetic photosignal is clean and the branching ratio of order 10^{-4} is accessible to currently planned experiments.
(5) We believe that by measuring $B \rightarrow K^*\gamma$ and $B \rightarrow K^{**}\gamma$ branching ratios the top-quark mass can finally be determined.

Our work[1,2] does have some other important impacts:

- tests of electoweak theory in one-loop are of interest in their own right, because they verify the gauge structure of the theory;
- if the fourth generation of quarks (b', t') does exist, it is clear that the heavy B system will also provide a good probe of new parameters involving the fourth generation;[3]
- if an enhancement of the supersymmetry (SUSY)[4] over the Standard Model contribution exists, it is clear that the B-meson radiative decays can be an interesting candidate directly affected by the SUSY contributions.

The dominant mechanism for the decays $B \longrightarrow K^*\gamma$ and $K^{**}\gamma$ is through the quark decay $b \longrightarrow s\gamma$. Namely, $b \longrightarrow s\gamma$ is the only contribution for B^0-decays, while for B^+ radiative decays we have the $u\bar{b} \longrightarrow \bar{s}u$ contribution, which is suppressed by KM angles[a] and by the wave function at the origin.

The flavor-changing vertex $b \longrightarrow s\gamma$, for real photons, proceeds in one loop through the exchange of u, c, t quarks, and the W boson is given by

$$J_\mu = G_2\{\bar{s}\sigma_{\mu\nu}q^\nu b_R m_b + \bar{s}\sigma_{\mu\nu}q^\nu b_L m_s\}\,. \tag{1}$$

The dominant contribution to G_2 comes from c, t quarks, and is

$$G_2 = \frac{G_F}{\sqrt{2}}\left(\frac{e}{4\pi^2}\right)(V_{bc}V_{cs}^+)\left(\frac{7}{12}\frac{m_c^2}{M_W^2} - F_2(m_t)\right), \tag{2}$$

where $F_2(m_t)$ is given in reference 5, table 2.

Vasanti[6] noted and Shifman *et al.*[7] showed explicitely (for $s \longrightarrow d\gamma$) that there is a large QCD correction for the G_2 term. Namely that including the gluon exchanges removes the power suppression, that is,

$$\frac{m_t^2 - m_c^2}{m_w^2} \longrightarrow \ln\frac{m_t^2}{m_c^2}, \tag{3}$$

and that the dominant part of the QCD corrected part of G_2 is given by

$$\tilde{G}_2 = \frac{G_F}{\sqrt{2}}\left(\frac{e}{3\pi^3}\right)\alpha_s\,(V_{bc}V_{cs}^+)\ln\frac{m_t^2}{m_c^2}\,. \tag{4}$$

We present our results ($B \longrightarrow K^*\gamma$ and $B \longrightarrow K^{**}\gamma$) with and without QCD correction for comparison. These results are expected to be valid for $m_t \leq m_w$, while for $m_t \gg m_w$ analysis is in progress.

It is straightforward to calculate the inclusive process $B \longrightarrow \gamma + x$, where x contains no charm, by equating it to $b \longrightarrow s\gamma$. The matrix element and the rate for this decay are

$$M(b \longrightarrow s\gamma) = \tilde{G}_2 m_b \bar{s}\sigma_{\mu\nu}q^\nu b_R \epsilon^\mu(q) \tag{5}$$

$$\Gamma(B \longrightarrow \gamma + x_{\text{no charm}}) = \Gamma(b \longrightarrow s\gamma) = \frac{m_b^5|\tilde{G}_2|^2}{16\pi}\left(1 - \frac{m_s^2}{m_b^2}\right)^3. \tag{6}$$

The inclusive process is, however, difficult if not impossible to detect. Still, we believe that exclusive decays $B \rightarrow K^*\gamma$ and $B \rightarrow K^{**}\gamma$ are expected[1,2] to be a measurable fraction of the inclusive process, as $B \rightarrow K\gamma$ is forbidden. We have computed the hadronic matrix elements using the constituent quark model (CQM) as a phenomenological model of QCD in the nonperturbative regime. That model was described and successfully applied to determine exactly the same type of mesonic matrix elements in recent papers by Grinstein *et al.*[8]

The general Lorentz structure of the matrix element of the operator $\bar{s}\sigma_{\mu\nu}q^\nu b_R$

[a]The suppression in angles is $(v_{ub}/v_{cb})^2 v_{us}^2$.

between the K^R and B states ($K^R = K^*, K^{**}$), and using current conservation, is

$$\sqrt{2E_R 2E_B}\,\langle K^R(k)|\,i\bar{s}\sigma_{\mu\nu}q^\nu b_R|B(p)\rangle = i\epsilon_{\mu\nu\lambda\sigma}\epsilon^\nu(k)p^\lambda q^\sigma f_1^R(q^2)$$
$$+ [(m_B^2 - m_{K^R}^2)\epsilon_\mu(k) - (p+k)_\mu(\epsilon(k)\cdot q)]\,f_2^R(q^2), \quad (7)$$

where $q = p - k$ (with $q^2 = 0$), $f_i^{K^*} = f_i$, and $f_i^{K^{**}} = \tilde{f}_i$.

We evaluate the left-hand side of equation 7 in the remaining system of the B-quark. Note that the zero-recoil of the K^R meson (i.e., $\mathbf{k} = 0$) corresponds to $q^2 = t_m = (m_B - m_{K^R})^2$, and is far from $q^2 = 0$ (actual physical point) where we need to evaluate the form factors. The constituent quark model permits one to evaluate the form factors with recoil. The results are[1,2]:

$$f_1(q^2) = \frac{1}{\sqrt{2}}\left(1 + \frac{m_{K^*}^2}{m_B^2}\right)^{1/2} F(q^2), \qquad f_2(q^2) = \frac{1}{2}f_1(q^2), \quad (8)$$

$$\tilde{f}_1(q^2) = \frac{1}{\sqrt{2}}\left(1 + \frac{m_{K^{**}}^2}{m_B^2}\right)^{1/2} \tilde{F}(q^2),$$

$$\frac{\tilde{f}_2(q^2)}{\tilde{f}_1(q^2)} = \frac{m_B}{m_B^2 - m_{K^{**}}^2}\sqrt{\frac{\tilde{m}_s}{\tilde{m}_b}(\tilde{t}_m - q^2)}, \quad (9)$$

$$F(q^2) = \int_{-\infty}^{+\infty} d^3p\,\phi_{K^*}\left(\mathbf{p} - \frac{m_d}{\tilde{m}_s}\mathbf{k}\right)$$
$$\cdot\,\phi_B(\mathbf{p})\alpha(\mathbf{k} - \mathbf{p}, \mathbf{p})\left(1 - \frac{\mathbf{p}\cdot(\mathbf{k} - \mathbf{p})}{6E_s m_b}\right), \quad (10)$$

$$\tilde{F}(q^2) = \int_{-\infty}^{+\infty} d^3p\,\phi_{K^{**}}\left(\mathbf{p} - \frac{m_d}{\tilde{m}_s}\mathbf{k}\right)\phi_B(\mathbf{p})\alpha(\mathbf{k} - \mathbf{p}, \mathbf{p})\frac{|\mathbf{k}|}{E_s} \quad (11)$$

$$\alpha(\mathbf{k} - \mathbf{p}, \mathbf{p}) = \sqrt{\frac{(E_s + m_s)(E_b + m_b)}{4E_s E_b}}; \qquad \tilde{m}_{s(b)} = m_{s(b)} + m_d, \quad (12)$$

$$E_s = \sqrt{m_s^2 + (\mathbf{k} - \mathbf{p})^2}, \qquad E_b = \sqrt{m_b^2 + \mathbf{p}^2}. \quad (13)$$

The ϕ_B, ϕ_{K^*}, and $\phi_{K^{**}}$ are the Gaussian wave functions for the B, K^*, and K^{**} mesons, respectively. From formulas 10 and 11 it is clear that we included recoil as well as the relativistic corrections.

Setting $q^2 = 0$ and using $\mathbf{k} \gg \mathbf{p}$, one can easily compute integrals 10 and 11. By comparing the left- and right-hand sides of equation 7 and evaluating integrals 10 and 11, we have found:

$$f_1(0) \simeq 0.23 \qquad f_2(0) \simeq 0.11 \quad (14)$$

$$\tilde{f}_1(0) \simeq 0.31 \qquad \tilde{f}_2(0) \simeq 0.10. \quad (15)$$

The small size of the form factor $f_1(0)$ is mainly due to the sharp exponential damping factor because of recoil. On the other hand, we have gotten surprisingly large $\tilde{f}_1(0)$ [compared to $f_1(0)$], which has been mainly due to its direct dependence on the large recoil momentum. Without recoil, that is, $\mathbf{k} = 0$, we have

$$f_1(\mathbf{k}=0) \simeq \frac{1}{\sqrt{2}}, \qquad \tilde{f}_1(\mathbf{k}=0) = \tilde{f}_2(\mathbf{k}=0) = 0. \tag{16}$$

The partial widths for $B \to K^R\gamma$ ($K^R = K^*, K^{**}$) are found to be

$$\Gamma(B \to K^R\gamma) = \frac{\tilde{G}_2^2}{32\pi} m_b^2 m_B^3 \left(1 + \frac{m_{K^R}^2}{m_B^2}\right)^3 [f_1^{R^2} + 4f_2^{R^2}]. \tag{17}$$

We found for the ratios of the exclusive to inclusive processes:[1,2]

$$\text{Ratio}^* = \frac{\Gamma(B \to K^*\gamma)}{\Gamma(b \to s\gamma)} \simeq 0.06 \tag{18}$$

$$\text{Ratio}^{**} = \frac{\Gamma(B \to K^{**}\gamma)}{\Gamma(b \to s\gamma)} \simeq 0.07 \tag{19}$$

and

$$\text{Ratio}^*: \text{Ratio}^{**} \simeq 1:1. \tag{20}$$

TABLE 1.

m_t (GeV)	Without QCD BR($B \to K^*\gamma$)	With QCD BR($B \to K^*\gamma$)
20	0.73×10^{-7}	0.45×10^{-4}
40	0.67×10^{-6}	0.72×10^{-4}
60	0.19×10^{-5}	0.92×10^{-4}
80	0.38×10^{-5}	0.11×10^{-3}
100	0.59×10^{-5}	0.12×10^{-3}

From the preceding results it is clear that the recoil momentum plays a very important role in both the $B \to K^*\gamma$ and $B \to K^{**}\gamma$ decays. The $B \to K^*\gamma$ decay is suppressed by a sharp exponential damping factor due to the recoil.

Note here that the resonances $K_{1^+}^{**}(1280)$ and $K_{0^+}^*(1350)$ do not contribute because of the spin. The contribution of $K_{1^+}^{**}(1280)$ is zero due to the spin as well as to K_{0^-} and $K_{0^+}^*(1350)$. That is, $K_{1^+}^{**}(1280)$ is the particle with the zero quark–antiquark spin and with an angular momentum equal to one, which then couples into the total spin one.

To evaluate the branching ratios, we calculate the total width for Γ_B in pure spectator approximation using an effective QCD-corrected Hamiltonian:

$$\Gamma_B = \frac{G_F^2 m_b^5 |V_{cb}|^2}{192\pi^3} \{(2c_+^2 + c_-^2)[r_c + r_{cc}] + [2r_c + r_{c\tau}]\}, \tag{21}$$

where $r_c = 0.447$, $r_{cc} = 0.119$, $r_{c\tau} = 0.062$ are phase space factors for one c, two c's, and $c\tau$ in their final states.

Finally we have the branching ratio for $B \to K^*\gamma$ as a function of m_t (TABLE 1).

Here we have used $N_f = 4$, $\Lambda_{QCD} = 300$ MeV, $\alpha_s(m_b/2) = 0.36$, and the QCD corrections $C_- = 1.35$, $C_+ = 0.86$.

It is clear that for lower values of m_t the QCD corrections turn out to be extremely important. For example, for $m_t \cong 40$ GeV the ratio BR (with QCD)/BR (without QCD) ~ 100.

The BR ($B \to K^*\gamma$) with QCD is within the current experimental range of $\sim 10^{-4}$. Note that the possibility of measuring the BR ($B \to K^*\gamma$) strongly depends on the lower experimental range. If this range is $\sim (10^{-4}$ to $5 \cdot 10^{-5})$, the $B \to K^*\gamma$ can be detected. Actually, according to formula 20 we predict that on each $K^{**}\gamma$ event there should be at least one $K^*\gamma$ event.

Besides the experimental value of the top-quark mass we have recent theoretical evidence of the importance of radiative B decays for the physics beyond the Standard Model.

Hou and Soni reported[3] the effects of the fourth family on the $b \to s(d)\gamma$ decays.

Bertolini *et al.*[4] pointed out that the mesonic radiative decays belong to the very small class of phenomena where SUSY contributions can be larger then the corresponding Standard Model ones. This gives the possibility of setting bound to the gluon–quark masses from BR($B \to K^*\gamma$) and BR($B \to K^{**}\gamma$).

In conclusion we think that the importance of understanding on top-quark mass would suggest a high priority for experiments on $B \to K^R\gamma$ or $B \to X\gamma$. So the main purpose of this paper is to encourage experimentalists to measure the radiative B-decays in forthcoming machines.

REFERENCES

1. DESHPANDE, N. G., P. LO, J. TRAMPETIC, G. EILAM & P. SINGER. 1987. Phys. Rev. Lett. **59:** 183.
2. DESHPANDE, N. G., P. LO & J. TRAMPETIC. 1987. Univ. of Oregon preprint OITS-352/1987 (revised).
3. HOU, W.-S., A. SONI & H. STEGER. 1987. PITT-87-01 and UCLA/87/TEP/1 preprints; HEWETT, J. L. 1987. DOE-ER40200-091 preprint; HOU, G. W.-S. & A. SONI. 1987. Loop-Induced Rare B-Decays and the Fourth Generation. This volume.
4. BERTOLINI, S., F. BORZUMATI & A. MASIERO. 1986. CMU-HEP 86-20 preprints NYU/TR13/86.
5. INAMI, T. & C. S. LIM. 1981. Prog. Theor. Phys. **65:** 297; DESHPANDE, N. G. & G. EILAM. 1982. Phys. Rev. **D26:** 2463; DESHPANDE, N. G. & M. NAZERIMONFARED. 1983. Nucl. Phys. **B213:** 390.
6. VASANTI, N. 1976. Phys. Rev. **D13:** 1889.
7. SHIFMAN, M. A., A. I. VAINSHTEIN & V. I. ZAKHAROV. 1978. Phys. Rev. **D18:** 2583; KOGAN, Y. I. & M. A. SHIFMAN. 1983. Sov. J. Nucl. Phys. **38:** 628.
8. GRINSTEIN, B., M. B. WISE & N. ISGUR. 1986. Phys. Rev. Lett. **56:** 298; 1985. CALT-68-1311/UTPT-85-37.

Theoretical Origin of Quark Mass Matrices[a]

RABINDRA N. MOHAPATRA

Department of Physics and Astronomy
University of Maryland
College Park, Maryland 20742

INTRODUCTION

Here we discuss the theoretical origin of Fritzsch- and Stech-type quark mass matrices in the grand unified theories. We then present recent work done in collaboration with Kagan and Pal, which presents the first natural derivation of the Stech-type mass matrix in unified gauge theories and also provides a solution to the strong *CP*-problem.

DISCUSSION

At the present time, there appears to be no fundamental understanding of the mixings and masses of the different quark generations. It is believed that quark mixings are related to the origin of the fermion generations and therefore, an understanding of the latter question is essential before any clear understanding of the former is possible. However, before we arrive at that happy state of affairs in quark–lepton physics,[1] we will have to remain content with taking small steps that will bring us closer to that goal.

One approach that appears promising at the moment is to discern the patterns in quark mixings from low-energy observations of weak processes such as *K*- and *B*-decays and attempts to derive them from ansatzes about quark mass matrices. Two mass matrices that appear to hold the greatest potential are the following.

Fritzsch[2]-type Mass Matrix

For *N*-generations, the Fritzsch-type mass matrix has the following form both in the up- and down-quark sectors:

$$M = \begin{pmatrix} 0 & m_{12} & & & \\ m_{12} & 0 & m_{23} & & \\ & m_{23} & 0 & & \\ & & & 0 & m_{N-1,N} \\ & & & m_{N-1,N} & m_{N,N} \end{pmatrix} \qquad (1)$$

[a]This work was supported by a grant from the National Science Foundation.

with the hierarchy $m_{N,N} \gg m_{N-1,N}$. This leads to the mixing pattern between generations of the form

$$\theta_{a,a+1} \approx \sqrt{\frac{m_a^{(-)}}{m_{a+1}^{(-)}}} - \sqrt{\frac{m_a^{(+)}}{m_{a+1}^{(+)}}}, \tag{2}$$

where $m_a^{\pm}$ are the masses of the up- and down-quarks.

Stech-type Mass Matrix

Another interesting ansatz is that due to Stech,[3] according to which

$$\begin{aligned} M_u &= M_u^{\dagger} \\ m_d &= \alpha M_u + iA \end{aligned} \tag{3}$$

with $A = -A^T$. This mass matrix, apart from predicting correct mixing angles[4] for V_{us} and V_{bc} has several additional interesting properties: (a) it leads to maximal *CP*-violation in weak currents, a feature that may be necessary if the Kobayashi–Maskawa (KM) model is to lead to a satisfactory understanding of *CP*-violating $K_L^0 \to 2\pi$ decays; (b) the mass matrices (equation 3) being Hermitian lead to a solution of the strong *CP*-problem.[5]

In this paper, we explore the theoretical origin of the mass matrices in equations 1 and 3. The main reason for this exercise is that since the Standard Model fails to provide any interesting pattern for quark mass matrices, our study may give us a clue to the new physics beyond the Standard Model. Another reason is that unless the preceding mass matrices can be derived in a "natural" manner (i.e., for an arbitrary range of parameters in a renormalizable field theory) in a gauge theory framework, radiative corrections will destabilize the pattern, making any prediction of mixing angles derived from them unreliable.

THEORETICAL FRAMEWORKS OF THE FRITZSCH MASS MATRIX

In this section, we discuss the theoretical frameworks that lead to the Fritzsch mass matrix. The minimal gauge group that is essential for this purpose is the left–right symmetric gauge group based on $SU(2)_L \times SU(2)_R \times U(1)_{B-L}$ symmetry.[6] One can then impose an additional Z_{2N_g} horizontal chiral symmetry with the quark and Higgs fields transforming as follows:[7]

$$\begin{aligned} Q_a &\to e^{(i\pi a/N_g)\gamma_5}\, Q_a \\ \Phi_{a,b} &\to e^{(-i\pi/N_g)(a+b)}\, \Phi_{a,b} \end{aligned} \tag{4}$$

where $a, b = 1, \ldots N_g$ = number of generations. The Fritzsch mass matrix follows if we admit into the theory only Higgs multiplets of type $\Phi_{a,a+1}$ and Φ_{N_g,N_g}. If we want to take this model seriously, we must try to understand the origin of the horizontal symmetry Z_{2N_g}. The fact that it is a chiral symmetry perhaps provides a clue that it is a remnant of a Peccei–Quinn-type (PQ) symmetry broken by instanton effects. The PQ-symmetry must be broken at a high scale leading to invisible axions.[8] This would then also solve the strong *CP*-problem, which would otherwise have made a Fritzsch-type model unattractive. One can embed these ideas into an $SO(10) \times U(1)_{PQ}$-type model.[9] One problem that remains unsolved is that we still have to put in by hand the hierarchical pattern for $m_{a,a+1}$.

DERIVATION OF THE STECH-TYPE MASS MATRIX

Let us now turn to the derivation of the Stech-type mass matrix. Before proceeding to the discussion of the model, we note two key aspects of the model: (a) Since the antisymmetric matrix A in equation 3 leads to small mixings, $|A_{ab}| \ll |M_{u,aa}|$, where a denotes the generation; (b) secondly, the proportionality constant α must be small ($\alpha \simeq 1/10$).

One attempt to derive equation 3 was made[10] in the context of the $SU(2)_L \times SU(2)_R \times U(1)_{B-L}$ theories, where it was found that the tree level derivation of the Stech-type mass matrix receives infinite corrections only at the four-loop level. Recently, an exact derivation of this mass matrix in the context of a supersymmetric $SO(10)$ model with softly broken supersymmetry has been carried out by Kagan *et al.*[11] The antisymmetric part of the down-quark mass matrix arises in this model at the one-loop level, which explains why it is small compared to the other part. The up-quark mass matrix remains unaltered by the one-loop corrections. The main idea of this paper relies on the dominance of the right-handed neutrino in the one-loop contribution to quark masses[12] and the special features of its supersymmetric extension.[13] It is important to point out that we do not make use of any horizontal symmetry in deriving this result.

We now present the elements of the supersymmetric $SO(10)$ model pertinent to the mass generation of the quarks. The left chiral superfields of our model include the matter fields $\psi_a\{\mathbf{16}\}$, a being a generation index, and the Higgs fields $\varphi\{\mathbf{10}\}$, $B\ \{\mathbf{120}\}$, $\Delta\{\mathbf{126}\}$, and $T\{\mathbf{210}\}$ (as well as some other field to make the model consistent), where in the curly brackets we have indicated the transformation properties of these fields under $SO(10)$. The superpotential of the model is given by

$$W = W_1 + W_2 \tag{5}$$

where

$$W_1 = h_{ab}\psi_a\psi_b\varphi + A_{ab}\psi_a\psi_b B + S_{ab}\psi_a\psi_b\Delta + \gamma_1 B\Delta T + \gamma_2\varphi\Delta T. \tag{6}$$

We do not display W_2 since it involves terms that are necessary to make the model consistent but not essential for the derivation of the mass matrix.

In equation 6, h and S are symmetric matrices in the generation space, whereas A is antisymmetric. Imposition of *CP* invariance on the superpotential implies that h and S are real and A is imaginary. This means that h, S, and A are all Hermitian matrices. We can always take h to be diagonal by an appropriate choice of basis. This will not alter the reality and the symmetry properties of S and A. *CP* invariance also implies that all the couplings γ_i are real.

The gauge symmetry breaking in the model is realized by the following vacuum expectation values:

$$\langle\varphi_0\rangle \equiv \langle(2, 2, 1)^{\mathbf{10}}_{[-1/2,+1/2,0,1]}\rangle = \kappa$$
$$\langle\varphi_0'\rangle \equiv \langle(2, 2, 1)^{\varphi\{\mathbf{10}\}}_{[+1/2,-1/2,0,1]}\rangle = \kappa'$$
$$\langle(1, 3, 10)^{\Delta\{\mathbf{126}\}}_{[0,1,-2,1]}\rangle = v$$
$$\langle(1, 1, 1)^{T\{\mathbf{126}\}}_{[0,0,0,1]}\rangle = M_G. \quad (7)$$

Here we have introduced a notation to indicate specific components of a multiplet. We make the $SO(10)$ multiplet is a superscript, whose submultiplet under $SU(2)_L \times SU(2)_R \times SU(4)_C$ is given in the parenthesis; the T_{3L}, T_{3R}, $B - L$ quantum numbers, as well as the $SU(3)_C$ transformation property of the component in question, are made subscripts. The VEV of $T\{\mathbf{210}\}$ breaks $SO(10)$ down to $SU(2)_L \times SU(2)_R \times SU(4)_C$, which is then broken down to $SU(2)_L \times U(1)_Y \times SU(3)_C$ by the VEVs of $\Delta\{\mathbf{126}\}$. The Weinberg–Salam stage of symmetry breaking is performed by the two VEVs in $\varphi\{\mathbf{10}\}$. The choice of $\kappa \neq \kappa'$ is consistent with the minimum of the potential due to the fact that φ is a complex multiplet. As we see later, there are other reasons for the splitting of κ and κ'.

The tree-level mass terms for the U-type and D-type quarks are given by

$$M_U^0 = h\kappa, \qquad M_D^0 = h\kappa'. \quad (8)$$

Next we look at the one-loop contribution to these mass matrices. In particular, we focus on the diagram in FIGURE 1. The various Yukawa and quartic couplings present in this diagram arise from the superpotential of equation 6. In FIGURE 1, we have explicitly put down the quantum numbers of D_L and $\overline{D}_R$ on the external fermion legs, which shows how it can contribute to M_D after Δ and φ develop VEVs. As far as M_U is concerned, there is no similar contribution. To see this, we just recall the argument of reference 13 and look at the conservation of T_{3R}. The internal fermion line is a $\bar{\nu}_R$, which has $T_{3R} = -1/2$. If we pick the left external fermion leg to be U_L and the right one to be $\overline{U}_R$, the internal scalar line on the left has to have $T_{3R} = +1/2$, whereas the one on the right should have $T_{3R} = +1$. The two scalar lines coming out at the top of the diagram must then carry $T_{3R} = 3/2$. However, they appear as the F-term of $T\{\mathbf{210}\}$, which does not have any component with $T_{3R} = 3/2$. That is why we get a contribution just for M_D. Note that supersymmetry is crucial for this result.

Another source of mass for the D-type quarks is a diagram that can be obtained by replacing the internal scalar lines on the left and right sides of FIGURE 1 by $(2, 2, 15)^{\Delta\{\mathbf{126}\}}_{[1/2,1/2,-4/5,3]}$ and $(1, 1, 10)^{B\{\mathbf{120}\}}_{[0,0,-2/3,3]}$, respectively. In the presence of these one-loop diagrams, the mass matrices for the U-type and D-type quarks at the one-loop level can be written as

$$M_U = h\kappa$$
$$M_D = M_D^{(0)} + \delta M_D = [h + \epsilon\,(AS^2 + S^2A)]\kappa' \quad (9)$$

where

$$\epsilon = \frac{\lambda_1\lambda_2 v^2}{8\pi^2 M^2}, \quad (10)$$

and M is the heaviest mass in the loop. The matrix $AS^2 + S^2A$ is antisymmetric and imaginary because of the reality and symmetry properties of A and S described before. Thus equation 9 gives exactly the Stech form of the mass matrices. However, in arriving at equation 9, it seems that we have assumed that the heaviest masses in the loop diagrams are equal. As has been shown by explicit construction,[11] this is indeed the case in realistic models. That completes our proof that the Stech-type mass matrix arises naturally within a supersymmetric $SO(10)$ model, where one-loop contributions play an important role. For further details of the model, we refer the reader to reference 11.

It is worth pointing out here that in this model, the strong CP-parameter θ is predictable and arises at the one-loop level and is of order $\simeq(\alpha/4\pi)(m_W/M_H)^2$ and is therefore small ($\leq 10^{-10}$ or so for $M_H \geq 10^7$ GeV).

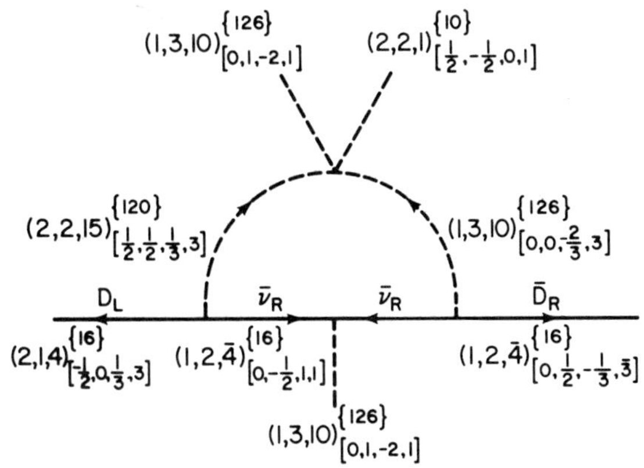

FIGURE 1. The typical one-loop graph that contributes exclusively to the down-quark mass matrix. There is another Feynman graph where the left dotted line carries the particle $(2, 2, 15)^{\Delta\{126\}}_{[1/2,1/2,-4/5,3]}$ and the right one carries $(1, 1, 10)^{B\{120\}}_{[0,0,-2/3,3]}$. The notation for designating the particles is explained in the text.

SUMMARY

In conclusion, we find that understanding the mixing patterns between quark generations usually implies an electroweak symmetry of the $SU(2)_L \times SU(2)_R \times U(1)_{B-L}$-type or its grand unifying group $SO(10)$. It is attractive that these models also provide a simultaneous solution of the strong CP-problem.

ACKNOWLEDGMENT

We wish to thank Alex Kagan and Palash Pal with whose collaboration the bulk of the work reported here was done.

REFERENCES

1. For a summary of the present status, see MOHAPATRA, R. N. 1986. Unification and Supersymmetry; the Frontiers of Quark-Lepton Physics. Springer-Verlag, New York.
2. FRITZSCH, H. 1977. Phys. Lett. **70B:** 436.
3. STECH, B. 1983. Phys. Lett. **130B:** 189; ECKER, G. 1984. Z. Phys. **C24:** 353.
4. For a thorough discussion, see GRONAU, M., R. JOHNSON & J. SCHECTER. 1986. Phys. Rev. **D33:** 264.
5. BÉG, M. A. B. & H. S. TSAO. 1978. Phys. Rev. Lett. **41:** 278; MOHPATRA, R. N. & G. SENJANOVIĆ. 1978. Phys. Lett. **79B:** 283; GEORGI, H. 1978. Had. J. **1:** 155.
6. PATI, J. C. & A. SALAM. 1974. Phys. Rev. **D10:** 275; MOHAPATRA, R. N. & J. C. PATI. 1975. Phys. Rev. **D11:** 566, 2558; SENJANOVIĆ, G. & R. N. MOHAPATRA. 1975. Phys. Rev. **D12:** 1502.
7. MOHAPATRA, R. N. 1985. Horizontal symmetries and flavor mixing in left-right symmetric models. *In* Proc Erice Workshop on Flavor Mixing in Weak Interactions. L. L. Chau, Ed. Plenum, New York.
8. For a review, see KIM, J. E. 1987. Phys. Rep. In press.
9. DAVIDSON, A., V. P. NAIR & K. C. WALI. 1984. Phys. Lett.
10. GRONAU, M. & R. N. MOHAPATRA. 1986. Phys. Lett. **168B:** 248; ECHER, G. & W. GRIMUS. 1985. Phys. Lett. **153B:** 279; For attempts in nonsupersymmetric $SO(10)$ models, see JOHNSON, R., S. RANFONE & J. SCHECTER. 1987. Phys. Rev. **D35:** 282.
11. KAGAN, A., R. N. MOHAPATRA & P. B. PAL. 1987. Phys. Rev. Lett. **59:** 2005.
12. CHANG, D., R. N. MOHAPATRA, P. B. PAL & J. C. PATI. 1985. Phys. Rev. Lett. **55:** 2756.
13. KAGAN, A. & P. B. PAL. Phys. Rev. **D36:** 269.

Status of AGS Experiment 780

WILLIAM M. MORSE

Brookhaven National Laboratory
Department of Physics
Upton, New York 11973

Experiment 780 is a search for the decays $K_L^0 \to \mu e$ and $K_L^0 \to e^+e^-$. The decay $K_L^0 \to \mu e$ violates lepton number and should not occur in the Standard Model. The decay $K_L^0 \to e^+e^-$ is helicity suppressed compared to $K_L^0 \to \mu^+\mu^-$, and is only expected to occur at about 3×10^{-12} in the Standard Model. We show the diagram for the decay $K^+ \to \mu^+\nu$ and a possible diagram for $K_L^0 \to \mu e$ in FIGURE 1. The decay $K_L^0 \to \mu e$ is mediated by a new heavy particle X with couplings g_{su} and g_{de}. The decay rate is proportional to the fourth power of the mass of the propagator:

$$B(K_L^0 \to \mu e) = 10^{-8} \times \frac{g_{su}^2 g_{de}^2}{g^4} \times \left(\frac{20\ \text{TeV}}{Mx}\right)^4.$$

The decay $K_L^0 \to \mu e$ is thus sensitive to lepton number violating interactions at very high mass scales. The decay $K_L^0 \to e^+e^-$ is sensitive to pseudoscalar currents if the couplings are not proportional to mass. We will normalize to the kinematically similar $K_L^0 \to \pi^+\pi^-$ and also measure $K_L^0 \to \mu^+\mu^-$.

The main background to $K_L^0 \to \mu e$ is the decay $K_L^0 \to \pi e \nu_e$ where the electron neutrino has very little energy in the K rest frame and the pion decays in flight. We have evaluated this background for our detector by Monte Carlo simulation. It should enter at about the 10^{-11} level. The main background to $K_L^0 \to e^+e^-$ is K_L^0 Dalitz decay $K_L^0 \to e^+e^-\gamma$, where the γ has very little energy in the K rest frame. This background should enter at about 10^{-12} for our detector.

A schematic representation of the detector is shown in FIGURE 2. The 3-meter decay region is in vacuum. Drift chamber modules $(xx'yy')$ are placed upstream and downstream of the 72D18 spectrometer magnet with a transverse momentum kick of 220 MeV/c. This is close to the Q value of the decays we hope to measure. The four modules are labeled A–D. Electrons are identified in the lead glass and hydrogen Cerenkov chamber. Muons are required to penetrate a distance into the range stock consistent with their momentum. Scintillation counter arrays are labeled E, G, H, and I. The fast trigger requires that an electron have a coincidence between a quadrant of the Cerenkov counter, an E scintillation counter in that quadrant, and at least 1.2-GeV pulse height in the appropriate lead glass quadrant with no hit in that H quadrant. The e^+e^- fast trigger then requires two such quadrants on opposite sides of the beam. The moun requirement is an E, G, H, and I quadrant in coincidence, and less than 1.2 GeV in that lead glass quadrant. The pion requirement is an E and G quadrant in coincidence with no Cerenkov hit or H hit in that quadrant. The four trigger types are $\pi\pi$, $\mu\mu$, μe, and ee.

Drift chamber information from the ORs of four wires is used in the trigger. We require at least one hit on either side of the beam in the A and B chambers (x and x'

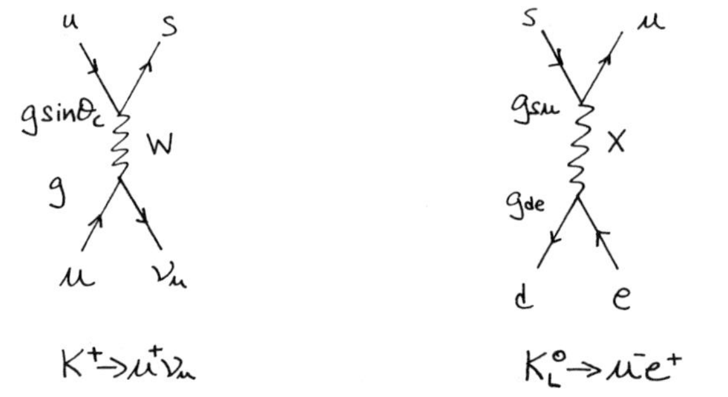

FIGURE 1. Diagram for $K^+ \rightarrow \mu^+\nu$ and possible diagram for $K^\circ \rightarrow \mu e$.

planes) and rough parellelism (± 25 mrad), after the spectrometer magnet in the C and D chambers.

This is the third experiment in which we have used Brookhaven Fastbus.[1] A schematic representation of the E780 seven-segment Brookhaven Fastbus system is shown in FIGURE 3. The segment interconnect and modules contain the arbitration logic for the lower segment and the broadcast register, besides connecting cable and crate segments. The lead glass analog-to-digital convertor (ADC) modules and the drift chamber TDC modules store their encoded data in the onboard first-in, first-out (FIFO) memory. The ADC modules subtract pedestals and only present hits above a settable threshold. The Fastbus list processor (FLIP) reads out the slaves and stores the data in the 2-megabyte Fastbus memory. This memory is large enough to contain all the events from one AGS spill. The host computer therefore analyzes events over the full AGS cycle. The FLIP takes about 200 nsec to read a data word from a slave and

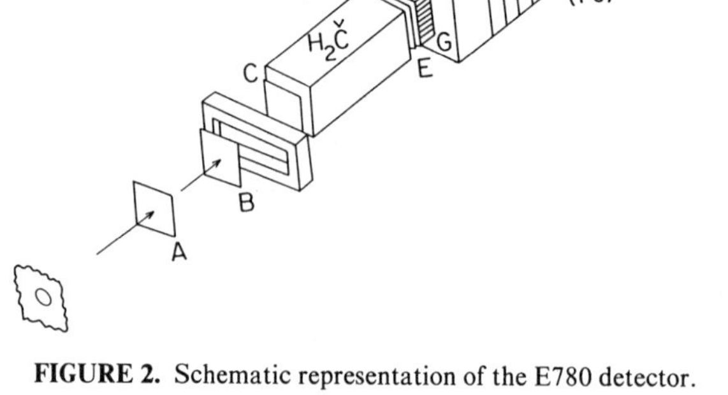

FIGURE 2. Schematic representation of the E780 detector.

store the data in memory. This memory is two-headed—the FIFO wire port is on the crate segment while the FIFO read port is on the cable segment. There is an additional address on the cable segment that causes the FIFO read counter to advance to the beginning of the next event when addressed. Thus the computer may quickly end processing one event and jump to the beginning of the next event.

The TDC spy resides on the lower cable segment. It examines the TDC data as the FLIP reads the TDC modules and signals the FLIP via a coaxial cable if there are less than two hits per horizontal plane or less than one hit per vertical half-plane or too

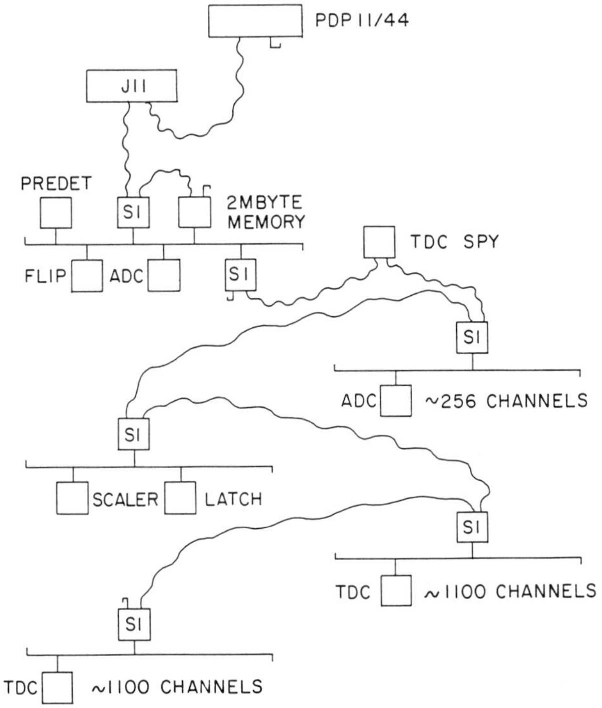

FIGURE 3. Schematic representation of the E780 data acquisition system.

many hits in the chambers. The FLIP then writes the abort word to the segment interconnect broadcast register, which issues a global abort broadcast. This clears the slave modules and the FIFO write pointer is reset to the beginning of the event. The spy also prevents TDC data with drift times outside of the programmable range from being written into memory, and requires hits in the vernier (say x') plane to have contiguous hits in the upstream (say x) plane to be recorded.

The AGS schedule for FY87 is to run slow extracted beam with protons from December 1, 1986, through February 1987, and the month of May. Unfortunately our experiment experienced an intense but limited fire in October 1986. We had rebuilt the experiment by late January and have been taking data since February.

By the end of February we were running with approximately 2.5×10^{11} protons on our target per AGS pulse. This generates a neutral beam of approximately 5×10^{8} neutrons and 2×10^{7} K_L^o. The detector experiences a rate of about 7 MHz in a square meter. We get about 2600 fast triggers going into Fastbus per AGS pulse. Most of these events are aborted. Only about 400 per pulse are buffered in the Fastbus memory. The host computer selects about twenty of these to write to tape for later off-line analysis.

We should have a sensitivity of better than 10^{-9} from the February run. In May we plan to implement the parallelism constraint in the trigger. This reduces the trigger rate by about a factor of 4. We should then have a sensitivity of approximately 10^{-10} from that run. This is at the interesting level of about 1% of the rate for $K_L^o \rightarrow \mu^+\mu^-$.

REFERENCES

1. LEIPUNER, L. B. *et al.* 1983. IEEE Trans. Nucl. Sci. **NS-30:** 3773; BENENSON, G. *et al.* 1983. IEEE Trans Nucl. Sci. **NS-30:** 3958; MORSE, W. M. *et al.* 1983. IEEE Trans Nucl. Sci. **NS-30:** 3964.

Status of Brookhaven Rare Kaon Decay Experiments[a]

ROBERT D. COUSINS

Physics Department
University of California, Los Angeles
Los Angeles, California 90024

INTRODUCTION

In this paper, we summarize the goals and status of the four rare kaon decay experiments at Brookhaven National Laboratory (BNL). In another paper W. Morse[1] has given the status of E780, one of the searches for the decay $K_L^0 \rightarrow \mu e$. We describe the other experiments after some introductory remarks, during which we reemphasize some of the points made by Dr. Morse. These remarks emphasize the status of the experiments. A more thorough discussion of the physics motivation and the background calculations can be found in the proposals for the experiments and in the papers to which we refer below.

DISCUSSION

The motivation for these experiments stems from the fact that we can learn about extremely massive particles by studying decays that they may mediate as virtual particles. As a historical example, we need only consider that the W-bosons were studied in their virtual state for nearly a century before they were produced and observed as real particles at the CERN collider in 1983. In fact, so much was learned from the virtual particles (including less than two decades of observations of neutral currents—virtual Z^0's), that there were few surprises when accelerators attained the energy needed to make the real particles. In an analogous fashion, it is possible now to search for kaon decays that may be mediated by particles of energy of about 100 TeV, which is higher than that of any planned accelerator.

Is there any reason to suspect such particles may exist? By now, the arguments are familiar, and the answer is yes. As the Standard Model has become accepted, the literature has become jammed with speculative papers on how to go "beyond the Standard Model" in order to provide a more fundamental explanation of its many parameters. Ironically, the impressive success of the Standard Model in accounting for all verified experimental data has made it difficult to choose among the many models that go beyond it (supersymmetry, technicolor, composite models, horizontal symmetries, extra Higgs, etc.). A promising path experimentally is to search for processes that are forbidden or extremely suppressed in the Standard Model. The mere observation of

[a]This work was supported in part by the U.S. Department of Energy, and in part by a Sloan Research Fellowship.

any of the processes at a level above that allowed by the Standard Model would indicate exciting new physics, with an opportunity to distinguish among the new speculative models.

The present BNL experiments are designed to be sensitive to branching ratios smaller than 10^{-10}, in one case attempting to push to 10^{-12}. To remind ourselves how small these numbers are, we need only look at the Particle Data Group listings for the K_L^0. Only four events have been seen from the decay $K_L^0 \to e^+e^-\gamma$, with a branching ratio of about 10^{-5}. A single candidate has been seen for $K_L^0 \to \mu^+\mu^-\gamma$, at about 10^{-7}. Perhaps the most sensitive experiment statistically was the pi-mu atom experiment at Fermilab, where over 350 events of $K_L^0 \to (\pi\mu)_{\text{atom}}\nu$ were observed while measuring the branching ratio to be about 10^{-7}. The rarest decay thus far observed is $K_L^0 \to \mu^+\mu^-$, where three experiments saw a total of about 27 events with a branching ratio of about 10^{-8}.

Thus, the present rare kaon decay experiments are quite difficult, and the ultimate limits will be achieved only after a series of iterative improvements in on-line rate-handling capability and off-line background suppression. (For example, a K_L^0 experiment sensitive to the 10^{-11} level must cope with over 10^8 Nobel Prize winning $K_L^0 \to \pi\pi$ decays.) These experiments have high priority at Brookhaven, but will nevertheless face a beam schedule that has been cut back due to funding difficulties. Proton beams were available during a ten-week run that ended on February 28, 1987, and there was a four-week run in May. The (very tentative) schedule for the next fiscal year shows only ten weeks of beam, perhaps beginning as late as January 1988. This is far from the ideal, which would have several lengthy runs (six to eight weeks long) throughout the year, interspersed with downtimes in which to analyze the data and upgrade the detectors.

The first class of experiments we discuss includes those looking for separate lepton number nonconservation. The first to be proposed was E777 (a BNL–SIN–Yale–University of Washington collaboration), a search for $K^+ \to \pi^+\mu^+e^-$. The next experiment to be proposed was E780, with which a BNL–Yale collaboration is searching for $K_L^0 \to \mu e$ and $K_L^0 \to e^+e^-$. The experiment on which we worked, E791, is designed to press further in the search for these modes, as well as some others. The decays $K^+ \to \pi^+\mu^+e^-$ and $K_L^0 \to \mu e$ do not conserve separate lepton number; the discovery of either decay would signal the presence of a new interaction. The sensitivity to new high-mass particles is traditionally illustrated by comparing, for example, the diagram for $K_L^0 \to \mu e$ mediated by a new heavy neutral particle H (with mass M_H) with the Standard Model decay $K^+ \to \mu^+\nu$ mediated by the W^+. We assign a coupling g_H to the vertices involving the H, though this coupling is in general not the same at the leptonic and hadronic vertices. The phase space of the two decays is nearly identical, and if we assume that helicity factors are similar, we have

$$\frac{\Gamma(K_L^0 \to \mu e)}{\Gamma(K^+ \to \mu^+\nu)} = \left(\frac{g_H^2 / M_H^2}{\sin\theta_C\, g^2 / M_W^2}\right)^2 .$$

Solving for the mass of the new particle and putting in numbers for known quantities gives

$$M_H = 20 \text{ TeV} \times \left(\frac{1 \times 10^{-8}}{B(K_L^0 \to \mu e)}\right)^{1/4} \left|\frac{g_H}{g}\right| .$$

As we discuss shortly, the current limit on the branching ratio $B(K_L^0 \rightarrow \mu e)$ is confusing, but we may say that the limit is no better than 10^{-8}. Experiment E791 attempts to push this down to 10^{-12}. Thus, the range of masses being explored is from 20 TeV to over 100 TeV (!) if the couplings are the same size as the electroweak coupling g.

The most stringent published limit on $K_L^0 \rightarrow \mu e$ comes from a 1971 Bevatron experiment. The same paper[2] reported $B(K_L^0 \rightarrow \mu e) < 1.57 \times 10^{-9}$ and $B(K_L^0 \rightarrow \mu^+\mu^-) < 1.82 \times 10^{-9}$, both at 90% confidence level. Three subsequent experiments saw $K_L^0 \rightarrow \mu^+\mu^-$ at a branching ratio of about 10^{-8} (more consistent with strong theoretical expectations); the reason for the Bevatron result remains a mystery. If one throws out the Bevatron limit on $K_L^0 \rightarrow \mu e$, as the Particle Data Group decided to do several years ago, then one must go back to a 1967 BNL experiment that set an upper limit of 6×10^{-6}. For the moment, many of us working on the $K_L^0 \rightarrow \mu e$ experiments would just say that the current limit for $K_L^0 \rightarrow \mu e$ is no better than the known branching ratio for $K_L^0 \rightarrow \mu^+\mu^-$, namely about 10^{-8}. Fortunately, we will soon have better limits from E780 and E791, so that model builders will not have to choose between the two published numbers that differ by a factor of 3000.

One must ask what limits can be placed on separate lepton number nonconservation from studies of related processes, such as $\mu \rightarrow e\gamma$, $\mu \rightarrow e^-e^-e^+$, μ–e conversion on nuclei ($\mu^- + A \rightarrow e^- + A$), and the $K_L^0 - K_S^0$ mass difference. In most of these cases, simple dimensional arguments of the type previously illustrated may easily be found. A useful classification scheme has been clearly described by Cahn and Harari[3] in the context of a simple model of a horizontal gauge symmetry. Because the mass limits from their model closely parallel the results of simple dimensional arguments, we highly recommend their paper as a starting point for interpreting the experimental limits.

The classification scheme invokes the concept of generation number G, whereby all first-generation fermions (u, d, ν_e, e) are assigned a G one unit less than second-generation fermions (c, s, ν_μ, μ) and two units less than the third generation. The processes may then be characterized by the net generation change, which is zero in $K_L^0 \rightarrow \mu e$ and $K^+ \rightarrow \pi^+\mu^+e^-$. In the multitude of models that have been considered in the literature, it is clear that the diverse processes will be beneficial in sorting out the character of any new physics that is discovered. Models are typically most severely constrained by the $K_L^0 - K_S^0$ mass difference unless they suppress the underlying $K^0 - \overline{K^0}$ mixing (a $\Delta G = 2$ diagram), and are also strongly challenged by the limits on μ–e conversion[4] and the other $\Delta G = 1$ processes. The decays $K_L^0 \rightarrow \mu e$ and $K^+ \rightarrow \pi^+\mu^+e^-$ are distinguished by having $\Delta G = 0$ and by involving the strange quark as well as leptons. Thus, in many models (though typically not in supersymmetry), these decays are a favored place to look for new physics. For more details, one may look at the numerous experimental and theoretical reviews that have been written in conjunction with workshops on future intense kaon beams.[5] These references also discuss a number of models having leptoquarks, which open up other possible diagrams for decays such as $K_L^0 \rightarrow \mu e$.

The current upper limit on the branching ratio for $K^+ \rightarrow \pi^+\mu^+e^-$ is 5×10^{-9}. Experiment E777 proposes to push this down to 10^{-11}. FIGURE 1 is a diagram of their apparatus, which looks at decays in flight of a 6-GeV K^+ beam. The design of the spectrometer was inspired by the strong need for particle identification in order to

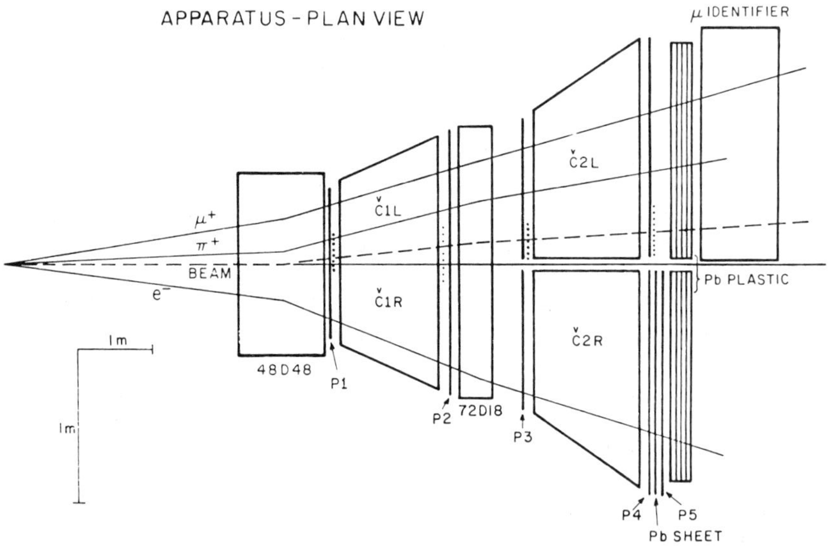

FIGURE 1. Plan view of the E777 detector (search for $K^+ \rightarrow \pi^+\mu^+e^-$). M1 and M2 are dipole magnets; P1–P4 are proportional chambers; the Čerenkov counters are described in the text. A BNL–SIN–Yale–University of Washington collaboration.

suppress background. A dominant background off-line is from $K^+ \rightarrow \pi^+\pi^+\pi^-$, where one π^+ decays to $\mu^+\nu$ in the detector, while the π^- is misidentified as an electron. Misidentification of pions as electrons is kept below 1 in 10^7 by a combination of two hydrogen Čerenkov counters (decoupled by an intervening magnet) followed by a lead-scintillator shower counter. The detector is asymmetric, accepting only negative tracks on that side. On the other side, the positive pion and muon are identified by Čerenkov counters with N_2 and CO_2 and by a muon filter. Experiment E777 wrote some physics data on tape during the recent run, and should have sensitivity below the old limit with the data in hand. As it was the first rare kaon experiment to push to a high intensity, E777 learned how the in-spill duty cycle of the beam strongly affects their accidental coincidence rates and their ability to trigger effectively. On behalf of all the experiments, it has urged machine studies in this area. After fixing problems with its collimators and its analyzing magnet, E777 ran again in May.

Dr. Morse has described the spectrometer used by his E780 collaboration to search for $K^0_L \rightarrow \mu e$ and $K^0_L \rightarrow e^+e^-$. The experiment employs traditional topological selection and particle identification techniques to single out two-body decays of the K^0_L. The experiments have recovered from a fire that damaged the spectrometer last October, and have some physics data on tape (at about the 10^{-9} level of sensitivity). They pushed on toward 10^{-10} in the May run.

The proponents of E791 (a UCLA–LANL–Penn–Stanford–Temple–William and Mary collaboration) have designed a new, dedicated spectrometer to search for $K^0_L \rightarrow \mu e$ and other rare kaon decays. Most of the apparatus was installed during the 1986–1987 winter and spring months, and some engineering data were written to tape during February 1987. An artist's view prior to installation is shown in FIGURE 2, with all shielding and portions of the magnets omitted. There are two large dipole magnets (48D48 and 96D40) used for precision measurement of each particle's momentum. This is an upgraded version of the method used by the most successful search for the decay $K_L^0 \to \mu^+\mu^-$ (at Argonne some years ago). A dominant background in E791 is the decay $K_L^0 \to \pi e \nu$, followed by the decay $\pi \to \mu\nu$ (or by misidentification of the pion as a muon). The detector sees an identified muon and electron, though they typically have an invariant mass less than that of the kaon. In a most pernicious case, the pion decay is in the middle of a dipole magnet, so that the decay kink confuses the bend used for momentum analysis. In E791, the bends in the two magnets can be checked for consistency in order to reject this background. (In E780, a muon range stack allows a similar check, albeit with less precision.)

The trajectories are measured with minidrift chambers and scintillation counters. Particle identification is accomplished with Čerenkov counters, lead glass, and muon filter and hodoscope. The left and right sides of the experiment have independent detectors in order to make reconstruction and triggering less vulnerable to double events. (A clean event has only one track in each side; even a fully contained double event only has two tracks on each side.) The magnets are run at equal and opposite field integrals of absolute magnitude corresponding to transverse momentum impulse of 300 MeV/c. This configuration results in approximate restoration of the original K_L^0 decay opening angle after the two particles have traversed the magnets and drift chambers. A scintillation counter system with 500 phototubes can then measure this angle, which is used in a memory lookup circuit in a fast trigger. This triggering technique frees us from the traditional constraint of 230-MeV/c transverse impulse used in spectrometers (like E780) to bend the tracks parallel for easy triggering.

Because E791 will have extremely high particle fluxes, we have developed a

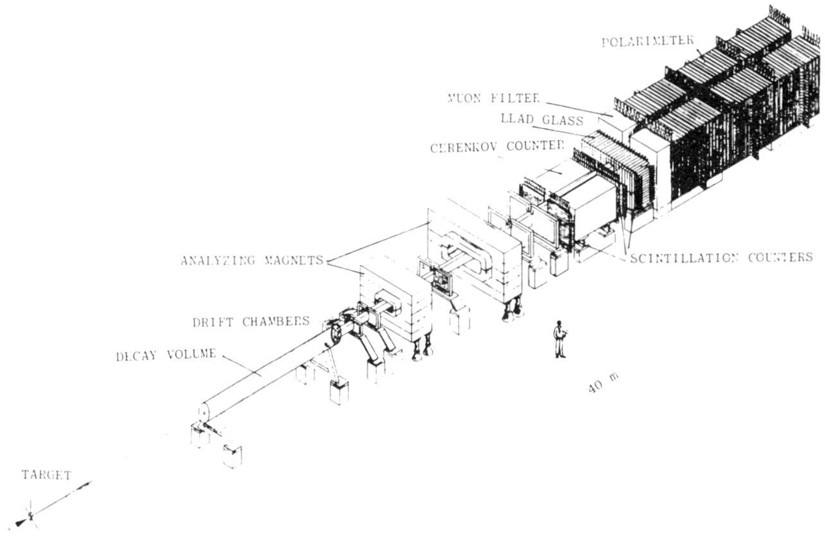

FIGURE 2. The E791 detector (search for $K_L^0 \to \mu e$, $K_L^0 \to e^+e^-$, $K_L^0 \to \pi^0 ee$, and other modes). A UCLA–LANL–Penn–Stanford–Temple–William and Mary collaboration.

number of special features in our beam line, spectrometer, and readout system. We briefly mention the beam line and the readout system. A new neutral beam line was built by Brookhaven, including a sophisticated beam dump for the primary protons and the subsequent hadronic showers. Primary protons (about 28 GeV/c) from the AGS impinge at 2.75 degrees on an 8-in.-long copper target that is followed by a 10-m-long system of collimators, sweeping magnets, and shielding. The neutral particles (mostly neutrons, about 3% K_L^0's, with most of the photons having been removed by a lead convertor) then travel through an 8-m-long decay volume, in which several percent of the kaons decay. Tests in June 1986 confirmed that muons and neutrons from the dump would not limit the experiment. The collimators for the experiment have precision tapered faces, positioned and monitored to 1/64 inch. After some initial settling of the shielding, the collimators were repositioned, and are now working. We envision a future upgrade of our vacuum window and an evacuated beam pipe through the spectrometer.

The completely custom-built readout system is virtually deadtimeless. All signals from all detectors are flash-digitized (in ADCs, TDCs, and latches) within a few hundred nanoseconds of the particles' passage through the spectrometer. Each channel of each front-end module can store two events, so that immediately after digitization of the first event, the system is live again, able to digitize a second event while reading out the first. The readout itself from the front end directly into custom dual-port memories in SLAC 3081 emulators takes only a few microseconds, thanks to sparse-data-scanning logic and a 48-byte-wide (!) dataway. The dataway is strobed every 100 ns, for an effective transfer rate of 0.5 Gbyte/sec, which is reduced slightly by fluctuations in the number of hits in different parts of the spectrometer. The system was used in February 1987 to log data from each part of the spectrometer, and was expanded for the May 1987 run.

Naturally, while investing in an experiment of such sensitivity, one considers what other measurements can be made with it. Our proposal discusses in detail one of the most interesting possibilities, a search for direct CP-violation in the decay $K_L^0 \rightarrow \pi^0 ee$. The discussion here is along the lines of that given by Gilman and Wise.[6] We recall that the K_L^0 is mostly the CP-odd state K_2, with a small admixture of the CP-even state K_1 at the level of $\epsilon = 2 \times 10^{-3}$. All measurements of CP-violation to date need only invoke "indirect" CP-violation, namely the use of the K_1 state as the source of CP-even final states in the decay of the K_L^0. It would be a fundamental breakthrough to observe "direct" CP-violation, namely the decay of the K_2 part of the K_L^0 into a CP-even state, as is the goal of experiments described at this conference to search for a nonzero value of the parameter ϵ'. There have been attempts to calculate the rate of $K_L^0 \rightarrow \pi^0 ee$ in the Standard Model from three different types of amplitudes. The first is indirect CP-violating through single-photon or Z exchange and yields by itself a rather reliable estimate of the branching ratio of 6×10^{-12}. The second amplitude is direct CP-violating through single-photon or Z exchange, and has been estimated to be about 10^{-11}. The third is CP-conserving through two virtual photons, and is estimated to be below 10^{-11}. While improvements to these calculations are still appearing in the literature, the exciting possibility exists that this decay may be dominated by direct CP-violation. The current limit on the branching ratio is only 2.3×10^{-6}, so in any case a large window exists within which to look for the effects of new physics. Experiment E791 plans to push to 10^{-10}, and possibly to 10^{-11} with future upgrades.

Other decays that can be looked at in the $K_L^0 \to \mu e$ experiments include $K_L^0 \to e^+e^-$, $K_L^0 \to \mu^+\mu^-$, $K_L^0 \to e^+e^-\gamma$, $K_L^0 \to \mu^+\mu^-\gamma$, $K_L^0 \to \pi^0\mu e$.

Finally, we discuss briefly experiment E787, a search for decays of the K^+ into a single observed π^+ plus unseen neutral particles. Referred to as $K^+ \to \pi^+ +$ "nothing", this class of decays includes a nonzero Standard Model prediction for the decay $K^+ \to \pi^+\nu_i\bar{\nu}_i$, where i stands for e, μ, or τ. The results of Standard Model calculations for this second-order effective strangeness-changing neutral current appear to be settling down at a branching ratio around 10^{-10}. Larger values are allowed in almost-standard models that have more than three generations of fermions. Experiment E787 (a BNL–Princeton–TRIUMF collaboration) currently has a design sensitivity of 2×10^{-10}, perhaps within reach of a Standard Model signal. The present

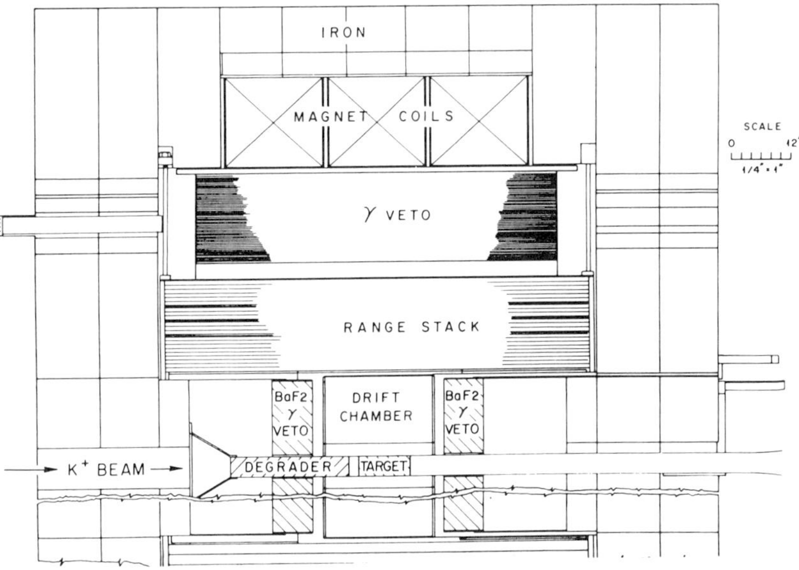

FIGURE 3. The E787 detector (search for $K^+ \to \pi^+ +$ "nothing"). A BNL–Princeton–TRIUMF collaboration.

limit is from a recent KEK experiment,[7] giving $B(K^+ \to \pi^+\nu\bar{\nu}) < 1.4 \times 10^{-7}$ (90% C.L.).

Since the neutrals are not observed, they might in fact not be neutrinos but speculative particles such as photinos or sneutrinos in some supersymmetric models. A particularly striking case experimentally would be the two-body decay $K^+ \to \pi^+a^0$, where a^0 is a new, nearly massless neutral particle. During the several years that E787 has been under construction, various candidates for the a^0 have gone in and out of fashion, such as the axion, the familon, and light Higgs. The same KEK experiment set a tighter upper limit on this two-body decay at 3.8×10^{-8}.

The E787 detector is represented in FIGURE 3. In contrast to the previously described experiments, E787 stops the kaons and looks at their decay at rest in the laboratory. The signal consists of a single pion whose energy is measured in drift chambers as it bends in a large solenoidal magnet not unlike those used in colliding-beams experiments. The pion itself is also stopped in a range stack, so that momentum and range can be compared to confirm the particle identification. The decay $\pi^+ \rightarrow \mu^+\nu$ is recorded, but ultimate background suppression requires that the electronics even remain sensitive to record the subsequent $\mu^+ \rightarrow e^+\nu\bar{\nu}$ decay.

Background suppression has required extensive preparation, since both particle and nuclear processes must be understood. An obvious example is the decay $K^+ \rightarrow \pi^+\pi^0$, where the photons from the π^0 escape detection. The π^+ is, in principle, monenergetic, so that these events can be cut out of the data. Of course, detector resolution makes this imperfect, and the photon veto must be as efficient as possible. In the final upgrade, barium floride detectors are expected to augment the initial lead-scintillator shower counters. Due to the lack of space, we must refer the interested reader to the literature[8] for a more complete discussion of the backgrounds and the detector.

The large magnet for E787 was built during the last months of 1986, and installation of some of the detector began in early 1987, with the first engineering data appearing in May 1987. First physics data will be taken during the proton running in fiscal year 1988.

SUMMARY

In summary, after several years' hard work, a number of experimenters are beginning to analyze or take their first physics data. We look forward to the exciting times in the months ahead, as we explore the high mass frontier (virtually).

ACKNOWLEDGMENTS

We would like to thank representatives from each of the BNL rare kaon decay experiments for informative conversations.

REFERENCES

1. MORSE, W. 1987. Status of AGS experiment 780. This volume.
2. CLARK, A. R. *et al.* 1971. Phys. Rev. Lett. **26:** 1667.
3. CAHN, R. N. & H. HARARI. 1980. Bounds on the masses of neutral generation-changing gauge bosons. Nucl. Phys. **B176:** 135.
4. BRYMAN, D. A. *et al.* 1985. Phys. Rev. Lett. **55:** 465.
5. See, for example, LUBATTI, H. J. 1982. Rare decays of the K-meson. *In* Proc. 2nd LAMPF II Workshop, LA-9572-C, vol. I, p. 115; KANE, G. L. & R. E. SHROCK. Rare *K* and μ decays: Theoretical and experimental status and prospects. *In* Intense Medium Energy Sources of Strangeness, AIP Conf. Proc. No. 102, Subseries on Particles and Fields No. 31, T. Goldman *et al.*, Eds. 123, and the accompanying article by J. Ellis; SMITH, G. A., Ed. 1984. Rep. of the AGS II Task Force, Brookhaven National Laboratory, Upton , N.Y.
6. GILMAN, F. & M. WISE. 1980. Phys. Rev. **D21:** 21.
7. ASANO, Y. *et al.* 1981. Phys. Lett. **107B:** 159; see also YAMAZAKI, T. *et al.* 1984. Phys. Rev. Lett. **52:** 1089.
8. See, for example, LITTENBERG, L. 1987. Nucl. Phys. **B279:** 171.

Rare B Decays: Experimental Prospects and Problems

JAMES D. BJORKEN

Fermi National Accelerator Laboratory
Batavia, Illinois 60510

INTRODUCTION

My interest in B-physics is conditioned by my presence at Fermilab. I carry a special bias toward the long-range potential contribution to this subject of experiments utilizing hadron–hadron collisions. Every minute that the Tevatron machine is delivering beam to experiments, more $b\bar{b}$ pairs are produced than will be produced in the history of SLC, LEP, CESR, and Doris combined. Regrettably these b's reside in the interior of the beamdumps, quite remote from direct access. But this still remains a spur toward working hard in exploiting this potential resource as much as possible.

In this survey of experimental prospects, I do not emphasize the electron–positron collider potentialities. This is not out of disinterest, but simply because the existing lore is quite extensive, and better covered by others.

WHAT WE WANT TO KNOW

We only briefly enumerate the obvious topics of interest in b-physics:

Production Dynamics

This is a more reliable quantum chromodynamics (QCD) test than charm photoproduction. The situation in e^+e^- collisions is already quite good. Charm production dynamics is reasonably described[1] in terms of QCD photon–gluon fusion mechanisms, although there is a normalization issue. This leaves hadroproduction as the most uncertain case. Important issues include the following.

(1) *Leading-particle effects:* Relative to charm hadroproduction, these effects diminish in importance as $(m_c/m_b)^{-1}$, but are important to establish (or rule out).

(2) *Normalization:* There is a rather big uncertainty[2] from choice of b-quark effective mass (5.0 ± 0.3 GeV seems to be the chosen range), choice of gluon distribution functions, and an unknown "K-factor."

(3) A *Dependence:* Are there any $A^{2/3}$ effects, especially for forward production?

(4) *Baryon yield:* The large yield of $\Xi_c^+(usc)$ baryons seen[3] by Fermilab experiment E400 (central production with 600-GeV incident *neutrons*) suggests

b-baryon production may likewise be quite substantial. We suspect everyone (even the e^+e^- community?) may underestimate the importance of baryons.

Spectroscopy

All species need independent study and identification. Lifetimes B_u and B_d should be separated. It is not satisfactory to claim a measurement of $B_s - \bar{B}_s$ mixing without having yet discovered the particles themselves. The various species of b-baryons are useful as probes of bound-state properties. An excited $B_d^* \rightarrow B_d + \pi^+$ would be an invaluable tag for mixing and CP studies. Or how about $B_s^* \rightarrow B_d + K^+$? or $\Sigma_b^* \rightarrow \bar{B}_d \times p$? Perhaps some species of B (i.e., $\bar{b}q$) may bind weakly to (or resonate with) some species of baryon à la deuterium. While these possibilities are somewhat unlikely, the search ought to be carried out (the same can, of course, be said for charm).

Decay Phenomena

To understand Cabibbo theory well, a large number of strange meson and baryon decay modes needed exhaustive study. The same is now true for heavy flavors: A large data base of charm and bottom decay channels is needed to reliably extract the Kobayashi–Maskawa (KM) matrix elements. The validity of the spectator/factorization model of B decays needs more testing. Optimal "signature" decay models of B-mesons and baryons need to be established. Final states containing $\tau\nu_\tau$ would be splendid to see. And final states containing ψ are, of course, extremely interesting.

Mixing

The recent Argus results[4] on B_d mixing, assuming they are confirmed, evidently have revolutionary consequences. Much of the theoretical work, especially the studies on CP-violation, now needs major revision and reconsideration. It is a good example of how acquisition of a few simple facts can greatly change experimental strategies for the future. For another example, our prejudices regarding m_t may now change, with a tilt toward a much heavier top quark.[5] If $m_t > m_w$, even the Higgs-search strategies for $m_H > 2m_w$ change. ($H^0 \rightarrow b\bar{b}$ dominates; $H^0 \rightarrow \mu^+\mu^-$ has a nontrivial branching ratio, etc.)

Also $B_s - \bar{B}_s$ mixing may be expected to be very rapid; several oscillations per mean life is quite reasonable. Measurement of $(\Delta m/\Gamma)$ for the B_s may therefore require very good time resolution and result in quite spectacular data.

Rare Decay Searches

The long lifetime and relatively large b-quark mass implies a special sensitivity to new physics in decay modes with eventually detectable branching ratio, for example, $B \rightarrow \mu e$ hadrons, $b \rightarrow s\ l\bar{l}$, etc. Other rare modes, such as $B_u \rightarrow \tau\nu_\tau$, are interesting in their own right.

CP-Violation

The large *CP*-violation effects in the *B*-system that are anticipated within the Standard Model[6] make this an irresistible experimental goal. The best experimental strategy for seeing *CP*-violation is as yet unclear. It should be emphasized that a firm experimental command of the preceding five topics is probably a prerequisite to succeeding in the detection of *CP*-violation in the *B*-system. I suspect we are a very long way from success in that endeavor.

Bottom-Quark Engineering

The long-range frontier of very high mass-scales is undoubtedly dominated by multijet spectroscopy. Detection of leptons, photons, and quark and gluon jets are therefore of highest priority. But close behind is identification of the flavor content of hadron jets. Quantum chromodynamics bremsstrahlung gluons will be a curse in this field, and sorting out interesting quark-induced jets (especially heavy-quark jets) from the large gluon-jet background will be a challenge of increasing importance as the energy scale continues to go up. In the long run, this may technically be the hardest problem for heavy flavor experimentation. Successful, efficient isolation of $W \to c\bar{b}$ and/or $Z^0 \to c\bar{c}$, $b\bar{b}$ signals would be a landmark accomplishment.

In any case, the study of generic *b*-physics has as its most important long-range goal the measurement of *CP*-violation in the *B*-system. Prior to the Argus results, my guess of the required number of produced $b\bar{b}$ pairs per experiment for doing this was at least 10^9. This is reasonably consistent with what is found[7,8] in the paper studies (mostly by theorists!). The back-of-the-envelope argument goes as follows.

(1) A specific nonleptonic decay chain appears to be needed:

$$B \to DX \quad (1\%)$$
$$\qquad \hookrightarrow Y(3\%)$$

The percentages measure typical branching ratios. Alternative options such as $B \to$ uncharmed, $B \to \psi X$, or $B \to$ charm–anticharm lead to net branching fractions no larger.

(2) The neutral *B* must be identified at birth as *B* or $\bar{B}$, and as strange or nonstrange. An efficiency of more than 10% per $b\bar{b}$ event seems optimistic.

(3) Good statistics is needed to be convincing even if the effect is gross. We take the number of events per decay channel of interest to be $\gtrsim 10^3$.

(4) Prior to the Argus result, either a first-forbidden B_s decay or a delicate effect in B_d decay (due to the small mixing expected) was required. We gave a factor 30 derating here, a factor perhaps no longer operative.

So our bottom line of $\geq 10^9$ $b\bar{b}$ produced per experiment might shrink to a mere 3×10^7 by the recent results. And there is of course a great uncertainty in this estimate. The number might be reduced by a clever, optimal strategy. On the other hand, little allowance has been made for the invariable losses from incomplete acceptance, detection inefficiencies, etc., that occur in real experiments as opposed to paper studies conducted by theorists. The real bottom line is, in order to have a chance of success with *CP*, to throw deep.

PROSPECTS AND LIMITATIONS OF $e^+e^- \rightarrow b\bar{b}$

Two distinct alternatives for $b\bar{b}$ physics are threshold e^+e^- machines, for example, CESR or Doris, or Z^0 factories. Other energies seem to me to be clearly disadvantageous.

Threshold Machines

At the $\Upsilon(4S)$, $\sigma(b\bar{b})$ is $\sim 10^{-33}$ cm^2. If we take the $\int L\,dt$/experiment to be 10^{39} cm^{-2}, we obtain 10^6 $b\bar{b}$ produced. The advantages of working at threshold are well-known; disadvantages for generic studies are that B_s and baryons are difficult, and no lifetime information can be gleaned. This might be mitigated somewhat if, as has been discussed, asymmetric collisions (2 × 12 GeV?) could be accommodated in some future machine.

Z^0 Factories

At the Z, $\sigma(b\bar{b}) \sim 5 \times 10^{-33}$ cm^2. With the $\int L\,dt$/experiment $\sim 2 \times 10^{38}$ cm^{-2}, one obtains again 10^6 $b\bar{b}$ per experiment. In this case, the *b*'s move relativistically and microvertex detection can be used. But technical problems of a big beam pipe and barrel geometry remain. The smaller SLC beam pipe and beam polarization are an advantage there; the bigger question at the Standard Linear Accelerator Center (SLAC) may be ultimate luminosity: Is 2×10^{38} cm^{-2}/experiment a reasonable estimate?

For generic studies of B_s and baryons, the Z^0 factories should have an advantage. Also LEP + SLD comprise an array of very sophisticated and expensive detectors; this ought somehow to be of importance.

Evidently the main limitation of e^+e^- colliders is simply luminosity; an order of magnitude improvement would produce a great deal of physics. But that may still fall short of what is needed for *CP* physics.

EXTERNAL PHOTON AND HADRON BEAMS

There are many initiatives for heavy flavor physics in this category that are underway. Before cataloging these, however, it may be worthwhile reviewing some of the necessary conditions for doing $b\bar{b}$-physics via this technique.

Spectrometer Properties

(1) Because the $b\bar{b}$ signal is so small ($\lesssim 10^{-6}$ of all events in hadron–nucleus collisions; $\leq 3 \times 10^{-5}$ of all hadron photoproduction events), high rates are

essential. For hadroproduction, one must deal with $10^{6\pm1}$ interacting beam particles/sec. Of these, $10^{1\pm1}$/sec can be expected to be recorded on tape, implying a selectivity of the data acquisition system of one in $\geq 10^{5\pm1}$ events.

(2) Isolation of the secondary decay vertex is essential. The success with silicon microstrip detectors is evidence that this can be done. Other techniques (e.g., high-resolution "straw" detectors, optical fibers, CCDs) are also candidates, as are "active targets," that is, detectors for which the secondary decay vertex is contained within. (These include high-resolution streamer chambers, emulsion, and scintillating glass fiber targets.)

(3) The usual criteria for a good spectrometer (broad acceptance, accurate charged particle resolution, good electromagnetic calorimetry and muon identification, Čerenkov identification, etc.) apply with special force because of the heavy demands of high rate and low signal/noise ratio.

Event Selection Criteria

It must still be learned how best to attain the large rejection factor of 10^5. The ideas extant include the following.

(1) *High* P_T *lepton* ($\gtrsim$*1–2 GeV*): While this strategy worked poorly for charm, there is good reason[9] to believe it is much better for bottom, because of the larger b-quark mass.

(2) ψ *Trigger:* The decay $b \rightarrow \psi s$ proceeds with a 1% branching ratio. The signature is evidently excellent, especially when combined with secondary-vertex information. It also is encouraging that CP violating effects are large in some channels, such as $B_d \rightarrow \psi K_s$. My favorite is the Cabibbo forbidden mode, $B_d \rightarrow \psi \pi^+ \pi^-$ (branching ratio $\sim 10^{-4}$), which has splendid signature and intrinsic properties, including CP violating effects at the level of ψK_s.

(3) *Secondary-vertex tag:* The goal here is to trigger on events with evidence of a secondary vertex, using only the data from the front-end microvertex tracking system. There are many ideas on doing this,[a] but as yet it seems not quite state of the art.

(4) *Charm tag:* Perhaps one might preselect $D^+ \rightarrow K^- \pi^+ \pi^+$ or $D^0 \rightarrow K^- \pi^+$ candidates by an on-line trigger processor, and thereby work back toward the B's. This seems hard.

(5) *Several high* p_T *hadrons:* This may be especially helpful in isolating low-multiplicity B-decays (the ones one is most interested in anyway). This approach will be tried[10] at CERN in the next year or two.

It seems probable that a *combination* of these strategies may be needed to get the desired rejection factor of $\gtrsim 10^5$.

[a] For example, CERN proposal WA82, scheduled to run this year.

Tagging the Initial B

The time dependence of neutral B-decays provides an especially sensitive measure of CP-violation. For example, for final states that are CP eigenstates

$$\Gamma(B \to f) = \Gamma_0[1 + (\text{Im}\,\lambda)\sin \Delta mt]e^{-\Gamma t}$$

$$\Gamma(\bar{B} \to f) = \Gamma_0[1 - (\text{Im}\,\lambda)\sin \Delta mt]e^{-\Gamma t},$$

where λ is of modulus unity and is real if CP is conserved. To see an effect, one must know whether the meson at birth was B or $\bar{B}$. Strategies for doing this[8] include the following.

(1) Finding an associated $B_u/\bar{B}_u$ or $\Lambda_b/\bar{\Lambda}_b$.
(2) Finding a resonance, for example, $B^* \to B\pi$ or BK, analogous to the often-used D^*. However, for the B-system, the B^* has to be a p-wave excitation.
(3) Finding an associated B^0 via, for example, semileptonic decay, and correcting for its mixing. This is tricky, especially if B^0 is a $B_s/\bar{B}_s$.
(4) Using only leading B's ($x_F \gtrsim$ 0.2–0.3?) where one may expect a $\bar{B}/B$ production ratio different from unity because of hadronization effects. This has the advantage of an inclusive measurement, but throws away a lot of B's.

The tagging problem and the related problem of resolving B_s and B_d final states is a heavy demand on experimental designs and needs careful attention.

What Is Happening Now?

(1) *CERN SPS:* To our knowledge only one directly observed hadroproduced $B\bar{B}$ pair has been reported;[11] more may be on the way. This was achieved via emulsion and muon trigger; a follow-up measurement in closed geometry[12] has yielded a coarse estimate of B production by pions at 300 GeV.

Other activities at CERN center around the Omega Spectrometer. This year saw an attempt[a] (experiment WA82) to trigger on secondary vertices with a silicon microstrip front end. Next year will feature[10] a scintillating glass fiber target, with fibers aligned along the beam (FIGURE 1). The readout is from upstream, using image-intensifiers and CCDs. The trigger will be three high p_T hadrons seen in the downstream spectrometer.

(2) *Fermilab:*[b] The very successful[13] charm photoproduction experiment E691 was followed by a charm hadroproduction experiment (E769) with incident pions and kaons. However, the low primary energy of ~250 GeV and loose trigger does not make this a strong $B\bar{B}$ experiment. But it does have the capability of seeing some $b\bar{b}$ events. This is also true of a new photoproduction experiment, E687. Also, an experiment

[b] The experiments cited below have the following spokespersons: J. Appel (E769), J. Butler and J. Cumulat (E687), N. Reay (E653), B. Knapp (E690), B. Cox (E771), P. Slattery (E706), and A. Zieminski (E672).

(E653) with emulsion target, a silicon microstrip tracker, followed by an open-geometry spectrometer and muon trigger, has had one run and continued last year with high-energy pions incident. One may expect some *B* events from this experiment also.

Another experiment (E690) scheduled to begin in 1989, will study charm production in target-dissociation processes. It features a very sophisticated pipeline processor capable of full reconstruction of complex events on-line at high rate. This system has been developed during a Brookhaven National Laboratory (BNL) experiment, and now appears to be essentially ready to go. This experiment has much to say

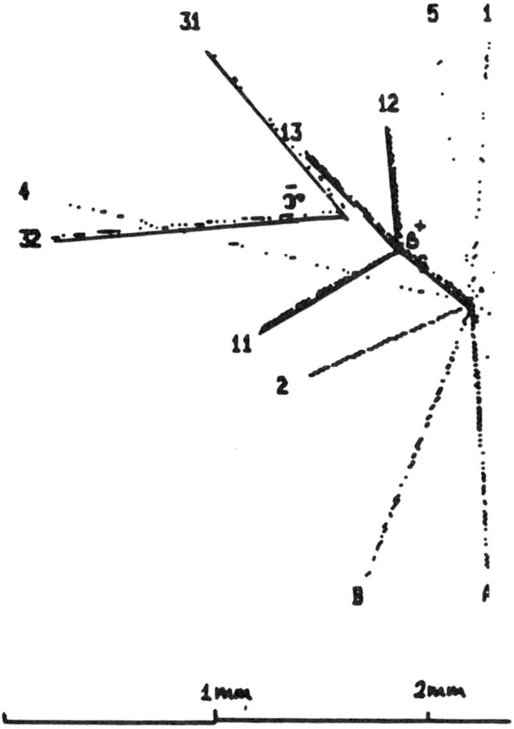

FIGURE 1. Simulation of a $B\bar{B}$ event as might be seen[10] from upstream via a scintillating glass fiber target. (Precision = 10 μm; hit density = 4/mm.)

about the capability of on-line event reconstruction, but it does not directly address *B*-physics in its present incarnation.

A new initiative with a fresh approval (E771), is a high-rate experiment that will trigger on $B \rightarrow \psi X \rightarrow \mu^+\mu^- X$ and observe the secondary decay vertex with a silicon microstrip system. It should have a preliminary run in 1989. Meanwhile an experiment (E706) designed for direct-photon physics has been merged with another (E672) designed for study of hadrons produced in association with Drell-Yan dileptons.

Because E706 has a silicon microstrip detector, this provides a serendipitous opportunity to try for this kind of physics already this year.

My view of all this is that at best the main result of this round of experiments will be to map out B-production characteristics. The yield will probably be too small for spectroscopy, mixing studies, etc. However, useful lifetime information may emerge, along with initial data on Λ_b and B_s. The $b \rightarrow \psi s$ approach is also very strong, and has a great deal of promise. However, it may well be that hadroproduction of $b\bar{b}$ today is where hadroproduction of $c\bar{c}$ was a decade ago. In any event, the aforementioned round of experiments should do a splendid job on charm physics, fully competitive with and complementary to what is being done in e^+e^- collisions. The topics should include rare decays, baryon physics, $D - \bar{D}$ mixing, and perhaps double-charm physics, such as, Ξ_{cc}^{++} (ccu) production.

GENERIC BOTTOM-PHYSICS IN HADRON–HADRON COLLIDERS

It is a difficult but not hopeless business to find secondary vertices in the high-rate environment characteristic of a hadron–hadron collider. The presence of a beam pipe, intense circulating beam, barrel geometry, and a large interaction-region volume exacerbates the problems. Furthermore, the typical "4π" detector that typically surrounds such collision regions is optimized for high-p_T jet/lepton/photon physics, not low-p_T "minimum-bias" final states characteristics of generic $b\bar{b}$ production. Initiatives and plans for UA1, CDF, and DØ can be expected to emphasize physics goals other than generic $b\bar{b}$ production.

Beyond these initiatives exists an effort[14] to look at the forward heavy-flavor production upstream/downstream of UA2. And at the Fermilab collider there are nascent ideas of generic $b\bar{b}$ experiments at 2 TeV in the center-of-mass. The $b\bar{b}$ yields

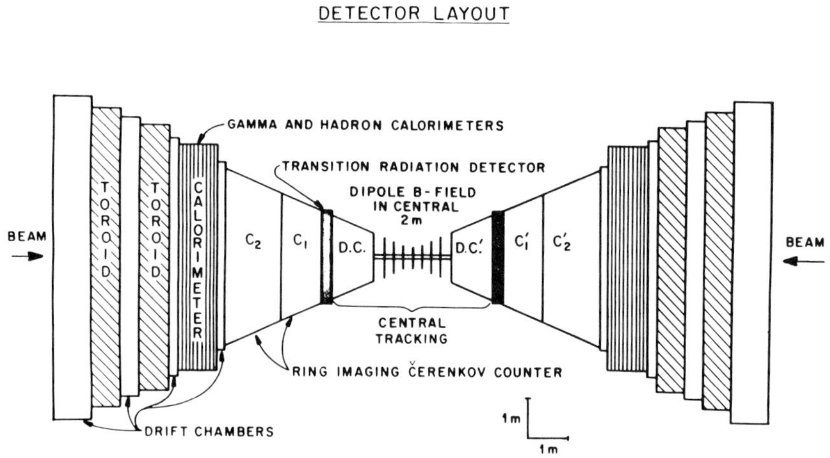

FIGURE 2. A layout[15] (Reay *et al.*) of a possible Tevatron collider experiment for generic B-physics.

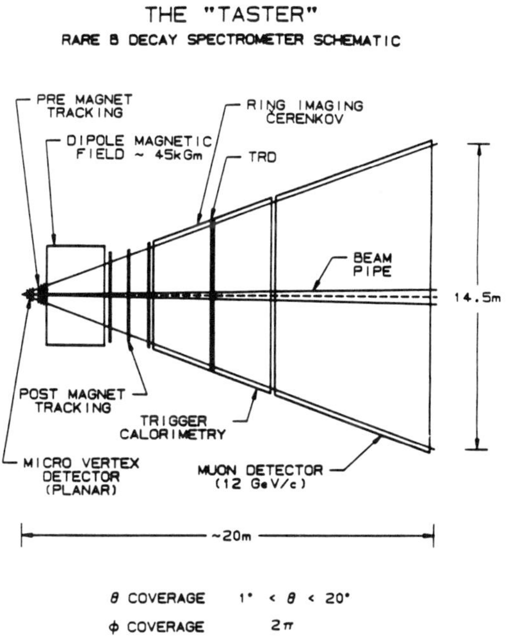

FIGURE 3. Layout[17] of an SSC experiment (the TASTER; Cox & Wagoner) for generic *B*-physics.

may be as high as 3×10^{-4} per collision, or $\gtrsim 20\ \text{sec}^{-1}$ at $L = 10^{30}\ \text{cm}^{-2}\text{sec}^{-1}$. Thus in principle the yields per experiment suffice to enter the domain of *CP* violating effects. However, the problems are daunting.

The products of *B*-decays are found over a large range of rapidities; the most useful *B*'s are those that move relativistically in the laboratory frame. I estimate a good angular acceptance to be 20 mrad $\lesssim |\theta| \lesssim$ 800 mrad, implying emphasis on "planar" detector geometry similar to fixed-target architectures. However, much is gained if the "barrel" region is also covered. One of the suggested designs[15] is shown in FIGURE 2; it is "planar" and covers angles up to ~300 mrad. The other,[16] produced by N. Lockyer and P. Karchin, puts more emphasis on the central "barrel" region. Both use a great deal of silicon tracking along with a transverse dipole magnetic field, and triggering via high-p_T electrons.

One of the biggest problems of doing this physics at Fermilab, other than support, is where to do it. Finding an independent collision region is problematic. At present the leading option is AØ, the locale of injection from the booster and of extraction to fixed-target experiments. Considerable work has to be done in order to determine whether this is a viable option.

The most detailed documentation[8] for generic $b\bar{b}$ collider experimentation exists for the superconducting super collider (SSC). At SSC energies, at least one event in 10^3 should contain $b\bar{b}$ pairs. Given a luminosity $\lesssim 10^{32}\ \text{cm}^{-2}\ \text{sec}^{-1}$, implying an event rate perhaps usable for open-geometry measurements, one finds a yield of at least 3×10^{10}

produced $b\bar{b}$ pairs per experiment. Again the problem of digging them out is a great challenge. A sample experiment (the TASTER) is described[17] by B. Cox and D. Wagoner in the 1986 Snowmass proceedings. Most $b\bar{b}$ pairs, even at the high SSC cms energy, are produced reasonably centrally (i.e., within an interval of ± 3 units of rapidity). Thus back-to-back detectors, each looking like typical fixed-target spectrometers, are a reasonable architecture. The TASTER (FIGURE 3) utilizes a high-p_T lepton and/or ψ trigger to select B candidates. Silicon planes up close are used to find the secondary vertices. It is remarkable how similar the TASTER is to Cox's fixed-target experiment E705/E771, and Reay's Tev I detector layout is to his experiment E653. This exhibits how similar the detection problems at the colliders are to those in external hadron beams. Or perhaps it is an indication of the advancing age of the proponents.

SUMMARY: SOME OPINIONS ON LONG-RANGE POSSIBILITIES

Doing generic $b\bar{b}$-physics via photoproduction and hadroproduction is, to say the least, in a primitive state. Nevertheless, the long-range possibilities are technology-limited, not rate-limited. The numbers of $b\bar{b}$ pairs that are produced per experiment in hadron–hadron collisions, both at colliders and even in external beams, can be far in excess of what can be anticipated from *any* e^+e^- collider. The technology centers on (a) precise event-by-event identification of secondary vertices via microvertex techniques, (b) high rate-capability, and (c) sophisticated on-line data processing and event selection. All these technologies are now progressing rapidly, and it is not easy to project where the ultimate limitations will be.

Probably the e^+e^- colliders will remain the principal source of information on the $b\bar{b}$ system for some time. For hadroproduction experiments to truly compete, one should strive for experiments in which the number of $b\bar{b}$ produced per experiment is greatly in excess of 10^6. This is not to say exploratory experiments are not important. They are needed as the first step. And $\gtrsim 10^5$ produced $b\bar{b}$ per experiment implies $\gtrsim 10^8$ $c\bar{c}$ produced, far in excess of e^+e^- capabilities. Nevertheless, I think the long-range goal should be kept clearly out in front.

I also suspect that the architecture and scale of detectors to do this physics may well be large compared to what is found in traditional fixed-target initiatives. A criterion of good acceptance for a fixed-target spectrometer is that it cover a solid angle of $\gtrsim 6\pi$. By this I mean ~ 5 units of rapidity should be well covered; a typical "4π" detector covers about 3 units. For the Tevatron fixed-target application my choice of angular range would be 5 mrad $< \theta <$ 500 mrad. This large angular range is important not only because produced b's are found in a range of rapidities, but because of the aforementioned tagging problem. This requires finding not only the products of the "trigger" b but the debris of its companion $\bar{b}$. The "triggering" b may be disciplined into falling within the optimal portion of the spectrometer acceptance. But that usually will not be the case for the spectator.

Another simpleminded way of judging fixed-target spectrometer architectures is to view them in a boosted reference-frame, where the b-hadron of interest is produced at rest. It is easy to see that the resolution of conventional charged-particle tracking elements is boost-invariant. Thus in this boosted frame, one might well demand a detector volume, acceptance, pattern-recognition capability, etc., at least as good as one would have in a generic 4π detector such as CLEO or Argus. The reader is invited to try this for existing detectors.

This line of argument can be reversed. If one takes CLEO or Argus as a suitable architecture and boosts it into the laboratory frame, it should be a suitable detector. The solenoid and barrel calorimetry is stretched by a factor $\gtrsim\gamma$ and the downstream endwall is a factor 2γ further away from the collision point.[c] For $\gamma \sim 20$, appropriate for Tevatron fixed-target applications, this makes a compact 2-by-2-meter collider-frame detector into a solenoid and barrel calorimeter $\gtrsim$40 m in length and $\gtrsim$2 m in diameter. To do just this would require doubling the world inventory of 4π-detector solenoids and barrel calorimeters. More efficient is to replace the calorimeter barrels with an endwall of diameter 6 m, plus two more upstream walls each of the same diameter and each with a 2-m-diameter hole. If supplemented with three 2-m-diameter magnets with transverse *B*-fields (dipole or quadrupole; my choice is quadrupole), this would provide excellent calorimetry and magnetic analysis over all of phase space, with perhaps ample space for nondestructive particle identification as well.

I would not claim this is anywhere near optimal. But even at this extravagant level, we are talking about an investment maybe twice that of a big collider detector. And no matter how one proposes to do quality *b*-physics, the threshold investment is probably large. The large acceptance and very large amount of information per event that *must* be acquired implies a big investment in readout and data processing electronics alone.

Whatever the fixed-target detector turns out to be, for collider applications one might contemplate a similar $12\pi = 2 \times 6\pi$ back-to-back pair to cover both forward and backward directions (we here assume that planar geometry dominates because most decay products of useful *b*-hadrons have $|\theta| < 45°$). An extra 2π or π of solid angle for a central barrel region would also be of use.

The similarity of each detector arm with the fixed-target detectors suggests that fixed-target initiatives can be a useful bridge to the ultimate, more difficult collider case. "Collider-compatibility" of fixed-target initiatives would imply no active target and transmission of the primary beam through the detector without traversal of detection elements.

But this is more than enough speculation about detectors. I ask the reader's forgiveness for such naive indulging in such experimental design fantasies. But the main point is that maybe it is already time to think big. I cannot imagine this kind of physics going into early obsolescence. Indeed, how likely is it that the new physics at the teraelectronvolt mass scale will shed significant light on the origin of *CP*-violation? The Argus results encourage the hope that in the long run *CP*-violation in the *b*-system can be observed. Maybe it is already time to throw deep.

ACKNOWLEDGMENTS

Among our many colleagues who have helped educate me in these matters, I especially want to thank B. Cox, N. Reay, D. Jovanovic, T. Nash, and J. Slaughter.

[c] The factor 2 originates from the fact that a fixed-target endwall boosted to the collider frame moves toward the collision point at the speed of light.

REFERENCES

1. For a review, see REUCROFT, S. 1986. *In* Proceedings of the 6th Int. Conf. on Physics in Collision. Chicago, Ill.
2. See, for example, ELLIS, S. & S. QUIGG. 1987. Fermilab preprint FN-445.
3. COTEUS, P. *et al.* 1986. Univ. of Colorado preprint COLO-HEP-140.
4. ALBRECHT, H. 1987. New results on B-$\bar{B}$ mixing from Argus. Presented at the 1st Int. Symp. for the Fourth Family of Quarks and Leptons. Los Angeles, Calif. February 26–28, 1987.
5. GILMAN, F. J. 1987. Overview of the Standard Model. This volume; ELLIS, J., J. HAGELIN & S. RUDAZ. 1987. CERN-TH4679/87; BARGER, V., T. HAN, D. NANOPOULOS, R. PHILLIPS. 1987. Univ. of Wisconsin preprint MAD/TH/87-12.
6. CARTER, A. & A. SANDA. 1980. Phys. Rev. Lett. **45:** 952.
7. ROSNER, J. 1987. Univ. of Chicago preprint EFI 87-9 and references therein.
8. CRONIN, J. *et al.* 1984. *In* Proceedings of the 1984 Summer Study on the Design and Utilization of the Superconducting Super Collider. Snowmass, Colo. R. Donaldson, Ed. COX, B. *et al. In* Proceedings of the 1986 Summer Study on the Design and Utilization of the Superconducting Super Collider. Snowmass, Colo. R. Donaldson & J. Marx, Eds.
9. See, for example, REAY, N. Ohio State Univ. Report (unpublished); also reference 8.
10. FISCHER, C., Spokesman. 1987. OMEGA-SCIFI proposal SPSC 87-2; SPSC/P266.
11. ALBANESE, J. *et al.* 1985. Phys. Lett. **15B:** 186.
12. CATANESI, M. *et al.* 1987. Phys. Lett. **187B:** 431.
13. ANJOS, J. *et al.* 1987. Phys. Rev. Lett. **58:** 311, 1818.
14. SCHLEIN, P. *et al.* 1986. Letter of intent to CERN SPSC, November 1986.
15. REAY, N., Spokesman. Fermilab letter of intent P783; REAY, N. reference 9.
16. LOCKYER, N. & P. KARCHIN. Fermilab letter of intent P784.
17. COX, B. & D. WAGONER. 1986. *In* Proceedings of the 1986 Summer Study on the Design and Utilization of the Superconducting Super Collider. Snowmass, Colo. R. Donaldson & J. Marx, Eds.

In Search for the Beyond

LING-LIE CHAU

Department of Physics
University of California
Davis, California 95616

INTRODUCTION

To search for the beyond it is best to begin with what we know for sure, which can serve as a base for making guesses at what is coming. This conference has seen many new results, including the existing $B^0 \leftrightarrow \bar{B}^0$ mixing, and the new measurements of $K_{L,S} \to 2\gamma$. Comments on these experimental results are included. In analyzing recently measured exclusive two-body decays of charm mesons, we demonstrate the uses of the model-dependent Quark Diagram Scheme. Predictions are made for future experimental measurements. We then point out specific beauty particle nonleptonic decays for experimental measurements to establish the nonzeroness of $|V_{ub}|$, and to measure $|V_{ub}/V_{cb}|$. Some interesting decays, which are from W-loop diagrams only and which are easy to detect experimentally, are also pointed out. General properties of partial decay rate differences in charm, beauty, and t particles are discussed, also using the Quark Diagram Scheme. For the search for CP noninvariance in the decay amplitudes, we give some interesting B' decays, with all charged particles in the final states. We point out some highlights in the event that more than three generations of quarks exist.

B-MESON MIXING REVISITED

The $B_d^0 \leftrightarrow \bar{B}_d^0$ mixing has been observed by the Argus group,[1] $r_{B_d} = 0.23 \pm 0.11$, and limits on the $B_s^0 \leftrightarrow \bar{B}_s^0$ mixing have been set.[2] Theoretical studies on the mixing in the B-meson system have been carried out for quite some time.[3–7,a] It was apparent quite early that large B_s^0-mixing was more likely than B_d^0-mixing, given reasonable $B_B f_B^2$, and t-quark mass m_t. The appreciable B_d^0-mixing observed by Argus came as a surprise, because recent theoretical analyses were biased toward choosing parameters such that B_d^0-mixing was negligibly small. Here we[8] analyze the experimental results according to some interesting features noted in references 5–7 and footnote a in the Kobayashi–Maskawa (KM) case of three generations of quarks.

Besides the obvious sensitive dependence of B^0-mixing on $f_B^2 B_B$ and on the t-quark mass m_t, it was demonstrated in references 5–7 and footnote a that the B_d^0-mixing is particularly sensitive to the value of the CP-invariance violating phase, and it was shown that the B_d^0-mixing can be substantial when the CP-invariance violating phase is

[a]Reference 6, fig. 23; reference 7, fig. 3.5; reference 8, fig. 4.

more than 90°.[a] The phase dependence of B_d^0 mixing is actually more dramatic than the dependence on m_t, given the current experimentally allowed range of $|V_{ub}/V_{cb}|$. When the value of $|V_{ub}/V_{cb}|$ is close to $|V_{us}|$, which is a realistic possibility from current experimental limits[9] of $|V_{ub}/V_{cb}| < 0.23$, the B_d^0-mixing increases drastically when the *CP*-invariance violating phase increases from 0° to beyond 90°. Such behavior can provide the scenario that, for some suitable range of values for $B_B f_B^2$ and m_t, B_d^0-mixing is substantial as given by recent experiments, yet B_s-mixing is still less than maximal. Actually for $|V_{ub}/V_{cb}| \approx 0.63$ and the *CP*-invariance violating phase being near 180°, $r_{B_d} \approx r_{B_s}$, independent of m_t, if we take $B_{B_d} f_{B_d}^2 \approx B_{B_s} f_{B_s}^2$; for $|V_{ub}/V_{cb}| > 0.63$, even $r_{B_d} > r_{B_s}$ is possible. On the other hand, if the observed B_d-mixing is not because of large ϕ near 180° and substantial $|V_{ub}/V_{cb}|$, but because of small ϕ, <90°, and large Bf_B^2 and/or large m_t, the B_s^0-mixing will be close to the maximum. Because of this sensitivity, the accurate measurements of B_d^0-mixing and B_s^0-mixing may provide information on the quark-mixing matrix and the *CP*-invariance violating phase, and on weak decay dynamics.

Here we give the formula for B^0-mixing

$$r_B \equiv \frac{P(B^0 \rightarrow l^-)}{P(B^0 \rightarrow l^+)} = \frac{P(e^+e^- \rightarrow B^0B^0 \rightarrow l^-l^-)}{P(e^+e^- \rightarrow B^0B^0 \rightarrow l^+l^-)} = \frac{(\Delta m/\tilde{\Gamma})^2 + (\Delta\Gamma/2\tilde{\Gamma})^2}{(\Delta m/\tilde{\Gamma})^2 - (\Delta\Gamma/2\tilde{\Gamma})^2 + 2}. \quad \textbf{(1.1)}$$

The large mixing came about either through $\Delta\Gamma/2\tilde{\Gamma} \approx 1$, which was the case in the *K*-meson case, that is, $\Delta\Gamma_K \approx \Gamma_{K_s} \approx 2\tilde{\Gamma}$, or through $\Delta m/\tilde{\Gamma} \gg 2$. In either case, the box graph that gives Δm and $\Delta\Gamma$ must compete with the decay graphs that give $\tilde{\Gamma}$. Thus mixing is more favorable in the *B*-meson-like systems, where the tree graphs are suppressed due to the quark-mixing suppression. In contrast, the mixings in the charm mesons and top mesons are less favorable, because the box graph is suppressed when compared with the tree graphs. These were clearly discussed in references 5–7.

For the B^0-mesons, it is reasonable to assume that Δm mainly comes from the box graph, the so-called short-distance contributions, which is dominated by the *t*-quark term

$$\Delta m \approx \frac{G_F^2 B_B f_B^2 m_b}{6\pi^2} m_t^2 F\left(\frac{m_t^2}{m_w^2}\right) |V|^2, \quad \textbf{(1.2)}$$

where $|V|^2 = |V_{td}V_{tb}|^2$, for B_d-mixing, $|V|^2 = |V_{ts}V_{tb}|^2$, for B_s-mixing, and

$$F(x) = 1 - \frac{(3/4)(x + x^2)}{(1 - x)^2} - \frac{(3/2)x^2 \ln x}{(1 - x)^3},$$

which is a smooth and slowly decreasing function. The $\Delta\Gamma$ comes from the absorptive part of the box diagram, and therefore the *t*-quark intermediate state cannot contribute. Thus $\Delta\Gamma \ll \Delta m$ is expected, and equation 1.1 simplifies to

$$r_B \approx \frac{(\Delta m/\tilde{\Gamma})^2}{(\Delta m/\tilde{\Gamma})^2 + 2}. \quad \textbf{(1.3)}$$

It is obvious that as $B_B f_B^2$ and m_t^2 increase, the mixing r_B increases. The reason that

B_s-mixing in general is bigger than B_d-mixing is because $|V_{ts}|^2 > |V_{td}|^2$. What was noted in references 5–7 is that $|V_{td}|$ is very sensitive to the CP-invariance violating phase.[a] This is most easily seen using the parametrization given in reference 10:

$$V = \begin{pmatrix} c_x & s_x & s_z e^{-i\phi} \\ -(s_x + s_y s_z e^{i\phi}) & c_y & s_y \\ s_y\left[s_x - \left(\frac{s_z}{s_y}\right)e^{i\phi}\right] & -s_y\left[1 + s_x\left(\frac{s_z}{s_y}\right)e^{i\phi}\right] & 1 \end{pmatrix}. \qquad (1.4)$$

The advantage of such parametrization is that each of s_x, s_y, s_z is the magnitude of the mixing matrix element, which is directly measurable: $s_x \approx |V_{us}|$, $s_y \approx |V_{cb}|$, $s_z \approx |V_{ub}|$, and the phase factor appears with small factors. Now we see that it is also particularly suitable for the discussions here.

From the expression 1.4 we see that $|V_{td}| = s_y[|s_x - (s_z/s_y)\cos\phi|^2 + |(s_z/s_y)\sin\phi|^2]^{1/2}$ is very sensitive to the sign of $\cos\phi$, and it increases as ϕ increases from 0° to 180°. Recently, experimental upper limits on $|V_{ub}/V_{cb}| = s_z/s_y$ have been loosened from 0.14 to 0.23.[9] As the value of $|V_{ub}/V_{cb}|$ becomes closer to the value of $|V_{us}|$, the dependence of $|V_{td}|$ on ϕ becomes more dramatic. The dependence of $|V_{ts}|$ on ϕ is the opposite, that is, it decreases as ϕ increases from 0° to 180°. However, since $s_x(s_z/s_y) < 1$, so the dependence of $|V_{ts}|$ on ϕ is less than that of $|V_{td}|$.

In FIGURE 1a and 1b we show some examples of B_d-mixing, B_s-mixing as a function of $\cos\phi$ for various values of $|V_{ub}/V_{cb}|$. We see that if $|V_{ub}/V_{cb}|$ is large enough, as allowed by the current experimental limit, we can have the interesting scenario that B_d-mixing is $r_{B_d} \approx 20\%$, as given by the Argus measurement, but B_s-mixing is still less than maximal, that is, $r_{B_s} < 1$. To emphasize the importance of the value of $|V_{ub}/V_{cb}|$, we also show in this figure that, if $|V_{ub}/V_{cb}| = 0.63$, $r_{B_d} \approx r_{B_s}$ near $\phi \simeq 180°$, as we assume $B_d f^2_{B_d} \approx B_s f^2_{B_s}$. For $|V_{ub}/V_{cb}| > 0.63$, r_{B_d} can become bigger than r_{B_s} as ϕ increases toward 180°. It is interesting to note from the two parts of this figure that the allowed value of m_t can be rather low, with Bf^2_B still within the reasonable range of current theoretical guesses. All these show that it is most important to make accurate measurements of r_{B_d} and r_{B_s}.

It is important to note that the scenario discussed here is allowed by the constraints on the quark-mixing parameters from CP-invariance violating effects $\epsilon \approx 2.3 \times 10^{-3}$ and $\epsilon'/\epsilon = 0.0035 \pm 0.003 \pm 0.002$.[11] As shown in reference 10, in the case of three generations of quarks, all CP noninvariance effects depend upon

$$X_{cp} = c_x c_y c_z^2 s_x s_y s_z s_\phi \approx s_x s_y s_z s_\phi,$$

which is the unique phase-convention independent CP-invariance violating parameter. So for given CP noninvariance effects ϵ,ϵ'/ϵ, only the product $s_z s_\phi$ is determined, since s_x,s_y are given.[12] Therefore the region of small s_ϕ, ϕ near 180°, and large s_z are allowed from the calculation of ϵ and ϵ'/ϵ. Eventually it is conceivable that the accurate measurements of B_d-mixing and B_s-mixing, together with better knowledge[13] of $|V_{ub}/V_{cb}|$ can help to determine s_ϕ,s_z separately for given Bf^2_B and m_t, or help to measure the dynamical parameter Bf^2_B and m_t.

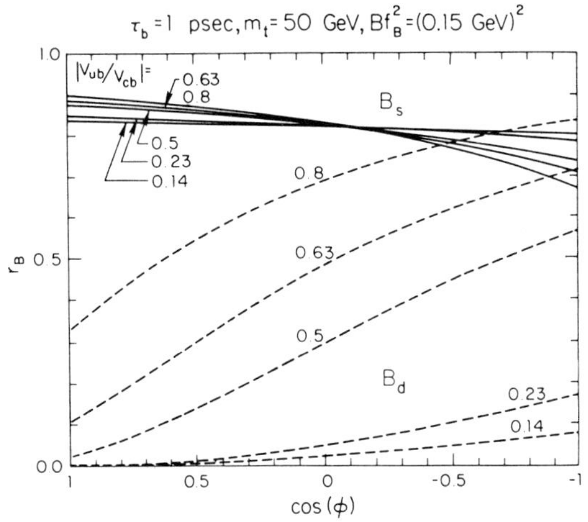

FIGURE 1. (a) The B_d-mixing r_{B_d} and the B_s-mixing r_{B_s} are plotted as a function of cos ϕ for various values of $|V_{ub}/V_{cb}|$ [note that 0.23 is the current experimental limit of $|V_{ub}/V_{cb}|$, see reference 10], given the input of the τ_b = 1 psec, t-quark mass m_t = 50 GeV, the dynamical-coupling parameter $B_d f_{B_d}^2 \approx B_s f_{B_s}^2 = (0.15\text{ GeV})^2$.

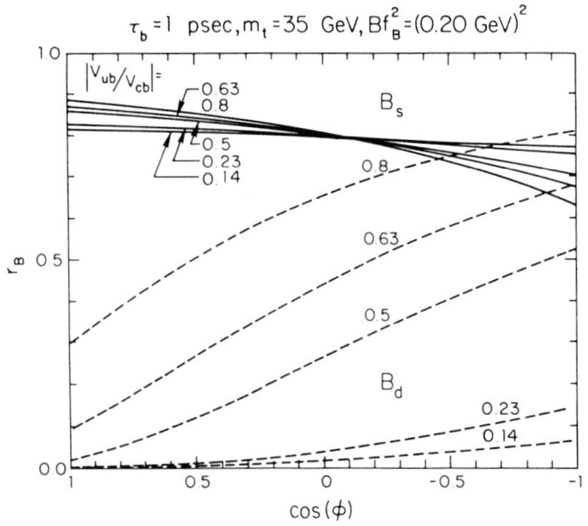

FIGURE 1. (b) The same as FIGURE 1a, except m_t = 35 GeV, and $B_d f_{B_d}^2 \approx B_s f_{B_s}^2 = (0.20\text{ GeV})^2$.

FURTHER COMMENTS ON $K^0_{L,S} \rightarrow 2\gamma$

Dr. Holder reported their new results:[14]

$$\frac{\Gamma(K_S \rightarrow 2\gamma)}{\Gamma(K_L \rightarrow 2\gamma)} \equiv R = 2.4 \pm 1.1$$

$$\Gamma(K_L \rightarrow 2\gamma) = (6.05 \pm 0.04 \pm 0.08) \times 10^{-4}$$

have confirmed early theoretical prediction of R. This agreement indicates that $K_S \rightarrow 2\gamma$ is dominated by 2π intermediate states (and no substraction in the π-loop calculations and $A(K \rightarrow 2\pi)$ is at its on-shell value in the π-loop). Further, even more interesting is the fact that $K_S \rightarrow 2\gamma$ does not overwhelm $K_L \rightarrow 2\gamma$ and allows the possibility of observing the existing new decay-amplitude CP noninvariance as proposed in reference 14.

As discussed in reference 14, since $K_S \rightarrow 2\gamma$ is dominated by 2π intermediate states, the decay-amplitude CP noninvariance is the same as in $K \rightarrow 2\pi$, that is, the hotly pursued ϵ'. Unfortunately ϵ'/ϵ decay-amplitude vs. mass-matrix CP noninvariance is inhibited to begin with by a factor of 20 due to the known $\Delta I = \frac{1}{2}$ rule. So we expect that CP noninvariance in $K_S \rightarrow 2\gamma$ is given by the mass-matrix CP noninvariance ϵ. However, the mechanism of $K_L \rightarrow 2\gamma$ is through the rather complex contributions from various poles: π, η, η', and 3π intermediate states. It was shown in reference 15 that decay-amplitude CP noninvariance in $K_L \rightarrow 2\gamma$ can be as big as the mass matrix CP noninvariance (unlike in ϵ'/ϵ, there is the $\Delta I = \frac{1}{2}$ suppression of $\frac{1}{20}$ to begin with). However, in order to observe this new kind of CP noninvariance in $K_L \rightarrow 2\gamma$, it must not be overwhelmed by $K_S \rightarrow 2\gamma$. It was shown in reference 14, for small enough R, for example, that $R = 2$, it is possible to see the new decay-amplitude CP noninvariance. Since unpolarized 2γ has both CP even and odd states, it was pointed out and emphasized in reference 14 that the appropriate way to measure the new decay-amplitude CP noninvariance is to measure partial decay rate differences of tagged K^0, $\bar{K}^0$, which is ideal for the current LEAR experiment. FIGURE 2 shows the CP-violating effects in partial rate differences.

THE QUARK DIAGRAM SCHEME, AND THE ANALYSIS OF EXCLUSIVE TWO-BODY DECAYS OF CHARM MESONS

It has been shown[16,17] that all meson decays can be expressed model independently in terms of six quark diagrams (shown in FIGURE 3): $\mathcal{A}$, the external W-emission diagram; $\mathcal{B}$, the internal W-emission diagram; $\mathcal{C}$, the W-exchange diagram; $\mathcal{D}$, the W-annihilation diagram, which we call the W tree diagrams; $\mathcal{E}$, the horizontal W-loop diagram; $\mathcal{F}$, the vertical W-loop diagram. These quark diagrams are specific, well-defined physical quantities. They are classified according to the topology of weak interactions, with only the quark and intermediate bosons explicitly shown. All quantum chromodynamics (QCD) strong-interaction effects are there though not explicitly shown. The QCD effects can also be treated perturbatively, for example, the one-gluon-exchange diagram of the $\mathcal{E}$ graph is the so-called penguin diagram. These diagrams keep their identities under all QCD effects. Only so much classification is useful. It has recently been successfully used[18] to analyze two-body charm decay data, to compare experimental data with various theoretical models to check their validity, and further to make predictions and to suggest new experimental measurements.

We begin with the *PV* decays because of the relative simplicity in presenting the discussion, though the data of *PV* decays are not yet as good as those for some of the *PP* decays. The simplicity in discussing the *PV* decays comes from the purity of the quark contents in ϕ and ω. Many *PV* decays are given by one type of amplitude, for example, $F^+ \longrightarrow \phi\pi^+(\alpha a')$, $D^+ \longrightarrow \phi\pi^+(\alpha b')$, $D^0 \longrightarrow \phi\bar{K}^0(\alpha\bar{c}')$, respectively. Thus from the decay rates, using the known lifetime measurements, we can determine their absolute values:

$$|a'| = (2.50 \pm 0.42) \times 10^{-6}, \qquad |b'| = (3.67 \pm 0.51) \times 10^{-6},$$
$$|\bar{c}'| = (1.68 - 2.10) \times 10^{-6}. \qquad \textbf{(3.1)}$$

The only theoretical assumption used here is that $|\exp(i\delta^{\phi\pi})| = 1 = |\exp(i\delta^{\phi\bar{K}})|$. Note here that neither amplitude b' nor amplitude $\bar{c}'$ is negligible, as preferred by some model calculations.

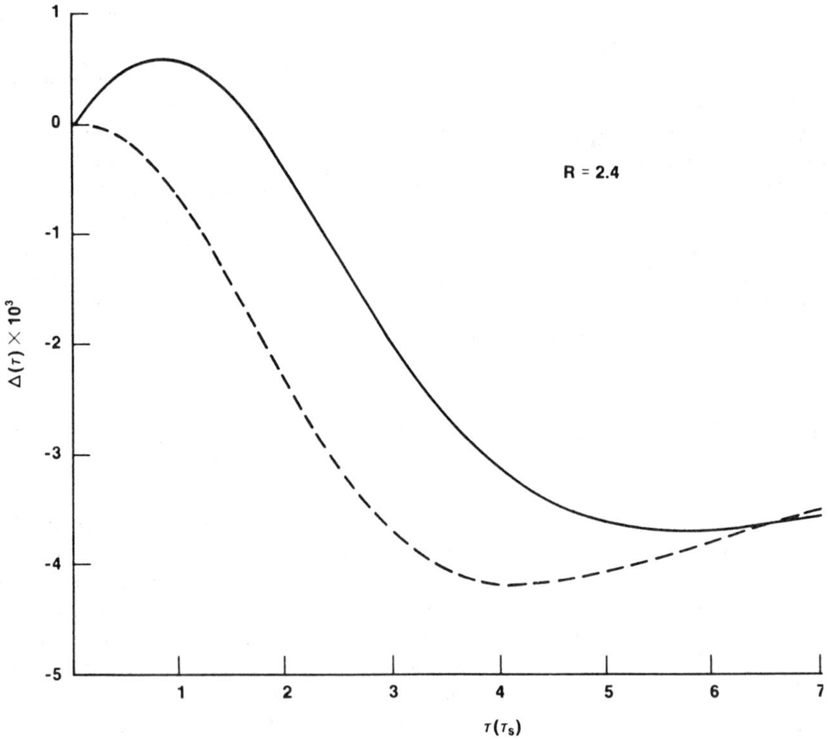

FIGURE 2. The time evolution for the partial rate difference $\Delta(\tau)$. The solid line is from the superweak model, and the broken curve corresponds to $\rho = 0.35$ and nonzero t_0 (for parameters see the text).

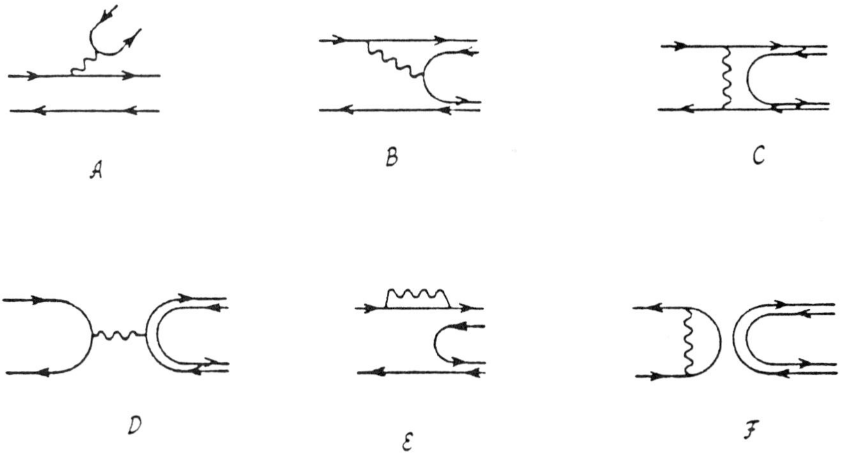

FIGURE 3. The six quark diagrams for a meson ⟶ two mesons.

The expression $D^+ \longrightarrow \bar{K}^{*0}\pi^+[\alpha(a' + b')]$ is an exotic channel that implies elastic and small $\delta^{\bar{K}^*\pi}_{3/2}$. Therefore, the rate gives

$$|a' + b'| = (1.16 \pm 0.37) \times 10^{-6}. \tag{3.2}$$

Combining equations 3.1 and 3.2, we obtain

$$a' = (2.50 \pm 0.42) \times 10^{-6}, \qquad b' = -(3.67 \pm 0.51) \times 10^{-6}. \tag{3.3}$$

From $D^0 \longrightarrow K^{*-}\pi^+, \bar{K}^{*0}\pi^0$ measurements we obtain the following two solutions:

$$\frac{(a' + c')}{(a' + b')} = 2.36 \pm 0.67 \longrightarrow c' = -(5.25 \pm 0.39) \times 10^{-6},$$

$$\Delta_{\bar{K}^*\pi} = (52^{+30}_{-52})^\circ, \tag{3.4a}$$

$$\frac{(a' + c')}{(a' + b')} = -1.70 \pm 0.67 \longrightarrow c' = -(0.53 \pm 0.39) \times 10^{-6},$$

$$\Delta_{\bar{K}^*\pi} = 180^\circ - (52^{+30}_{-52})^\circ, \tag{3.4b}$$

We note that the errors on $\Delta_{\bar{K}^*\pi}$ are so large that the data can be accommodated by real amplitudes without final-state interactions. The amplitude c' is quite different from $\bar{c}'$ in equation 3.1. (Amplitudes with tildes involve strange-quark and -antiquark pair production.) To make definite conclusions, we need better measurements.

One nice prediction from this analysis is $B(D^0 \longrightarrow \phi\pi^0) \simeq 0.21\%$, from $D^0 \longrightarrow \phi\pi^0(\alpha b')$. This will be an important measurement if we are to test this scheme.

Next we proceed to determine the unprimed amplitudes from $D \longrightarrow p\bar{K}$ and $D^0 \longrightarrow \omega\bar{K}$ decays. The reactions $D^+ \longrightarrow \rho^+\bar{K}^0$, $D^0 \longrightarrow \omega\bar{K}^0$ determine

$$|a + b| = (2.18 \pm 0.25) \times 10^{-6}, \qquad |b + c| = (2.57 \pm 0.51) \times 10^{-6}. \tag{3.5}$$

From the measurements of $D^0 \to \rho^0\bar{K}^0$, ρ^+K^-, we find two solutions for $(a + c)/(a + b)$. Combining all those results we obtain the following three possible solutions for amplitudes a, b, and c: two for $\Delta_{\rho\bar{K}} = (24^{+25}_{-24})°$,

$$a = (4.06 \pm 0.38) \times 10^{-6}, \qquad b = -(1.88 \pm 0.28) \times 10^{-6},$$

$$c = -(0.68 \pm 0.28) \times 10^{-6}; \tag{3.6a}$$

$$a = (1.50 \pm 0.38) \times 10^{-6}, \qquad b = (0.68 \pm 0.28) \times 10^{-6},$$

$$c = (1.89 \pm 0.28) \times 10^{-6}, \tag{3.6b}$$

and one for $\Delta_{\rho\bar{K}} = 180° - (24^{+21}_{-24})°$,

$$a = (1.41 \pm 0.38) \times 10^{-6}, \qquad b = (0.77 \pm 0.28) \times 10^{-6},$$

$$c = -(3.34 \pm 0.28) \times 10^{-6}, \tag{3.6c}$$

The measurement of $D^+ \to \bar{K}^{*0}K^+$, gives us $|a' - \bar{d} + \delta_e| = (2.70 \pm 0.61) \times 10^{-6}$.

Future measurements of $\bar{K}^{*0}\eta_8[\alpha(b' + c' - 2\bar{c})]$ and $\bar{K}^{*0}\eta_0[\alpha(b' + c' + \bar{c})]$ can help to determine amplitudes $\bar{c}$ and $(b' + c')$, and then c', since b' is known. From future measurements of $D^0 \to \bar{K}^{*0}K^0$, $\bar{K}^{*0}\bar{K}^0[\alpha(c - c')]$, and the known information on c' we can determine the amplitude c. Then we can check which solution of equation 3.6 will be picked, and thus determine a and b individually. From $F^+ \to \rho^+\pi^0[\alpha(d - d')]$, $F^+ \to \omega\pi^+[\alpha(d + d')]$, we can determine d and d'. We then know all the amplitudes a, b, c, d, and a', b', c', d', and their relative signs. The rest of the PV decays are predictable, up to $SU(3)$ breaking and final-state interactions. These results must be conformed to by any theoretical calculations.

For results of charm meson decays into two pseudoscalar mesons, see reference 18.

QUARK MIXING MATRIX AND THE SEARCH FOR $|V_{ub}|$ AND $|V_{ub}/V_{cb}|$ FROM THE QUARK DIAGRAM SCHEME

Since the proposal of the Kobayashi–Maskawa (KM) scheme of quark mixing for CP noninvariance[19] and the discovery of the beauty particles,[20] much progress has been made in measuring the quark mixing matrix in the case of three generations of quarks. Besides the urgently needed observation of the t-quark and its decays to measure V_{tb}, V_{ts}, V_{td}, the measurement of V_{ub} and the verification of its nonzeroness is currently most urgent. If $|V_{ub}|$ is too small or zero, higher than three generations of quarks will be needed if we still want to describe CP noninvariance through the quark mixing phenomena. Theoretically the neatest way to measure $|V_{ub}|$ is to observe $B_u \to \tau\nu_\tau$. Unfortunately the branching ratio of such decay is very small $\mathrm{BR}(B_u \to \tau\nu_\tau) \lesssim 10^{-5}$, using the current bound on $|V_{ub}/V_{cb}|^2 < (1/5)^2$. It is therefore important to find other decay channels that are nonzero only when $|V_{ub}|$ is not zero. Certainly the observation of charmless semileptonic decay $b \to l^-\bar{\nu}_l X_{c=0}$ inclusively or exclusively will establish $|V_{ub}| \neq 0$. Here we point out some pure nonleptic decays that are nonzero only if $|V_{ub}| \neq 0$, and some ratios of nonleptonic decays that are proportional to $|V_{ub}/V_{cb}|$.

Such a selection is possible based upon the general model-independent Quark Diagram Scheme.

This scheme can also provide a model-independent way to pick out nonleptonic decays of beauty particles, which are nonzero only if $|V_{ub}|$ is nonzero. The result is that the following decay channels are purely proportional to $|V_{ub}|$:[21] $B_{d,u} \to F^+X_{c=0,s=0}$ or $DX_{c=0,s=1}$, which include exclusive channels like:

$$\begin{aligned} B_u^+ &\to F^+\pi^0, F^+\rho^0, F^+\pi^+\pi^-, F^+\phi, F^+K^+K^-, F^+\eta_8, F^+\eta_0, F^+\eta_c, F^+J/\psi, \\ &\quad \ldots; D^0K^+, D^+K^0, D^{*0}K^+, D^0K^{*+}, D^{*0}K^{*+}, \ldots \\ B_d^0 &\to F^+\pi^-, F^+\rho^-, F^+\pi^-\pi^0, \ldots; D^0K^0, D^{*0}K^0, \\ &\quad D^0K^+\pi^-, D^{*0}K_d^{0*}, D^{*0}K^{0*}, \ldots . \end{aligned} \tag{4.1}$$

$B_s \to FX_{c=0,s=-1}$, or $DX_{c=0,s=0}$, which include exclusive channels like:

$$\begin{aligned} B_s^0 \to\ & F^+K^-, F^+K^{*-}, F^+\overline{K}^0\pi^-, \ldots; D^0\pi^0, D^0\eta_8, D^0\eta_0, D^0\phi, D^{*0}\phi, \\ & D^{*0}\eta_8, D^{*0}\eta_0, D^0\eta_c, D^0J/\psi, \ldots; D^+\pi^-, D^+\rho^-, D^+\pi^-\pi^0, \\ & D^{*+}\pi^-, D^{*+}\rho^-, \ldots . \end{aligned}$$

They typically have a branching ratio of 10^{-4}, which is within the reach of current experimental capability. Note that all these channels have a charm particle in the final states.

Recently, bounds on charmless nonleptonic decays like $B^0 \to \pi^+\pi^-$ have been used as direct measurement of bound on $|V_{ub}|$. Caution must be given in such a procedure. Observation of charmless beauty particle decays does not necessarily mean $|V_{ub}| \neq 0$. The important point is that those charmless decays do not have to be zero if $|V_{ub}| = 0$ according to the general Quark Diagram Scheme, since they have W-loop-diagram contributions with the dominant coefficient V_{cb}. Only if we can establish experimentally that the W-loop-diagram contributions are negligibly small compared to the tree-diagram contributions can charmless decays give indications on $|V_{ub}|$. The Quark-Diagram Scheme is again useful in finding ways to study the contributions of the W-loop diagram. It can be shown that the following decays can have only W-loop-diagram contributions, even with the dominant coefficient $V_{cs}V_{cb}^*$:

$$\begin{aligned} &B_d^0 \to K^0\phi, K^{*0}\phi; \qquad B_s^0 \to \overline{K}^0\phi, \overline{K}^{*0}\phi; \\ &B_{d,s}^0 \to \phi\phi, K^0K^0, K^{*0}\overline{K}^0, K^0\overline{K}^{*0}, K^{*0}\overline{K}^{*0}. \end{aligned} \tag{4.2}$$

Experimentally studying these decays can give direct measurements of the size of the W-loop diagrams.

Once $|V_{ub}|$ is established to be nonzero, we ought to determine its precise value. It is then worthwhile to measure the dynamically simplest $B \to \tau\nu_\tau$ in a dedicated experiment. Another way is to find other b particle decays that have the same Quark-Diagram Scheme amplitudes as previously listed, except with known mixing matrix elements. Then the ratios of such decays can give ratios of $|V_{ub}/V_{cb}|$. The idea here is similar to a previous well-known case in charm decays[6] $\tilde{\Gamma}(D^+ \to \pi^0\pi^+)/$

$\tilde{\Gamma}(D^+ \rightarrow \overline{K}^0\pi^+) = \frac{1}{2}|V_{cd}/V_{cs}|^2$. The ratios of decay widths that give $|V_{ub}/V_{cb}|$ are, for example,

$$\frac{2\tilde{\Gamma}(B_u^+ \rightarrow F^+\pi^0)}{\tilde{\Gamma}(B_d^0 \rightarrow D^-K^+)} = \frac{\tilde{\Gamma}(B_d^0 \rightarrow D^0K^0)}{\tilde{\Gamma}(B_d^0 \rightarrow \overline{D}^0K^0)} = \frac{\tilde{\Gamma}(B_s^0 \rightarrow F^+K^-)}{\tilde{\Gamma}(B_s^0 \rightarrow F^-K^+)}$$

$$= \frac{2\tilde{\Gamma}(B_s^0 \rightarrow D^0\pi^0)}{\tilde{\Gamma}(B_s^0 \rightarrow D^-\pi^+)} = \left|\frac{V_{cs}}{V_{us}}\right|^2 \cdot \left|\frac{V_{ub}}{V_{cb}}\right|^2. \tag{4.3}$$

Here we ought to be careful about final-state interactions. It is, however, very likely that final-state interactions are negligible, since the B-masses are far beyond known resonances.

CP NONINVARIANCE OF PARTIAL DECAY RATE DIFFERENCES FROM THE QUARK-DIAGRAM SCHEME

Despite many recent dedicated searches, the original observation made in 1964 of the minute *CP* noninvariance effects of $K_L \rightarrow 2\pi$,[22] and in $K_L \rightarrow \pi^\pm e^\mp \nu_e$ stays still as the only system and the sole kind of *CP* noninvariance. Yet *CP* noninvariance makes up one of the essential ingradiences in explaining the matter dominance of our universe.[23] Further, the origin of *CP* noninvariance seems to reside at the same level of ignorance of our fundamental understanding as the now triumphant electroweak unified theories, that is, the mass generating mechanism. It is natural that the searching for new kind of *CP* nonvariance in reactions other than the K_L system continue to be dedicated efforts at various laboratories, and to be the central theme in the discussions of possible searches in future experiments.

In weak decays there are basically two types of mechanisms for making *CP* noninvariance: the mass-matrix type (also called superweak type), originated in the particle–antiparticle mixing of neutral mesons, and are therefore limited to neutral mesons; and the decay-amplitude type, originated directly in the weak decay amplitudes. The ϵ from $K_L \rightarrow 2\pi$ belongs to the mass-matrix type. The expression $\epsilon' \equiv |\eta_{+-}/\eta_{00}| - 1$ is of the decay amplitude type.[24] Unfortunately ϵ' is suppressed relative to ϵ due to the $\Delta I = \frac{1}{2}$ rule, that is, $\Gamma(K^+ \rightarrow \pi^+\pi^0)/\Gamma(K^0 \rightarrow \pi^+\pi^-) \approx 1/670$. Ironically such suppression was vital to the discovery of parity violation in the 1950s. So far ϵ' has still eluded experimental observation.[25] Dedicated experiments are being carried out to search for its possible presence.[b] And to avoid the $\Delta I = \frac{1}{2}$ rule suppression of decay-amplitude *CP* noninvariance ϵ', it was pointed out[26] that the time-integrated partial decay rate difference in $K^0,\overline{K}^0 \rightarrow 2\gamma$ can be as large as ϵ. It is therefore important to do an experiment on this decay.

A general convenient way to search for such decay-amplitude *CP* noninvariance is to measure the partial decay rate differences between a particle and its antiparticle. It was shown that the Quark Diagram Scheme can provide a model-independent general

[b] Current experiments measuring ϵ'/ϵ: CERN NA31; Fermilab 631; and the CERN LEAR experiment, p. 82.

survey of whether decay-amplitude *CP* noninvariance can exist in a particular exclusive decay in the KM way of violating *CP* invariance, and such a general survey has been carried out in reference 27 (see also references 28–30). Many channels of charm and beauty particle decays were shown to possibly have partial decay rate differences via basic interferences between the quark diagrams, belonging to the following major categories:[27] interferences among the tree graphs alone, between the tree graphs and the loop graphs, and between the loop graphs themselves. How big these partial decay rate differences are depends upon specific model calculations. A series of investigations has been carried out based on this Quark Diagram Scheme,[31,32] using the currently available models. A general picture has emerged that the partial decay rate differences can be very large, 10%, in beauty particle decays. Such large partial decay rate differences are very encouraging.

The decay of $B_d^0 \rightarrow K^+\pi^-$ calculated in reference 32 has been singled out in the physics studies for the superconducting super collider (SSC) because of its ease in detecting its final particles K^- and π^+, which are all charged. Here some clarifying discussions are in order to compare the advantages of searching partial decay rate differences in the charged $B^\pm$ decays over the neutral B^0 decays. It was pointed out in reference 32 that the partial decay rate differences in neutral B^0 decays, like $B_d^0 \rightarrow K^+\pi^-$, D^-F^+, etc., that is, $B_d^0 \rightarrow$ total $s = 0$, $c = 0$ states, which are not *CP* self-conjugate states and only B^0 (not $\bar{B}^0$) can decay into it, are purely of the decay-amplitude type of *CP* noninvariance. There are no $B^0 \leftrightarrows \bar{B}^0$ mixing effects nor mass-matrix *CP* noninvariance involved. Therefore the interpretations of any observations of such partial decay rate differences are straightforward. There are neutral channels that both B^0 and $\bar{B}^0$ can decay into, for example, $B_d^0 \rightarrow \pi^+\rho^-$, $\pi^-\rho^+$, $\pi^+\pi^-$, D^+D^-. Their time-integrated partial decay rate differences are results of all three distinct effects: decay-amplitude *CP* noninvariance, mass-matrix *CP* noninvariance, and $B^0 \leftrightarrows \bar{B}^0$ mixing.[32,33] Therefore the interpretation of observed results of partial decay rate differences in these decays are more complicated. To untangle these three effects, two other types of experiments must be supplemented. Here we confine our discussion to the decays in which partial decay rate differences are purely from the decay-amplitude *CP* noninvariance. The charged $B^\pm$ decays for studying partial decay rate differences have a further advantage over the neutral B^0 decays, which we must follow experimentally in order to avoid the contamination from $B^0 \leftrightarrows \bar{B}^0$ mixing in the measurement of integrated partial decay rate differences.

In reference 21, we elaborate these points and then point out a few more such beauty particle decays, which have large partial decay rate differences, yet are easily observed experimentally, for example

$$B_u^\pm \rightarrow K^\pm\rho^0, \bar{K}^{*0}\pi^\pm, K^\pm\pi^+\pi^-, \ldots. \qquad \textbf{(5.1)}$$

These decays are from similar kinds of quark diagrams as in $B_d^0 \rightarrow K^-\pi^+$, so we expect similar partial decay rate differences ($\sim$10%) and branching ratios ($\sim 10^{-4}$ to 10^{-5}).[34]

Also a simple explanation is given in reference 21 as to why the partial decay rate differences are naturally large in beauty particle decays, but rather small in charm decays. The partial decay rate differences in *t*-particle decays can also be large, as in the beauty particle decays. Unfortunately, the branching ratios of two or three particle decays of *t*-particles are very much suppressed due to the weightiness of the *t*-particles.

To enhance the branching ratio, high-multiplicity final states must be studied. This certainly complicates the experimental search. In charm particle decays, the difficulty in the search for *CP* noninvariance of partial decay rate differences is quite different. In general, the branching ratios of two-body exclusive decays of charm particles are bigger than in beauty particle decays, but the partial decay rate differences are in general much smaller than in beauty particle decays.

BEYOND THE THREE GENERATIONS OF QUARKS

First we give a brief account of the status of the KM matrix for three generations of quarks. Recent results show that the quark mixing matrix can be parameterized most conveniently in the following form:[35]

$$V = \begin{bmatrix} 1 & 0 & 0 \\ 0 & c_y & s_y \\ 0 & -s_y & c_y \end{bmatrix} \begin{bmatrix} c_z & 0 & s_z e^{-i\phi} \\ 0 & 0 & 0 \\ -s_z e^{i\phi} & 0 & c_z \end{bmatrix} \begin{bmatrix} c_x & s_x & 0 \\ -s_x & c_x & 0 \\ 0 & 0 & 1 \end{bmatrix}$$

$$\begin{array}{cccc} & d & s & b \\ s_i \approx 0 & \left[\begin{array}{c} 1 \\ -s_x - s_y s_z e^{i\phi} \\ s_x s_y - s_z e^{i\phi} \end{array}\right. & \begin{array}{c} s_x \\ 1 - s_x s_y s_z e^{i\phi} \\ -s_y - s_x s_z e^{i\phi} \end{array} & \left.\begin{array}{c} s_z e^{-i\phi} \\ s_y \\ 1 \end{array}\right] \cdot \begin{array}{c} u \\ c \\ t \end{array} \end{array} \qquad \textbf{(6.1)}$$

An important feature of the three-generation quark mixing matrix as first shown in reference 35 is that all *CP* noninvariance in decays of those three-generation quark particles are described by one universal phase-convention-independent parameter

$$\begin{aligned} X_{CP} &= s_x s_y s_z s_\phi c_x c_y c_z^2, \\ &\approx 10^{-5} \qquad \text{if } m_t \approx 50 \text{ GeV is used in fitting } \epsilon. \end{aligned} \qquad \textbf{(6.2)}$$

Because of this property we can put the phase factor at the most convenient place, that is, where the matrix element is the smallest.[36] This has the advantage that for all practical purposes, without involving *CP* noninvariance, the matrix can be considered to be real. Further, this parameterization is also the simplest in considering *CP* noninvariance of partial decay rate differences.

Another important feature of this parameterization is that it takes advantage of all the experimental information: Each sine is directly related to one type of experiment[37]

$$s_x \approx 0.22, \text{ determined from strange-particle decays,} \qquad \textbf{(6.3a)}$$

$$s_y \approx 0.05, \text{ from } b\text{-particle lifetime,} \qquad \textbf{(6.3b)}$$

$$s_z \lesssim 0.01, \text{ from bounds on } (b \rightarrow u)/(b \rightarrow c). \qquad \textbf{(6.3c)}$$

This convenience is inherited from the original Maiani parametrization—the rotation order was found originally in reference 35—and has its final form in the real angles agreeing with Maiani's.[38] We pick the subscript x, y, z because that was the order of experimental measurements and easiest to remember. Such parameterization via sequences of rotations, which put phases at the furthest corners, can be generalized to cases that generate higher numbers of quarks.[39,40]

When there are more than three generations of quarks, we can generalize the sequential rotations as given in equation 6.1—for example, for four generations of quarks—so that the quark mixing matrix can be written as follows:

$$V_4 = \begin{bmatrix} 1 & 0 & 0 & 0 \\ 0 & 1 & 0 & 0 \\ 0 & 0 & c_u & s_u \\ 0 & 0 & -s_u & c_u \end{bmatrix} \begin{bmatrix} 1 & 0 & 0 & 0 \\ 0 & c_v & 0 & s_v e^{-i\phi_3} \\ 0 & 0 & 1 & 0 \\ 0 & -s_v e^{i\phi_3} & 0 & c_v \end{bmatrix} \begin{bmatrix} c_w & 0 & 0 & s_w e^{-i\phi_2} \\ 0 & 1 & 0 & 0 \\ 0 & 0 & 1 & 0 \\ -s_w e^{i\phi_2} & 0 & 0 & c_w \end{bmatrix} \begin{bmatrix} & V_3 & & 0 \\ & & & 0 \\ & & & 0 \\ 0 & 0 & 0 & 1 \end{bmatrix}, \tag{6.4}$$

where V_3 is the 3×3 matrix given in equation 6.1. The reason we put the additional phase ϕ_2,ϕ_3 where they are in equation 6.4 is to anticipate that the widening of the generation gap will keep its pace. Of course, physical consequences are independent of ways of parameterization. The striking new phenomena are the proliferation of the number of X_{CP}

$$(X_{CP})_{ij}^{\alpha\beta} \equiv \text{Im } T_{ij}^{\alpha\beta} = (V_{\alpha j}^* V_{\alpha i} V_{\beta i}^* V_{\beta j}), \tag{6.5}$$

where α, β, i, j run from 1 to $(N - 1)$, with $\alpha < \beta, i < j$. So there are $[(N - 1)(N - 2)/2]^2$ of $(X_{CP})_{ij}^{\alpha\beta}$; however, there are a total of only $(N - 1)^2$ real parameters. Thus the X_{CP} saturate the whole parameter space of the quark mixing matrix! We see that the three-generation case is a rather singular case. Therefore the estimates of partial decay rates can be very different in the presence of higher than three generations of quarks. One important effect is that the partial decay rate differences in charm particle decays can be large too. Of special interest is that some of these large effects involving only tree graphs are independent of the masses of the fourth or higher generations of quarks, and are dependent solely on the existence of higher generations of quarks. The partial decay rate difference between

$$F^+ \to \overline{K}^0\pi^+ \qquad F^- \to K^0\pi^- \tag{6.6}$$

is such an example.

With this scenario of the possible existence of higher than three generations of quarks, it is worthwhile to be on the lookout for *CP* noninvariant decays in experiments where charm particles are or will be copiously produced.

CONCLUDING REMARKS

The observation and precise measurement of the $B_d^0\bar{B}_d^0$ mixing, and the $B_s^0\bar{B}_s^0$ mixing, can provide us with vitally needed information—for example, separate determination of the mixing angle and *CP* noninvariant phase s_z, ϕ—and shed light on m_t, Bf_B^2.

The measurement of $\Gamma(K_s \rightarrow 2\gamma)/\Gamma(K_L - 2\gamma) = 2.4 \pm 1.1$, that is, $K_s \rightarrow 2\gamma$ does not overwhelm $K_L \rightarrow 2\gamma$, encourages the future measurement of the new decay-amplitude *CP* noninvariance effects in $K_L \rightarrow 2\gamma$ through tagged K^0, $\bar{K}^0$ partial decay rate differences.

We see that the Quark Diagram Scheme has been very useful in our studies on weak decays and *CP* noninvariance. It provides a framework within which to advance our knowledge on nonleptonic decays: from experimental measurements to predict new experimental results; for example, to test and give guidance to various theoretical models, and to provide experimental values for various amplitudes and final-state interaction phases that any theoretical calculations must agree to.

For the urgently needed measurements of $|V_{ub}|$, $|V_{ub}/V_{cb}|$, the Quark Diagram Scheme has provided a useful model-independent framework to select nonleptonic decays, and decay ratios, as given in equations 4.1 and 4.3, which are directly proportional to $|V_{ub}|$ and $|V_{ub}/V_{cb}|$, respectively. Dedicated experimental measurements of these decays are very important for our understanding of the quark mixing phenomena and whether there are more than three generations of quarks. Charmless decays are useful for the search for $|V_{ub}|$ only if the *W*-loop-diagram contributions with dominant mixing-matrix coefficients are negligible. To directly measure the size of the *W*-loop-diagram contribution, it is useful to study the decays that are purely from the *W*-loop diagrams as given in equation 4.2.

In our search for *CP* noninvariance, the Quark Diagram Scheme has been very useful, providing guidance on what decay channels in heavy quark decays can possibly have nonzero partial decay rate differences. Using the scheme, we can easily see the qualitative reason why the beauty-particle decays provide favorable reactions for the *CP* noninvariance search. Here a few more good channels for experimental studies are pointed out, see, for example, equation 5.1.

In the presence of higher than three generations of quarks, the number of *CP* nonvariance parameters increases from a unique X_{CP} to saturate the whole parameter space: for example, nine X_{CP} for four generations of quarks. This gives the possibility that the *CP* noinvariance of partial decay rate differences can also be appreciable in charm particle decays (equation 6.6). It is therefore worthwhile to be on the lookout for *CP* noninvariance decays in experiments where charm particles are or will be copiously produced.

ACKNOWLEDGMENTS

We would like to thank F. Botella, H.-Y. Cheng, and W.-Y. Keung for many collaborations on the subject, and especially H.-Y. Cheng for many collaborations in carrying out the program using the Quark Diagram Scheme.

We would also like to thank the organizers of the conference for their nice conference and for their warm hospitality.

REFERENCES

1. ALBRECHT, H. 1987. New results on $B - \bar{B}$ mixing from Argus. Presented at the 1st Int. Symp. for the Fourth Family of Quarks and Leptons. Los Angeles, Calif. February 26–28, 1987.
2. SCHADD, T. 1985. Mark II collaboration. Phys. Lett. **160B:** 188; ALBAJAR, C. *et al.* 1986. UA1 collaboration. CERN preprint EP-86-209; HURST, R. 1987. MAC collaboration. Talk referred to in reference 1.
3. ELLIS, J., M. K. GAILLARD, D. N. NANOPOULOS & S. RUDAZ. 1977. Nucl. Phys. **B131:** 1285.
4. ALI, A. & Z. Z. AYDIN. 1978. Nucl. Phys. **B148:** 165; see also more recent work by ALI, A., B. VAN EIJK & I. TEN HAVES. 1986. DESY preprint #108, Oct. 1986; see also FRIDMAN, A. & A. SCHWARZ. 1985. Phys. Rev. **D32:** 1650.
5. CHAU, L.-L., W.-Y. KEUNG & M. D. TRAN. 1983. Phys. Rev. 2145.
6. CHAU, L.-L. 1983. Phys. Rep. **95:** 1.
7. CHAU, L.-L. & W.-Y. KEUNG. 1984. Phys. Rev. **29:** 592.
8. CHAU, L.-L. & W.-Y. KEUNG. 1987. The *B* meson mixing revisited. UCD-87-02; see also GILMAN, F. J. 1987. Overview of the Standard Model. This volume.
9. STONE, S. 1987. CLEO collaboration. Talk at the APS Meeting, Salt Lake City, Utah, January 14–17, 1987; WACHS, K. 1987. X-ball collaboration. Talk referred to in reference 1.
10. CHAU, L.-L. & W.-Y. KEUNG. 1984. Phys. Rev. Lett. **53:** 1802.
11. BOCK, G. J. 1987. Status of *CP* violation studies at Fermilab. This volume.
12. CHAU, L.-L., H.-Y. CHENG & W.-Y. KEUNG. 1985. Phys. Rev. **D. 32:** 1837; CHAU, L.-L., H.-Y. CHENG & W.-Y. KEUNG. 1986. CP violation in the kaon system. BNL preprint.
13. CHAU, L.-L. & H.-Y. CHENG. 1986. Ways to measure $|V_{ub}|$, $|V_{ub}/V_{cb}|$ from the quark-diagram scheme. Univ. of California at Davis preprint UCD-86-35.
14. HOLDER, M. 1987. Status of *CP* violation at CERN. This volume.
15. CHAU, L.-L. & H.-Y. CHENG. 1984. Phys. Rev. Lett. **54:** 1768; CHAU, L.-L. & H.-Y. CHENG. 1987. Univ. of California at Davis preprint UCD-87-09.
16. For reviews, see CHAU, L.-L. 1983. Phys. Rep. **95:** 1; CHAU, L.-L. 1985. *In* A Passion for Physics, Essays in Honor of Geoffrey Chew, C. De Tar, Finkelstein & C.-I. Tan, Eds.: 148. World Scientific. Singapore; CHAU, L.-L. 1985. *In* Proceedings of Kyoto Int. Symp. on the Jubilee of the Meson Theory. Kyoto, Japan. August 15–17, 1985; CHAU, L.-L. 1985. Prog. Theor. Phys. Suppl. **85:** 147.
17. CHAU, L.-L. 1981. *In* Proceedings of the 1980 Guangzhou Conf. on Theoretical Particle Physics. Van Nostrand–Reinhold, New York, and Science Press, People's Republic of China; CHAU, L.-L. 1983. Phys. Rep. **95:** 1.
18. CHAU, L.-L. & H.-Y. CHENG. 1986. Phys. Rev. Lett. **56:** 1655; CHAU, L.-L. & H.-Y. CHENG. 1986. Univ. of California at Davis preprint UCD-86-32. Phys. Rev. In press.
19. KOBAYASHI, M. & T. MASKAWA. 1973. Prog. Theor. Phys. **49:** 652.
20. BEHRENDS, S. *et al.* 1983. Phys. Rev. Lett. **50:** 881; for the *b* lifetime, FERNANDEZ, E. 1984. Phys. Rev. Lett. **51:** 1022; LOCKYER, N. S. *et al.* 1983. Phys. Rev. Lett. **51:** 1316; for $\Gamma(b \rightarrow u)/\Gamma(b \rightarrow c)$, KLOPFENSTEIN, C. *et al.* 1983. Phys. Lett. **103B:** 444; CHEN, A. *et al.* 1984. Phys. Rev. Lett. **111B:** 1084.
21. CHAU, L.-L. 1986. In search of V_{ub} and CP non invariance. Univ. of California at Davis preprint UCD-86-33-R. Proceedings of the 1986 Snowmass Summer Study on the Design and Utilization of the Superconducting Super Collider. June 23–July 13, 1986. In press; papers by CHAU, L.-L. & H.-Y. CHENG. 1986. Univ. of California at Davis preprints UCD-86-34 and UCD-86-35.
22. CHRISTENSON, J. H., J. W. CRONIN, V. L. FITCH & R. TURLAY. 1964. Phys. Rev. Lett.

13: 138; the classical theoretical discussions on *CP* violation: LEE, T. D., R. OEHME & C. N. YANG. 1957. Phys. Rev. **106:** 340; WU, T. T. & C. N. YANG. 1984. Phys. Lett. **13:** 380.
23. SAKHAROV, A. 1967. Zh. Eksp. Teor. Fiz. Pisma Red. **5:** 32; for a review, see GOLDHABER, M., P. LANGACKER & R. SLANSKY. 1980. Science **210:** 851.
24. GILMAN, F. J. & M. B. WISE. 1979. Phys. Lett. **83B:** 83.
25. BLACK, J. K. *et al.* 1985. Phys. Rev. Lett. **54:** 1628; BERNSTEIN, R. H. *et al.* 1985. Phys. Rev. Lett. **54:** 1631.
26. CHAU, L.-L. & H.-Y. CHENG. 1985. Phys. Rev. Lett. **54:** 1768.
27. For general discussions on the KM scheme using the Quark Diagram Scheme to map out decays that have partial decay rate differences, CHAU, L.-L. 1980. AIP Conf. Proc. No. 72, Particle and Fields, Subseries No. 23, Virginia Polytechnic Institute and State Univ., Blacksburg. G. B. COLLINS, L. N. CHANG, & J. R. FICENEC, Eds.; for other developments, see references 14–16.
28. For earlier discussions, see SEHGAL, L. M. & L. WOLFENSTEIN. 1976. Phys. Rev. **162:** 1362; OVERSETH, O. E. & S. PAKVASA. 1969. Phys. Rev. **189:** 1663; PAIS, A. & S. B. TREIMAN. 1975. Phys. Rev. **D12:** 2744; OKUN, L. B., V. I. ZAKHAROV & B. M. PONTECORVO. 1975. Lett. Nuovo Cimento **13:** 218.
29. For a timelike penguin diagram contribution of partial decay rate difference in the KM scheme, see BARSHAY & J. GERIS. 1979. Phys. Lett. **84B:** 319; BANDER, M., D. SILVERMAN & A. SONI. 1979. Phys. Rev. Lett. **43:** 242; for tree graph contributions, see BARNABEU, J. & C. JARLSKOG. 1981. Z. Phys. **C8:** 233.
30. For *CP* noninvariance in neutral *B* decays into the *CP* self-conjugate state, see BIGI, I. I. & A. I. SANDA. 1981. Nucl. Phys. **B193:** 81.
31. For general estimates of *CP* noninvariance in charge $B^{\pm}$ decays, see CHAU, L.-L. & H.-Y. CHENG. 1984. Phys. Rev. Lett. **53:** 1037.
32. For estimates of *CP* noninvariance in B^0 decays, including *CP* non-self-conjugate states, like $B_d^0 \rightarrow K^+\pi^-$, see CHAU, L.-L. & H.-Y. CHENG. 1985. Phys. Lett. **165B:** 429.
33. DU, D. S., I. DUNIETZ & D. D. WU. 1986. Enrico Fermi Institute Rep. EFI 86-9; DUNITEZ, I. & J. L. ROSNER. 1986. Phys. Rev. **D34:** 1404.
34. For detailed calculations, see CHAU, L.-L. & H.-Y. CHENG. 1987. Univ. of California at Davis preprint UCD-87-9.
35. CHAU, L.-L. & W.-Y. KEUNG. 1984. Phys. Rev. Lett. **53:** 1802.
36. WOLFENSTEIN, L. 1984. Phys. Rev. Lett. **51:** 1945.
37. For a recent thorough review on determination of the mixing matrix, see STONE, S. 1986., Experimental determination of the KM matrix. Cornell Univ. preprint CLNS-86/753.
38. MAIANI, L. 1976. Phys. Lett. **62B:** 183; MIGNAMI, R. 1980. Lett. Nuovo Cimento **28:** 529.
39. BOTELLA, F. J. & L.-L. CHAU. 1986. Phys. Lett. **168B:** 97.
40. See also HE, X.-G. & S. PAKVASA. 1985. Phys. Lett. **156B:** 236; ANSELM, A. A. *et al.* 1985. Phys. Lett. **156B:** 102; GRONAU, M. & J. SCHECHTER. 1985. Phys. Rev. **D31:** 1668; HARARI, H. & M. LEURER. 1986. Weizmann Institute of Science. preprint, W15-86/24/May-Ph; FRITZSCH, H. & J. PLANKL. 1986. Univ. of Munich and Max-Planck-Institute preprint MPI-PAE/PTH 20/86.

Searching for Fourth-Generation Fermions at High-Energy e^+e^- Machines[a]

JONATHAN DORFAN AND RICHARD J. VAN KOOTEN

Stanford Linear Accelerator Center
Stanford University
Stanford, California 94305

INTRODUCTION

This writeup summarizes Monte Carlo studies on the feasibility of searching for fourth-family fermions at high-energy e^+e^- machines. By high energy we mean energies beyond the Z^0, the Z^0 searches having been studied extensively by the SLC and LEP collaborations. We have chosen the specific examples of $\sqrt{s} = 200$ GeV (namely LEP200) and $\sqrt{s} = 600$ GeV (in line with work being done at SLAC to investigate the potential for a future linear collider). The studies have been performed with the four vectors produced by the models and no attempt has been made to account for the effects of finite resolution and inefficiencies due to a real detector environment.

We will begin by considering a canonical fourth family in the Standard Model:

$$\begin{pmatrix} u \\ d \end{pmatrix} \begin{pmatrix} c \\ s \end{pmatrix} \begin{pmatrix} t \\ b \end{pmatrix} \begin{pmatrix} h \\ l \end{pmatrix}$$

$$\begin{pmatrix} e^\pm \\ \nu_e \end{pmatrix} \begin{pmatrix} \mu^\pm \\ \nu_\mu \end{pmatrix} \begin{pmatrix} \tau^\pm \\ \nu_\tau \end{pmatrix} \begin{pmatrix} L^\pm \\ \nu_L \end{pmatrix}$$

and will return later to the possibility that the fourth-generation neutral lepton is in fact massive.

PRODUCTION, DECAY, AND BACKGROUNDS

The attractiveness of the e^+e^- environment is the direct and relatively background-free production of the fundamental fermions; in all cases studied here the fourth-family fermions $(f, \bar{f})$ are produced in pairs via the intermediate exchange of a virtual photon or Z^0. The production cross section for this process $e^+e^- \rightarrow \gamma, Z^0 \rightarrow f\bar{f}$ is given by the

[a]This work was supported by the Department of Energy under Contract DE-AC03-76SF00515.

electroweak model:

$$\frac{d\sigma}{d\cos\theta} = \frac{3\pi\alpha^2 Q_f^2}{2s}(1 + \cos^2\theta)$$
$$- \frac{\alpha D G_F Q_f M_z^2(s - M_z^2)}{8\sqrt{2}[(s - M_z^2)^2 + \Gamma_z^2 M_z^2]} \cdot [4v_e v_f(1 + \cos^2\theta) + 8a_e a_f \cos\theta]$$
$$+ \frac{D G_F^2 M_z^4 s}{64\pi[(s - M_z^2)^2 + \Gamma_z^2 M_z^2]} \cdot [4(a_e^2 + v_e^2)(a_f^2 + v_f^2) + 32 a_e v_e a_f v_f \cos\theta]$$

where θ is the production angle of the fermion, a and v are the weak couplings, s is the center-of-mass energy squared, M_z and Γ_z are the mass and width of the Z^0, Q_f the fermion charge, G_F the Fermi constant, and $D = 1$ for leptons, 3 for quarks.

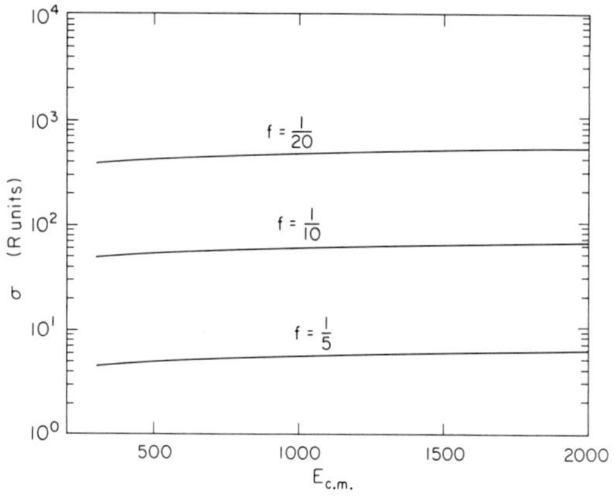

FIGURE 1. R for $e^+e^- \rightarrow e^+e^-$ + hadrons via the two photon process. The f is the fraction of energy carried by the hadronic system.

To good approximation, this cross section falls like s^{-1}, which implies that as s increases we need correspondingly larger machine luminosities. The appropriate threshold factor that moderates this cross section is $\beta_f(3 - \beta_f^2)$, where β_f is the fermion velocity.

This covers the production mechanism for h, l, and $L^{\pm}$. They will decay via the standard weak interaction, namely

$$h \rightarrow lW^+ \quad \hookrightarrow f\bar{f}$$

$$l \rightarrow tW^- \quad \hookrightarrow f\bar{f}$$

$$L^{\pm} \rightarrow \nu_L W^{\pm} \quad \hookrightarrow f\bar{f}$$

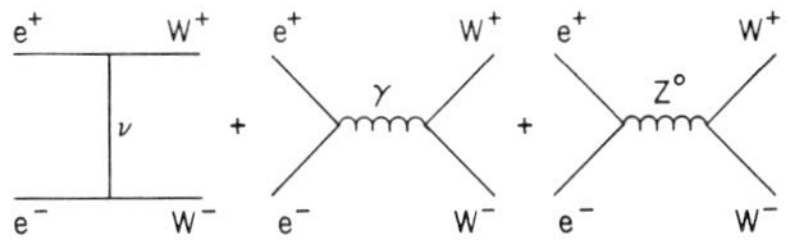

FIGURE 2. Feynman graphs for the process $e^+e^- \rightarrow W^+W^-$.

where now $f(\bar{f})$ are first-, second-, and third-generation fermions. Depending on the mass differences involved, the W can be both real or virtual. For all the studies presented here we take the top-quark mass to be 40 GeV/c^2. None of the conclusions changes dramatically if M_t is in fact heavier.

What are the background processes that could inhibit the search for fourth-generation fermions? There are three obvious sources of background:

(1) Conventional processes from the three light generations. This background will prove to be large for searches done using $R = \sigma_{\text{had}}/\sigma_{\mu\mu}$, while $e^+e^- \rightarrow t\bar{t}$ will be a contributing background in topological searches.

(2) Two-photon production of hadrons. This process has a very large cross section. However, if the hadronic system carries sufficient energy to be a serious background, the effective cross section becomes small enough to ignore. This is shown in FIGURE 1, where the two-photon cross section is shown as a function of $\sqrt{s}$. In this figure, the fraction of energy carried by the hadronic system is denoted by f.

(3) We encounter a background at these energies not yet seen in e^+e^- machines, namely $e^+e^- \rightarrow W^+W^-$. This is in fact the main background, since it has a large cross section and since it can easily mimic the decays of the h, l, and $L^\pm$.

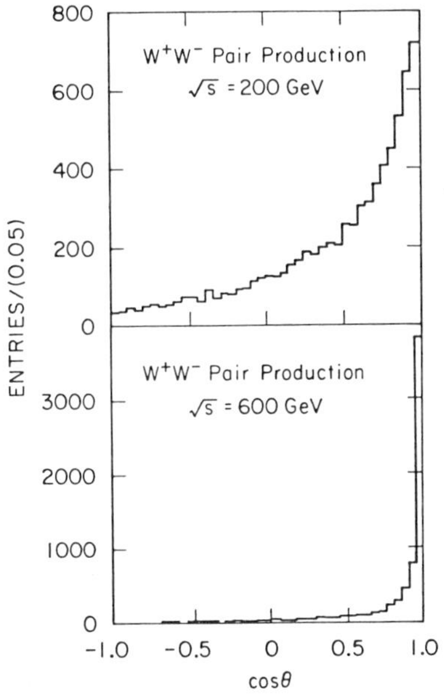

FIGURE 3. Angular distribution of W^+ with respect to the e^- beam direction at $\sqrt{s} = 200$ GeV and $\sqrt{s} = 600$ GeV.

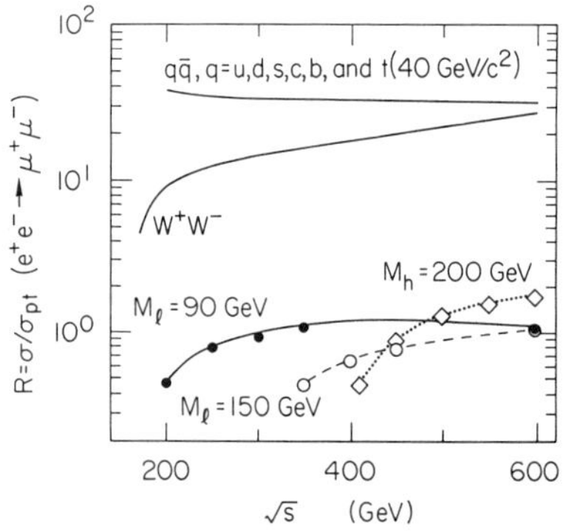

FIGURE 4. Radiatively corrected R values of fourth-family quarks and the two major backgrounds.

The W pair production occurs via the three Feynman graphs shown in FIGURE 2, and will be studied at LEP200. We have assumed here the Standard Model for calculating cross sections and kinematic distributions; $\sigma_{W^+W^-}$ grows rapidly with $\sqrt{s}$, mainly due to the t channel ν exchange graph in FIGURE 2. This growth, however, is mostly in the forward direction (see FIGURE 3), so this background can be reduced by requiring that the event thrust axis not point in the forward direction.

The cross section for the production of the fourth-family fermions and the two major backgrounds are shown in FIGURES 4 and 5. These cross sections have been corrected for initial state electromagnetic radiative effects, which is a particularly large, positive correction for the conventional light quarks. This comes about when the radiation is sufficiently large to generate the e^+e^- collision at M_Z, where the cross section is enormous. As we will see, these events are easily removed with simple kinematic cuts. FIGURES 4 and 5 set the magnitude of the rejection required to extract the fourth-family signals.

ESTABLISHING THE EXISTENCE OF h, l, OR $L^{\pm}$

The first thing that comes to mind at a new e^+e^- collider is to look for an increase in R. By looking at FIGURES 4 and 5 it becomes clear that even if one were above

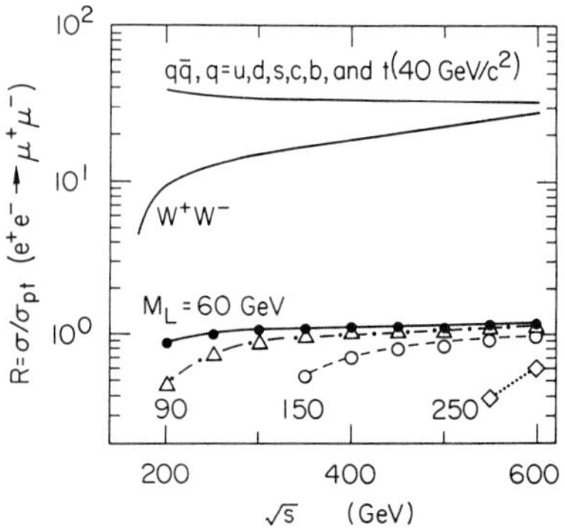

FIGURE 5. Radiatively corrected R values of fourth-family charged leptons and the two major backgrounds.

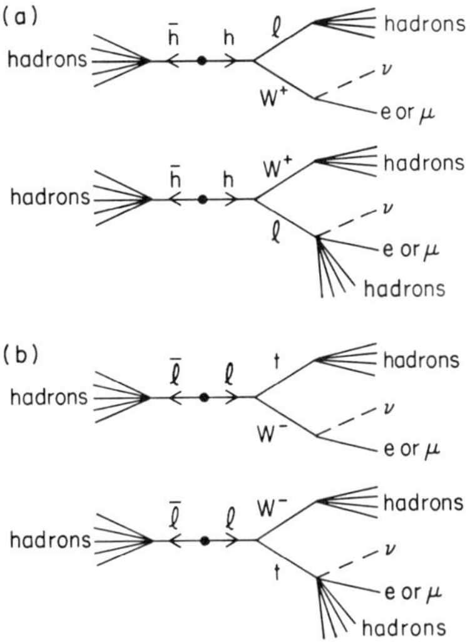

FIGURE 6. Signal topologies used to search for (a) h and (b) l.

threshold, the increase in R from fourth-family charged fermions is at most 3%. Since the PEP and PETRA experiments achieve systematic errors no better than 3% for R measurements, this procedure will not work.

It proves much more fruitful to look for distinctive topologies. By making relatively simple kinematic cuts one could greatly reduce the contribution to R from the light quarks and the W^+W^- backgrounds. Under these circumstances one could probably turn this into a useful search procedure. However, by adding the additional requirement of an isolated lepton, signal to background improves even more. Since leptonic decays are reasonably copious, this is not a large price to pay.

For all the charged fermion searches discussed here, Monte Carlo simulations were done for both the signal and background channels. In all cases the Standard Model was assumed and initial-state radiative processes are included. High-energy linear colliders will not have the well-defined $\sqrt{s}$ that e^+e^- storage rings provide. The machine parameters chosen to achieve high luminosity result in large ($\simeq$30%) uncertainties in

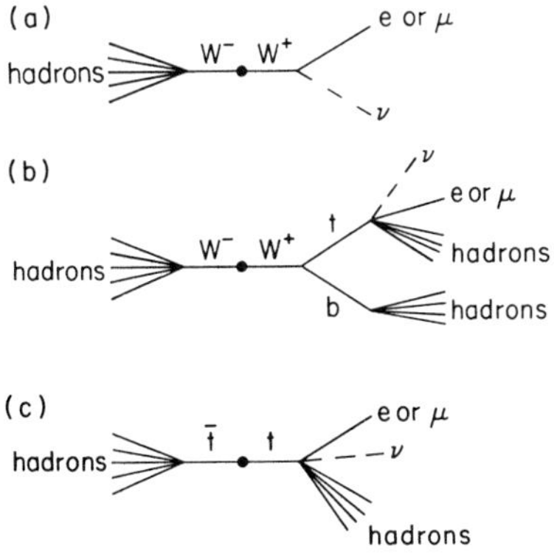

FIGURE 7. Background sources for the search for h and l.

TABLE 1. $\sqrt{s} = 600$ GeV

Channel	⟨Aplanarity⟩	⟨Charged Multiplicity⟩
$h\bar{h}$	0.26	77.1
$l\bar{l}$	0.20	62.2
W^+W^-	0.08	31.2
Light quarks	0.07	29.7
	$\sqrt{s} = 200$ GeV	
$l\bar{l}$	0.28	44.2
W^+W^-	0.19	29.8

the beam energy because of the intense radiation emitted when the beams pass through each other (beamsstrahlung). Hence, at energies above LEP200 we have to avoid analysis techniques that rely on a precise knowledge of $\sqrt{s}$. These beamsstrahlung effects have not been included in the heavy quark searches considered here because the search technique uses kinematic selection cuts that are relatively immune to the beam spread. However, this is more crucial for the $L^{\pm}$ search, and in this case we have included the beamsstrahlung effects.

Search for the Fourth-Generation Quarks **h**(2/3), **l**(−1/3)

These searches rely heavily on the fact that the decays of the *h*- and *l*-quarks produce isolated leptons with very large transverse momentum. In addition their kinematics differ considerably from the background processes, particularly at the higher energies. The signal topologies are shown in FIGURE 6a and 6b and are characterized by isolated leptons (e or μ), high charged-particle multiplicity, and high aplanarity. (Aplanarity is a measure of the amount of momentum that is perpendicular to the event plane as defined by a sphericity analysis.) The background topologies shown in FIGURE 7 are characterized by relatively low multiplicity and aplanarity and mostly lower transverse momentum for the e and μ. Based on the actual kinematic distributions, the following set of cuts were used:

(a) aplanarity > 0.2 for $\sqrt{s} = 200$ GeV, > 0.15 for $\sqrt{s} = 600$ GeV,
(b) charged-particle multiplicity > 35,
(c) isolated lepton; < 100 MeV of energy within a 10° cone centered around the lepton direction.

The means of the distributions for (a) and (b) are given in TABLE 1 for the signal and

TABLE 2. $\sqrt{s} = 600$ GeV

Channel	Number Produced	Number Isolated Leptons
$h\bar{h}$	3,000	2,873
$l\bar{l}$	2,060	1,017
W^+W^-	37,500	212
Light quarks	60,000	147
Misidentification of hadrons	97,500	60

TABLE 3. $\sqrt{s}$ = 200 GeV

Channel	Number Produced	Number Isolated Leptons
$l\bar{l}$	825	495
W^+W^-	14,850	685
Light quarks	63,000	124

background channels. The parameters for the simulations were:

$$\sqrt{s} = 200 \text{ GeV} \qquad M_l = 90 \text{ GeV}/c^2 \qquad (\text{no } h \text{ is reasonable})$$

$$\langle \mathcal{L} \rangle = \frac{1}{2} \times 10^{32} \text{ cm}^{-2} \text{ sec}^{-1},$$

$$\sqrt{s} = 600 \text{ GeV} \qquad M_l = 150 \text{ GeV}/c^2 \qquad M_h = 200 \text{ GeV}/c^2$$

$$\langle \mathcal{L} \rangle = \frac{1}{2} \times 10^{33} \text{ cm}^{-2} \text{ sec}^{-1}.$$

The results of this study are summarized in TABLES 2 and 3, where we have assumed a six-month run. The background from misidentifying hadrons as leptons has been included, assuming the misidentification probability is ½%. This is modest, since these leptons are truly isolated, and most detectors will do much better. The lepton transverse momentum spectra are shown in FIGURES 8 and 9. One concludes from this study that both at $\sqrt{s}$ = 200 GeV and at higher energies ($\sqrt{s}$ = 600 GeV being typical) one has a relatively clean signal for h and l, and that the efficiencies of such searches are large enough to yield sizable signals in a period of $\lesssim$ 6 months.

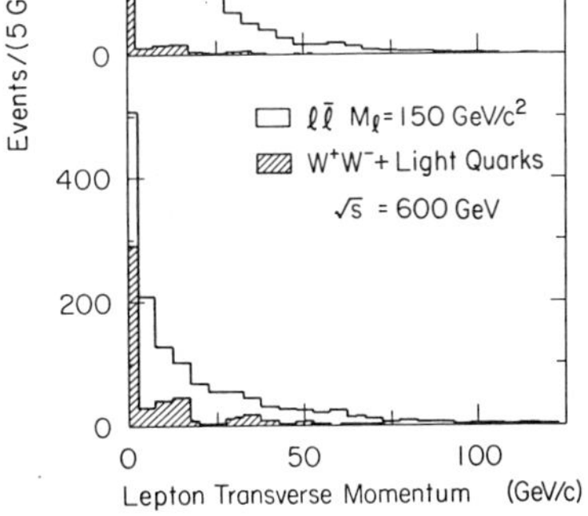

FIGURE 8. Lepton transverse momentum for the events selected for the h and l search and for the contributions from the background ($\sqrt{s}$ = 600 GeV).

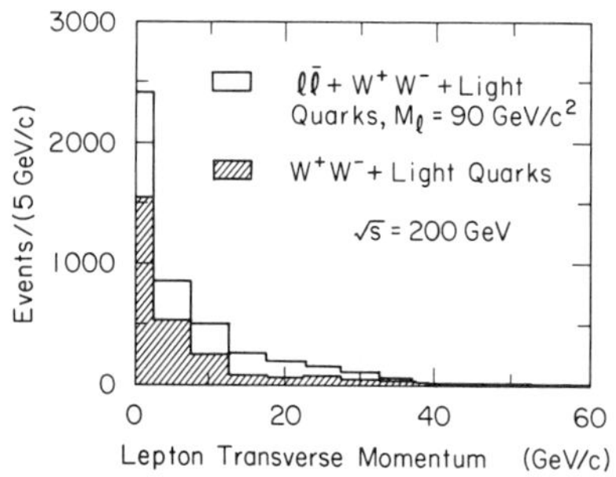

FIGURE 9. Lepton transverse momentum for events passing the selection criteria at $\sqrt{s}$ = 200 GeV. The shaded region shows the contribution from the background, the unshaded the contribution from the signal plus background.

Search for the Fourth-Generation L$^\pm$

FIGURE 10a shows the signal topology used for the $L^\pm$ search, while FIGURE 10b shows the main background topology. We have concentrated here on the leptonic decay of the $L^\pm$, where one pays a fairly large price in terms of their branching fraction ($B(L^\pm \rightarrow e\nu_e\nu_L) \approx 8\%$). This has been done in order to generate a clean signature. The alternative would be to use the hadronic decays of the $L^\pm$ that have larger branching fractions. This will probably work, but has not been tried here. The characteristics that allow one to separate the signal and background topologies are the facts that the background tends to be peaked in the forward direction, there is small missing energy on the hadronic side of the event, and there is an isolated lepton on the other side of the event that is accompanied by large missing transverse momentum that is carried off by ν_L. For LEP200 the beam energy will be accurately known, and this can be used to take maximum advantage of these characteristics. However, for the 600-GeV linear collider where the beam energy is poorly known, a less optimal set of cuts is possible.

Analysis at $\sqrt{s}$ = *200 GeV*

Cuts can be made that take advantage of the preceding characteristics of signal and background. In addition, if one *assumes* a W^+W^- hypothesis, $\bar{P}_{\nu_l}$ can be reconstructed. In the W^+W^- case, the invariant mass ($M_{\nu_l l}$) of this ν_l and the isolated

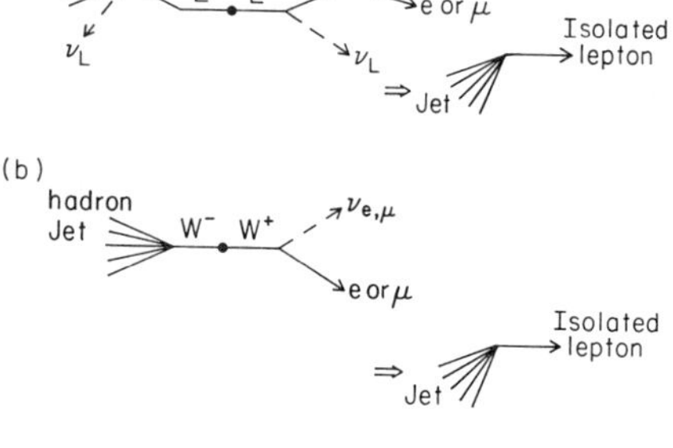

FIGURE 10. (a) Signal topology selected for $L^\pm$ search. (b) Major background topology (W^+W^-).

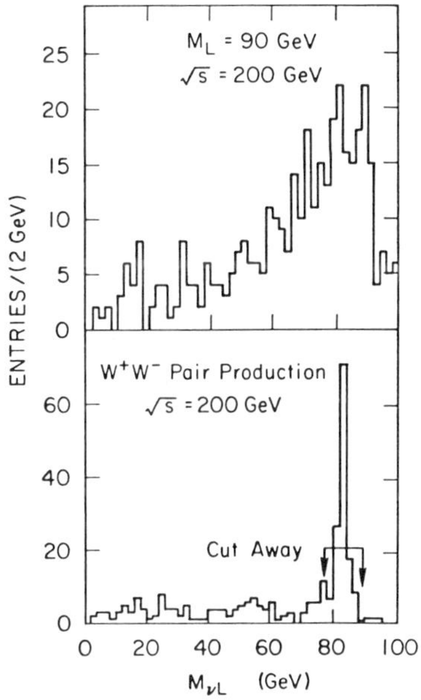

FIGURE 11. After *assuming* a W^+W^- event and reconstructing $\bar{P}_{\nu_l}$, the invariant mass of the isolated lepton and ν_l. Top: L^+L^- events. Bottom: W^+W^- events.

lepton shows a distinct mass peak. The extra neutrinos in the L^+L^- topology make the assumption of W^+W^- kinematics incorrect, $\bar{P}_{\nu_l}$ is reconstructed incorrectly, and $M_{\nu_l l}$ does not show any peaks, as shown in FIGURE 11. Therefore, the following set of cuts was used.

(a) $|\cos\theta_{\text{thrust}}| < 0.85$; thrust is that of the hadronic jet.
(b) Isolated $e^{\pm}$ or $\mu^{\pm}$. Require < 1 GeV within a 30° cone around the lepton.
(c) E_{vis} = measured energy in hadronic hemisphere,

$$E_{\text{vis}} < 0.8\, E_{\text{beam}}.$$

(d) Assume W^+W^- event. Reconstruct $\bar{P}_{\nu_l}$ and calculate $M_{\nu_l l}$. Demand $M_{\nu_l l} < 78$ GeV; $M_{\nu_l l} > 86$ GeV (not near M_W).

We used a luminosity of $\langle \mathcal{L} \rangle = \frac{1}{2} \times 10^{32}\ \text{cm}^{-2}\ \text{sec}^{-1}$ over a six-month run, a lepton mass of $M_{L^{\pm}} = 90$ GeV, and a massless ν_L. Results of this study are summarized in TABLE 4.

TABLE 4. $\sqrt{s} = 200$ GeV, $M_{L^{\pm}} = 90$ GeV

Channel	Number Produced	Number Passing Cuts
L^+L^-	1,100	39
W^+W^-	14,850	10
Light quarks	63,000	3

Analysis of $\sqrt{s} = 600$ *GeV*

The analysis techniques for $\sqrt{s}$ of 200 GeV and 600 GeV differ because the higher luminosity required for the 600-GeV machine will result in synchrotron radiation in the collision region (beamsstrahlung) that leads to substantial smearing of the center-of-mass energy of annihilation. Therefore, we will not know $\sqrt{s}$ (or E_{beam}) with any degree of precision, and should develop techniques that do not rely on these parameters and that are not sensitive to arbitrary boosts along the axis of the beam (z-axis).

Looking in the transverse plane of the event, a W^+W^- hypothesis is again assumed, as shown in FIGURE 12. The invariant mass of the leptonic decay is insensitive to boosts along the z-axis, and we have

$$M_W^2 = (E_l + E_\nu)^2 - (P_{lx} + P_{\nu x})^2 - (P_{ly} + P_{\nu y})^2 - (P_{lz} + P_{\nu z})^2.$$

Using this equation, and knowing that the transverse momentum of the hadronic side must be balanced by transverse momentum on the leptonic side, a quadratic equation in $P_{\nu z}$ is derived. In the case of L^+L^-, the assumption of W^+W^- kinematics is incorrect, and quite often imaginary solutions for $P_{\nu z}$ are obtained. Cuts can then be made on η, the discriminant of the solution to the quadratic, where $\eta < 0$ indicates imaginary solutions.

Event samples were generated by Monte Carlo from distributions of electron and positron energies computed using a beam–beam simulation code written by K. Yokoya[1] based on typical sets of machine parameters for a 600-GeV collider. Cuts used were:

(a) $|\cos\theta_{\text{thrust}}| < 0.6$; thrust is that of the hadronic jet,
(b) isolated $e^\pm$ or $\mu^\pm$ with the same requirements as for the $\sqrt{s} = 200$-GeV case,
(c) η = discriminant of solution to quadratic in $P_{\nu z}$,

$$\frac{\eta}{M_W^6} < 0.1.$$

We used $\langle \mathcal{L} \rangle = \frac{1}{2} \times 10^{33}\ \text{cm}^{-2}\,\text{sec}^{-1}$ over a six-month run. The results of this study for $M_L = 150$ and 250 GeV with $M_{\nu_L} = 0$ are summarized in TABLE 5.

One concludes that at $\sqrt{s} = 200$ GeV, one has a relatively clear signal with high

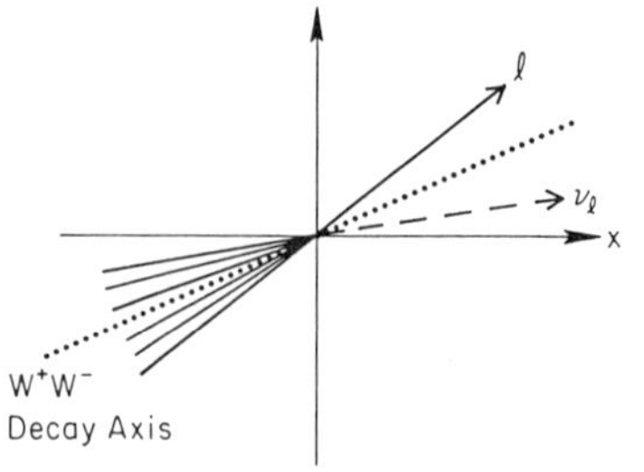

FIGURE 12. Transverse plane of W^+W^- background event. P_x, P_y of the hadronic jet hemisphere must be balanced on the lepton hemisphere.

TABLE 5. $\sqrt{s} = 600$ GeV

Channel	Number Produced	Number Passing Cuts
L^+L^-		
$M_L = 150$ GeV	1,860	74
$M_L = 250$ GeV	1,200	69
W^+W^-	37,500	54
Light quarks	60,000	6

enough yield in a six-month run. At $\sqrt{s} = 600$ GeV, the signal is less clear because of problems caused by beamsstrahlung, but is still evident. In both cases, after cuts are made, the momentum spectrum of the isolated lepton can be examined. The two-body decay of the $W^\pm$ results in a lepton momentum spectrum peaked at high momentum, whereas the lepton from $L^\pm$ decays has a much softer momentum spectrum. Fits can then be made to determine the presence of an L^+L^- signal.

SEARCHING FOR NEUTRAL HEAVY LEPTONS

We consider now the possibility that the fourth family comprises $L^\pm$, L^0, where L^0 is a massive Dirac neutrino.

The production is via $e^+e^- \rightarrow L^0\bar{L}^0$ with a cross section given by

$$\frac{d\sigma}{d\cos\theta} = \frac{1}{64\pi}\frac{G_F^2 s}{((1 - s/M_z^2)^2 + \Gamma_z^2/M_z^2)}\{(1 - 4\sin^2\theta_w + 8\sin^4\theta_w)$$

$$+ \beta(1 + \beta^2\cos^2\theta) + 2(1 - 4\sin^2\theta_w)\beta^2\cos\theta\}$$

where β is the L^0 velocity and θ_w is the weak mixing angle. For $s \gg M_Z^2$,

$$\sigma \sim \frac{0.025}{s(\text{TeV})}\text{ pb.}$$

This gives $\sigma = 1$ pb (R = 0.43) at $\sqrt{s} = 200$ GeV and $\sigma = 0.07$ (R = 0.27) for $\sqrt{s} = 600$ GeV. Hence, there is substantial production of $L^0\bar{L}^0$. There are several modes for decay, examples of which are given below.

Within the Standard Model one can have the following.

(1) $L^0 \rightarrow L^-W^+ \qquad (M_{L^0} > M_{L^-})$

(2) Mixing with a lighter generation (presumably the τ, but e and μ are not excluded):

$$L^0 \rightarrow \tau^- W^+.$$

For a mixing angle ϵ, the weak isodoublets are

$$\begin{pmatrix} \nu_\tau\cos\epsilon + L^0\sin\epsilon \\ \tau^- \end{pmatrix}_L \begin{pmatrix} L^0\cos\epsilon - \nu_\tau\sin\epsilon \\ L^- \end{pmatrix}_L.$$

(3) One can also imagine decays that violate the Standard Model like the flavor-changing neutral current decay where the L^0 mixes with ν_τ:

$$L^0 \longrightarrow Z^0 \nu_\tau .$$

All of these decay scenarios lead to striking signatures with no obvious backgrounds. The distinctive features are low multiplicity and multiple, mixed leptons. For the purposes of illustration we choose Model 2 in which the L^0 couples to the τ. We can define two classes of search topologies.

(1) Four charged-particle events in which both the τ and intermediate W decay leptonically. Using standard branching fractions, this topology occurs in 2% of the $L^0\bar{L}^0$ events and results in events of the type $llll$, $ll\pi\pi$, $lll\pi$ (where $l = e$ or μ) with a ratio of $e{:}\mu{:}\pi$ of 1:1:2.5. We label these 2 + 2 events.

(2) Events with two charged particles in one hemisphere and a high-multiplicity jet in the other hemisphere. Again the low-multiplicity hemisphere is populated with mixtures of e, μ, and π. This process occurs in 20% of the $L^0\bar{L}^0$ events and we label it the $2 + N$ topology.

If we assume as before $\langle \mathcal{L} \rangle = ½ \times 10^{32}$ cm^{-2} sec^{-1} at $\sqrt{s} = 200$ GeV, $\langle \mathcal{L} \rangle = ½ \times 10^{33}$ cm^{-2} sec^{-1} at $\sqrt{s} = 600$ GeV, and a six-month run, one finds the number of

TABLE 6.

Topology	Number of Events (600 GeV)	Number of Events (200 GeV)
2 + 2	9	14
$2 + N$	100	160

events shown in TABLE 6 for the 2 + 2 topology and the $2 + N$ topology. This would be sufficient to establish the signal, since there appear to be no competing backgrounds.

CONCLUSIONS

We have studied the possibilities for the discovery of fourth-family fermions in e^+e^- machines at energies beyond the Z^0. Two specific energies were chosen: $\sqrt{s} = 200$ GeV, compatible with the highest LEP200 energy, and $\sqrt{s} = 600$ GeV, which serves as a reasonable example for machines in the 500–1000 GeV region. These Monte Carlo studies are by no means exhaustive, nor have they included the effects of detector inefficiencies and resolutions. However, it appears that sufficiently clean topologies exist for the discovery of all the fourth-family charged fermions and the neutral if it is a massive Dirac particle. Production cross sections and topological search efficiencies are sufficiently large that with the LEP200 design luminosity or a 10^{33} cm^{-2} sec^{-1} luminosity at $\sqrt{s} = 600$ GeV, sizable signals are obtained in six-month running periods. In short, the outlook for the discovery of fourth-family fermions in these higher energy e^+e^- machines looks very promising. More detailed studies,[2] including the effects of detector smearing and acceptance, are underway at SLAC for $\sqrt{s} = 600$–1000 GeV. A report on this work should be forthcoming soon.

ACKNOWLEDGMENTS

The authors wish to thank Alfred Peterson for providing the Monte Carlo implementation and tapes for this study. Useful discussions with Gary Feldman are also acknowledged.

REFERENCES

1. Yokoya, K. 1987. SLAC-AAS-27.
2. See upcoming report, Peskin, M. E. *et al.* Opportunities and requirements for experimentation at a 500 GeV–1 TeV e^+e^- collider.

Are Fourth-Generation Quarks Integrally Charged?[a]

X.-G. HE,[b,d] S. PAKVASA,[b] AND H. SUGAWARA[c]

[b] *Department of Physics and Astronomy*
University of Hawaii at Manoa
Honolulu, Hawaii 96822

[c] *KEK, National Laboratory for High Energy Physics*
Tsukuba, Ibaraki 305, Japan

INTRODUCTION

In this paper we review a novel way of implementing generation symmetry before grand unification in which the unification scale can be as high as the Planck scale (and the correct $\sin^2 \theta_W$ prediction is preserved). In the simplest model of this kind, corresponding to a low-energy gauge group $SU(2)_L \times SU(2)_R \times SU(3)_C \times SU(4)_G$, there are striking experimental signatures from production and decay of integer-charged fourth-generation quarks. We also describe the consequences of a scheme in which the gauge is enlarged to $SU(2)_L \times SU(2)_R \times SU(3)_C \times SU(5)_G$ and the fourth-generation quarks are normal (while the fifth-generation quarks carry charges $(-1/6, -7/6)$).

DISCUSSION

How many generations of quarks and leptons are there in nature? is one of the most frequently asked questions in particle physics. The Standard Model[1] has no answer for this. One of the solutions to this problem is to have generation symmetry, but there are still many different ways to implement generation symmetry. Here we describe a model of generation symmetry in which the generation symmetry is restored before grand unification, the unification scale is large enough $\mu_0 > 10^{17}$ GeV (as high as the Planck scale) so that proton stability should not present a problem, and the correct prediction of $\sin^2 \theta_W$ is preserved.[2] The generation gauge bosons would transform different generations into each other and, in order to satisfy the experimental limit on flavor-changing decays,[3] the generation symmetry scale is, depending on other parameters, bounded by $\mu > 1$ TeV.

The group we discuss is $SU(2)_L \times SU(2)_R \times SU(3)_C \times SU(N)_G$. This model is anomaly-free due to left–right symmetry. The generalized Nishijima–Gell-Mann

[a] This work was supported in part by the U.S. Department of Energy, under Contract DE-AM03-76SF00235, and in part by the Japan Society for the Promotion of Science.

[d] To whom correspondence should be addressed.

formula is

$$Q = I_L + I_R + \frac{C}{2}, \tag{1}$$

where $C = 1/3$ for usual quarks and -1 for usual leptons.

The quarks transform as

$$\begin{aligned} q_{\mathrm{L}}&: (2, 1, 3, N) \\ q_{\mathrm{R}}&: (1, 2, 3, N) \end{aligned} \tag{2}$$

where the first, second, third, and fourth numbers in the bracket correspond to $SU(2)_L$, $SU(2)_R$, $SU(3)_C$, and $SU(N)_G$, respectively. As is known, there are at least three generations of quarks (with usual charge assignments) in the nature, and the smallest generation symmetry is $SU(4)_G$. In this case, from the tracelessness condition on the $SU(4)_G$ generators, the C operator must be

$$\begin{pmatrix} 1/3 & 0 & 0 & 0 \\ 0 & 1/3 & 0 & 0 \\ 0 & 0 & 1/3 & 0 \\ 0 & 0 & 0 & -1 \end{pmatrix}. \tag{3}$$

This implies that fourth-generation quarks have charges $(0, -1)$ and that the fourth generation is integrally charged! The leptons transform as

$$\begin{aligned} L_{\mathrm{L}}&: (2, 1, 3, 20) \\ L_{\mathrm{R}}&: (1, 2, 3, 20), \end{aligned} \tag{4}$$

where 20 of $SU(4)_G$ decomposes under $SU(3)$ as

$$20 = 3\left(\frac{1}{3}\right) + \bar{3}\left(\frac{5}{3}\right) + \bar{6}\left(\frac{1}{3}\right) + 8(-1), \tag{5}$$

where the numbers in the brackets are the eigenvalues of C. We expect five additional ordinary leptons and twelve extra fractionally charged leptons.

Now let us consider unification and the prediction for $\sin^2\theta_W$. If we assume dynamical symmetry breaking, there should be no Higgs. In order to satisfy all requirements previously discussed, additional fermion multiplets transform, such as $(1, 1, 3, X)$, are needed.

The evolution of the gauge coupling constant is described by[4]

$$\beta(\mu)_i = \frac{\mu \partial g_i}{\partial \mu} = K_i g^3, \tag{6}$$

where $K_i = -(11C(G)/3 - 2T(R)_{\mathrm{fermion}}/3 - T(R)_{\mathrm{Higgs}}/6)/16\pi^2$, $C(G)\delta^{ad} = f^{abc}f^{dbc}$, and $T(R)$ is the index number of representation R. Matching coupling constants at different symmetry breaking scales, we have

$$\sin^2\theta_W = 1 - [(1 + 4C'^2 - \chi)(1 - 8\pi K_L \alpha(\bar{\mu}) \ln(\mu/\bar{\mu}))$$
$$+ \chi\alpha(\bar{\mu})/\alpha_s(\bar{\mu})(1 - 8\pi K_C \alpha_s(\bar{\mu})) \ln(\mu/\bar{\mu})$$
$$+ 32\pi C'^2 K_1 \alpha(\bar{\mu}) \ln(\mu/\bar{\mu})]/[1 + 8C'^2 - \chi]$$
$$\mu_0 = \frac{\mu \exp[\sin^2\theta_W/\alpha(\bar{\mu}) - 1/\alpha(\bar{\mu}) - (K_L - K_C)\ln(\mu/\bar{\mu})]}{8\pi(K_L^0 - K_C^0)}, \tag{7}$$

where K_i and K_i^0 are the K factor defined by equation 6 for coupling constant g_i below and above the generation symmetry scale μ, respectively, $\bar{\mu}$ is the electroweak breaking scale, $\chi = (K_L^0 - K_4^0)/(K_L^0 - K_C^0)$, and $C'^2 = 9/2N(N-1)$ (in our case $N = 4$). In order to have $\mu_0 > \mu$, it is necessary to have $K_L^0 - K_C^0 > 0$. Allowing for all possible (1, 1, 3, X) fermions in the theory, the condition $K_L^0 - K_C^0 > 0$ allows only 20 possible ways to have (1, 1, 3, X) fermions, among which only two cases are satisfactory for our requirements. These are (1) one (1, 1, 3, 4) and one (1, 1, 3, 6), both with masses below μ; and (2) two (1, 1, 3, 4) with masses below μ.

If the fourth generation of quarks should have the same charge assignments as the first three generations, the minimal generation symmetry group is $SU(5)_G$. In this case quarks transform as

$$q_L: (2, 1, 3, 5)$$
$$q_R: (1, 2, 3, 5) \tag{8}$$

and the C operator is:

$$\begin{pmatrix} 1/3 & 0 & 0 & 0 & 0 \\ 0 & 1/3 & 0 & 0 & 0 \\ 0 & 0 & 1/3 & 0 & 0 \\ 0 & 0 & 0 & 1/3 & 0 \\ 0 & 0 & 0 & 0 & -4/3 \end{pmatrix} \tag{9}$$

The fifth generation of quarks have charges $(-1/6, -7/6)$. It is interesting to note that in this case it is possible to have the number of standard quark generations equal to the number of standard lepton generations. The leptons in this case transform as

$$L_L: (2, 1, 1, 10)$$
$$L_R: (1, 2, 1, 10) \tag{10}$$

The 10 of $SU(5)_G$ decomposes under $SU(4)$ as

$$10 = 4(-1) + 6(2/3). \tag{11}$$

We expect six additional lepton generations with charges $(5/6, -1/6)$. It is not possible to satisfy our requirements if dynamic symmetry breaking is assumed but only if Higgs are introduced to break symmetries. It is possible that the following Higgs break symmetries and give masses to gauge bosons and fermions: $H(2, 2, 1, 1)$, $H(2, 2, 1, 24)$, $H(1, 2, 2, 75)$, and $H(2, 1, 1, 10)$. These Higgs give fermions masses of order $\bar{\mu}$ and break the electroweak symmetry: $H(1, 2, 1, 10)$ and $H(1, 1, 1, 24)$. They also break $SU(2)_R$ and generation symmetries. Again additional $(1, 1, 3, X)$ fermions are needed to achieve our requirements. We found by renormalization analysis of coupling constants that only two cases give the correct value of $\sin^2 \theta_W$. These two cases are: (1) four $(1, 1, 3, 5)$ with masses below μ and two $(1, 1, 3, 5)$ with masses above μ; and (2) three $(1, 1, 3, 5)$ with masses below μ and three $(1, 1, 3, 5)$ with masses above μ.

There are several experimental implications. We only discuss the $SU(4)_G$ model in the following, but the discussion of the $SU(5)_G$ model is very similar.

The generation gauge bosons mediate some rare decays, such as $K^{\pm} \to \pi^{\pm} e^{\mp} \mu^{\pm}$ and $\Sigma^{+} \to p e^{+} \mu^{-}$, with branching ratios 10^{-12} and 10^{-14}, respectively. Decays $K_L \to \pi^{\pm} e^{\pm} \mu^{\mp}$ and $\Sigma^{+} \to p e^{-} \mu^{+}$ are forbidden, and the contribution to $\mu \to e\gamma$ or $\mu \to 3e$ are negligible.

There are several new fermions with masses of order 100 GeV to a few TeV. These are (a) one doublet (U, D) (the fourth generation of quark) with charges $(0, -1)$; (b) nine lepton doublets $(L_{2/3}, L_{-1/3})$ of charges $(2/3, -1/3)$; (c) three lepton doublets $(L_{4/3}, L_{1/3})$ of charges $(4/3, 1/3)$; and (d) two quark singlets $(Q_{1/6}, Q_{1/6}, Q_{1/6}, Q_{-1/2})$ of charges $(1/6, 1/6, 1/6, -1/2)$ [or one $(Q_{1/6}, Q_{1/6}, Q_{1/6}, Q_{-1/2})$ and one $(Q_{1/3}, Q_{1/3}, Q_{1/3}, Q_{-1/3}, Q_{-1/3}, Q_{-1/3})$ of charges $(1/3, 1/3, 1/3, -1/3, -1/3, -1/3)$]. The fermion doublets will contribute to $\delta\rho$. From experimental bounds $\delta\rho < 0.02$,[5] the masses of quarks and leptons must be nearly degenerate in each doublet $m_U^2 - m_D^2 < (200\ \text{GeV})^2$ and $m_E^2 - m_N^2 < (340\ \text{GeV})^2$, respectively. If the integer-charged quarks are heavier than fractionally charged leptons, (U, D) will decay into $(u, d) + \bar{L}_{2/3} + \nu_e$, $(u, d) + L_{-1/3} + e^-$, etc., by $SU(4)_G$ gauge interactions. The leptons of charge $(4/3, 1/3)$ can decay into those of $(2/3, -1/3)$, for example, $L_{4/3} \to L_{2/3} + L_{2/3} + \bar{\nu}$, $L_{1/3} \to L_{-1/3} + L_{-1/3} + e^+$. In the quark singlets $(1/6, 1/6, 1/6, -1/2)$, the lightest quark (most likely of charge $1/6$) would be stable and the others decay as $Q_{-1/2} \to Q_{1/6} + \bar{L}_{2/3} + \nu$, etc. The lifetimes are expected to be in the range 10^{-9} to 10^{-16} sec for mass differences in the 50- to 100-GeV range. Finally, one of the leptons $L_{2/3}$ or $L_{-1/3}$ will decay into the other by β-decay and one will be stable.

There are spectacular signatures from the production and decay of the integer charge quarks at

$$P\bar{P} \to \begin{pmatrix} D & \bar{D} \\ U & \bar{U} \end{pmatrix} \to \left\{ \begin{matrix} \begin{pmatrix} d \\ u \end{pmatrix} + \bar{L}_{2/3} + \nu, \begin{pmatrix} \bar{d} \\ \bar{u} \end{pmatrix} + L_{2/3} + \bar{\nu} \\ \begin{pmatrix} d \\ u \end{pmatrix} + \bar{L}_{-1/3} + e^-, \begin{pmatrix} \bar{d} \\ \bar{u} \end{pmatrix} + L_{-1/3} + e^+ \end{matrix} \right\}, \tag{12}$$

leading to events with two jets plus (a) $e^+ + e^- + \bar{L}_{-1/3} + L_{-1/3}$; (b) $\not{P}_T + e^- + \bar{L}_{-1/3} + L_{2/3}$; (c) $\not{P}_T + e^+ + L_{-1/3} + \bar{L}_{2/3}$; and (d) $\not{P}_T + L_{2/3} + \bar{L}_{2/3}$, all with equal rates. Here L_q refers to a stable massive lepton of charge q where $q = -1/3$ or $2/3$. Similarly from the production and decay of the quarks of charges $(1/6, -1/2)$, one can expect signatures like $P\bar{P} \to q + \bar{q} + q + \bar{q} + Q_{1/6} + \bar{Q}_{1/6} \to 4$ jets $+ Q_{1/6} + \bar{Q}_{1/6}$, where $Q_{1/6}$ is expected to be stable.

SUMMARY

As have seen, there are some heavy stable fermions, namely $L_{2/3}$, $L_{-1/3}$, and $Q_{1/6}$. If these particles do indeed exist, they should have been produced in the early universe and should have been found in the experiment.[6] Experiments have set an upper limit $<10^{-18}$ (stable fermions/nucleon).[7]

REFERENCES

1. GLASHOW, S. L. 1961. Nucl. Phys. **B22:** 579; WEINBERG, S. 1967. Phys. Rev. Lett. **19:** 1264; SALAM, A. 1968. *In* Proceedings of the 8th Nobel Symp. Wiley. New York.
2. PAKVASA, S. & H. SUGAWARA. 1986. Prog. Theor. Phys. Lett. **75:** 1471.
3. PARTICLE DATA GROUP. 1986. Phys. Lett. **170B**: 1.
4. GEORGI, H., H. R. QUINN & S. WEINBERG. 1974. Phys. Rev. Lett. **33:** 451.
5. VELTMAN, M. 1977. Nucl. Phys. **B123:** 89; KIM, J., P. LANGACKER, M. LEVINE & H. WILLIAMS. 1980. Rev. Mod. Phys. **53:** 211.
6. DOVER, C. B., T. K. GAISSER & G. STEIGMAN. 1979. Phys. Rev. Lett. **42:** 1117.
7. LYONS, L. 1985. Phys. Rep. **129:** 225.

Probing the Four-Generation Kobayashi–Maskawa Matrix with Supergravity Proton Decay

R. ARNOWITT[a,b] AND PRAN NATH[c,d,e]

[a]*Center for Theoretical Physics*
Department of Physics
Texas A&M University
College Station, Texas 77843

[c]*Lyman Laboratory of Physics*
Harvard University
Cambridge, Massachusetts 02138

INTRODUCTION

It may at first sight seem paradoxical to use a decay that has not been seen to test for the existance of a fourth generation that has not been observed. However, proton decay has always represented a strong constraint on any grand unified theory (GUT). Thus in the standard $SU(5)$ GUT model, one has

$$\tau_{th}(p \rightarrow e^+\pi^0) \leq 2.3 \times 10^{31}\text{ y}, \quad (1)$$

while the experimental lower bound[1]

$$\tau_{\exp}(p \rightarrow e^+\pi^0) \geq 3.6 \times 10^{32}\text{ y} \quad (2)$$

is considerably larger.

In supergravity GUT models, the situation is more complicated. Since M_{GUT} is larger than in the Standard Model,

$$M_{\text{GUT}} \simeq 1 \times 10^{16}\text{ GeV} \quad (3)$$

proton decay proceeds through the exchange of, the superheavy Higgsino triplet, followed by gaugino dressing[2] (FIGURE 1). Each vertex contains Kobayashi–Maskawa (KM) matrix elements. Thus in supersymmetry, proton decay measures the strength of the KM matrix elements, and thus gives information on the number of generations.

[b]To whom correspondence should be addressed.
[d]Visiting Scholar.
[e]Permanent address: Department of Physics, Northeastern University, Boston, Massachusetts 02115.

DECAY AMPLITUDES

One may calculate the effective dimension six decay amplitudes of FIGURE 1 for arbitrary decay channels.[3] The full result is quite complicated, and we record here only the dominant terms. For an arbitrary number of generations, one has, for the $\bar{\nu}_i K$ mode, the effective nucleon (N) decay Lagrangian

$$L_6(N \rightarrow \bar{\nu}_i K) = \left(\frac{\alpha_2}{M_H}\right)^2 m_{di}\, V_{ui}^*\, A_2^*\, C_2 \left[\left\{1 + \sum_{j=3}\left(\frac{m_{uj}}{m_c}\right)\left(\frac{A_j^*}{A_2^*}\right) P_j\right\} \alpha_i^L\right] + \Delta_i, \quad \textbf{(4)}$$

where m_{di} and m_{ui} are the u and d quark masses, A_i is the combination $A_i = V_{id} V_{is}^*$ of KM matrix elements V_{ij}, and α_i^L are the quark–lepton four-field interactions $\alpha_i^L =$

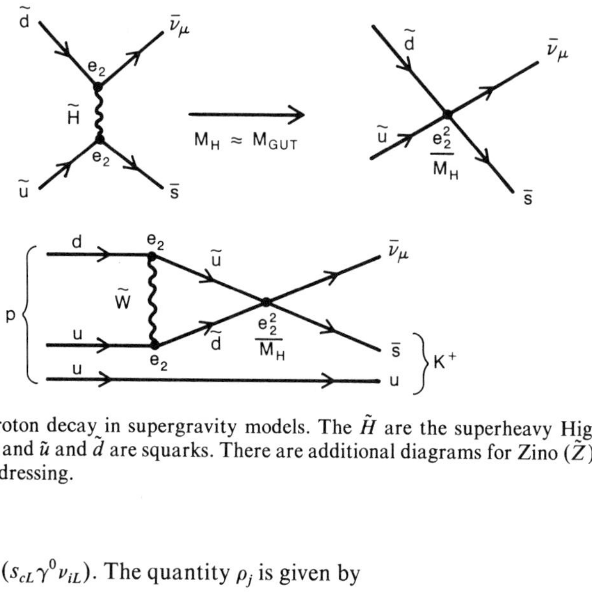

FIGURE 1. Proton decay in supergravity models. The $\tilde{H}$ are the superheavy Higgsino triplet, $\tilde{W}$ is the Wino, and $\tilde{u}$ and $\tilde{d}$ are squarks. There are additional diagrams for Zino ($\tilde{Z}$), photino ($\tilde{\gamma}$) and gluino ($\tilde{g}$) dressing.

$\epsilon_{abc}(d_{aL}\gamma^0 u_{bL})(s_{cL}\gamma^0 \nu_{iL})$. The quantity ρ_j is given by

$$\rho_j = \frac{P_j F_j}{P_2 F_2}, \quad \textbf{(5)}$$

where F_j is the form factor arising from the Wino triangle loop of FIGURE 1 for the jth generation, and P_j are the additional (PC-violating) phases of nucleon decay. The Δ_i represent additional, generally small, contributions, and C_2 is an overall normalization factor.

The KM elements that govern the rate of proton decay are the front factor V_{ui} and the combination A_i. This latter quantity obeys the unitarity relation

$$\Sigma A_i = 0. \quad \textbf{(6)}$$

The A_i also enter the rare K decays. Thus the branching ratio for $K^+ \rightarrow \pi^+ \nu\bar{\nu}$ is given by[4]

$$\mathrm{BR}(K^+ \rightarrow \pi^+ \nu\bar{\nu}) \simeq 1.5\, N_\nu \times 10^{-5} \left| \sum A_i\, D\left(\frac{m_{ui}2}{M_w2}\right)\right|^2 \qquad \textbf{(7a)}$$

and from the bound on $K_L \rightarrow \bar{\mu}\mu$ one has[4,5]

$$Q(K_L) \equiv \left| \mathrm{Re} \sum A_i C\left(\frac{m_{ui}2}{M_w2}\right)\right| \lesssim 2 \times 10^{-3}, \qquad \textbf{(7b)}$$

where C and D are corresponding loop integrals and N_ν is the number of neutrinos (i.e., generations).

We see, therefore, that it is possible to relate proton decay to the rare K decay modes.

TWO-GENERATION MODEL

In order to see the significance of the preceding formulas, it is convenient to consider successively models based on two generations, three generations, and four generations. If we consider only the contributions from the first two generations,[6] the dominant decay modes are $p \rightarrow \bar{\nu}_i K^+$ and $n \rightarrow \bar{\nu}_i K^0$. The decay rate depends on the Higgs triplet mass M_H, for example,

$$\Gamma(p \rightarrow \bar{\nu}_i K^+) = (\alpha_2)^4 (M_H)^{-2} |m_{di} V^*_{ui}|^2 |B_2|^2, \qquad \textbf{(8)}$$

where B_2 is an overall normalization factor depending on the masses in the loop of FIGURE 1. While M_H depends upon physics at the GUT scale and cannot be determined theoretically, Enqvist *et al.*[7] have pointed out that within the framework of $N = 1$ supergravity models that preserve the gauge hierarchy, one has the bound

$$M_H \lesssim M_{\mathrm{GUT}} \simeq 1 \times 10^{16}\,\mathrm{GeV}. \qquad \textbf{(9)}$$

Experimentally,[1] one has a lower bound on the partial lifetime for the $p \rightarrow \bar{\nu} K^+$ decay of 5.0×10^{31} y. For squarks with mass $\simeq$ 200 GeV this implies

$$M_H \gtrsim (10 - 70) \times 10^{16}, \qquad \textbf{(10)}$$

which grossly violates the bound of equation 9.

THREE-GENERATION MODEL

One can, of course, eliminate the preceding problem by making the squarks sufficiently heavy so that the loop integral in B_2 becomes small. However, the inconsistency still remains for squarks with masses $\lesssim$350 GeV (which is in the range where they would be detectable at Tevatron and/or SSC). Fortunately, however, when the third generation is included, an alternate way of satisfying equation 9 arises, that

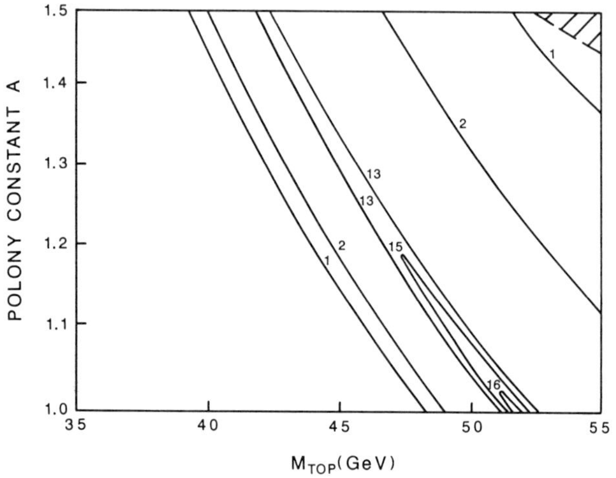

FIGURE 2. Contours of $\Gamma(p \rightarrow \bar{\nu}\pi^+)/\Gamma(p \rightarrow \bar{\nu}K^+)$ for gravitino $m_{3/2} = 178$ GeV.[8]

is, it is possible that the third- and second-generation effects roughly cancel.[3] This requires that the brace in equation 4 be small, that is,

$$\left(\frac{m_t A_3}{m_c A_2}\right)\rho_3 \approx -1. \tag{11}$$

Equation 11 is satisfied for the usual three-generation KM matrix if there is significant left–right mixing in the squark mass matrix, that is, if the Polonyi constant $A_P \approx 1$ (e.g.,[3] if $m_t \simeq 50$ GeV, one finds $\rho_3 \simeq -4$ for $P_3/P_2 = 1$). All the $N \rightarrow \bar{\nu}_i K$ modes are universally suppressed, while the $N \rightarrow \bar{\nu}_i \pi$ modes are enhanced. One finds generally $M_H \approx M_{\text{GUT}}$ and while existing experiments would not yet be able to see proton decay, several modes are reasonably close to observability.[3] Further, equation 11 can be satisfied for a wide range of parameters, as can be seen from FIGURE 2. Thus equation 11 holds when the $p \rightarrow \bar{\nu}_i \pi$ rate is comparable or greater than the $p \rightarrow \bar{\nu}_i K$ rate.

We next consider the K decay modes of equation 7. [Since the cancellation (equation 11) requires an approximately real proton decay amplitude, we neglect the imaginary parts of A_i in the following discussion.] One has

$$\text{BR}(K^+ \rightarrow \pi^+ \nu\bar{\nu}) \simeq 1.5\, N_\nu \times 10^{-5} \left| A_2 D_2 - \left(\frac{m_c A_2}{m_t \rho_3}\right) D_3 \right|^2 \tag{12a}$$

$$Q(K_L) = \left| A_2 C_2 - \left(\frac{m_c C_3}{m_t \rho_3}\right) A_2 \right|. \tag{12b}$$

For $m_t = 50$ GeV one finds $\text{BR}(K^+ \rightarrow \pi^+\nu\bar{\nu}) = 2 \times 10^{-10}$ and $Q(K_L) = 5 \times 10^{-4}$. In general, the experimental constraint $Q(K_L) \lesssim 2 \times 10^{-3}$ is automatically satisfied for top-quark mass $m_t < 130$ GeV, and for all m_t obeying the constraint on $Q(K_L)$ one has $\text{BR}(K^+ \rightarrow \pi^+\nu\bar{\nu}) < 8 \times 10^{-10}$.

FOUR-GENERATION MODEL

For four generations, the situation is again different. From equation 4 we see that the condition that the $\bar{\nu}_i K$ proton decay modes remain suppressed now reads

$$m_t A_3 \rho_3 + m_t' A_4 \rho_4 \simeq - m_c A_2, \tag{13}$$

where m_t and m_t' are the t and t' masses. Using equation 13 and the unitarity condition (equation 6), one may solve for A_3 and A_4:

$$A_3 = \frac{m_t' (A_1 + A_2) \rho_4 - m_c A_2}{(m_t \rho_3 - m_t' \rho_4)}$$

$$A_4 = \frac{-m_t (A_1 + A_2) \rho_3 + m_c A_2}{(m_t \rho_3 - m_t' \rho_4)}. \tag{14}$$

The result depends sensitively on the combination $A_1 + A_2$, which is not too well determined experimentally. One has[9]

$$A_1 + A_2 = 0.0246 \pm 0.0315. \tag{15}$$

There are also UA1 experimental lower bounds on the t and t' quark masses and the τ' lepton mass[10]

$$m_t, m_t' \gtrsim 40 \text{ GeV}$$

$$m_\tau' > 41 \text{ GeV}, \tag{16}$$

while the renormalization group analysis gives upper bounds on the t', b', and τ' masses[11]

$$m_t' < 140 \text{ GeV} \qquad m_b' < 135 \text{ GeV} \qquad m_\tau' < 70 \text{ GeV}. \tag{17}$$

One may consider two possible cases. (1) "Superstring-inspired models," where there is no left–right mixing in the squark mass matrices ($A_P \simeq 0$). Here $|\rho_3| = 1 = |\rho_4|$. (2) "Supergravity models" with large left–right mixing ($A_p \approx 1$). Here $|\rho_3| \neq |\rho_4|$ and both can be large. Here we only discuss case (1). In general, the region allowed by the $K \rightarrow \mu\bar{\mu}$ constraint (equation 7b) for masses in the range of equations 16 and 17 is a

TABLE 1. Examples of $Q(K_L)$ and $BR(K^+ \rightarrow \pi^+ \nu\bar{\nu})$ of Equations 7 for Various t and t' Masses (in GeV) for the Case $\rho_3 = -1 = \rho_4$

m_t	m_t'	$A_1 + A_2$	$Q(K_L)$	$BR(K^+)$
40	140	0.010	0.8×10^{-3}	1.6×10^{-9}
40	100	0.010	0.9×10^{-3}	1.6×10^{-9}
80	140	0.010	0.9×10^{-3}	1.5×10^{-9}
40	60	0.019	0.6×10^{-3}	0.8×10^{-9}
40	140	0.037	14.4×10^{-3}	0.02×10^{-9}
40	60	0.037	8.0×10^{-3}	0.01×10^{-9}

Note: Large values of $A_1 + A_2$ cause $Q(K_L)$ to exceed the experimental limit of 2×10^{-3}.

complicated function of m_t, m_t', A_2, and the signs of ρ_3 and ρ_4. In general, we find[9] acceptable solutions only when $A_1 + A_2$ is within the range

$$-0.025 \lesssim A_1 + A_2 \lesssim 0.022, \qquad \textbf{(18)}$$

which lies within but is considerably narrower than the current experimentally allowed range of equation 15. Further, the $\mathrm{BR}(K^+ \rightarrow \pi^+\nu\bar{\nu})$ falls below the maximum value allowed for three generations only for a very narrow band $-0.006 \lesssim A_1 + A_2 \lesssim 0.002$, where there is an accidental cancellation of third- and fourth-generation effects. Otherwise, one expects the $\mathrm{BR}(K^+ \rightarrow \pi^+\nu\bar{\nu})$ to be ≈ 5 times the maximum allowed three-generation bound, as can be seen from the examples of TABLE 1. Hence there would almost always be a clear distinction between three and four generations in supersymmetry (SUSY) models for experiments that can measure $\mathrm{BR}(K^+ \rightarrow \pi^+\nu\bar{\nu})$ values to an accuracy of $\approx 5 \times 10^{-10}$.

The approximate cancellation (equation 13) suppresses the $N \rightarrow \bar{\nu}_\mu K$ modes so that equation 9 is correctly satisfied. From equation 4 one sees that the $\bar{\nu}_\tau K$ and $\bar{\nu}_\tau' K$ modes will also be correctly suppressed, provided V_{ub} and V_{ub}' are not too large. This leads to the following bounds on these matrix elements:

$$|V_{ub}| \lesssim \frac{m_s\,|V_{us}|}{m_b} = 6 \times 10^{-3} \qquad \textbf{(19a)}$$

$$|V_{ub}'| \lesssim \frac{m_s\,|V_{us}|}{m_b'} \lesssim \frac{m_s\,|V_{us}|}{2\,m_\tau'} \lesssim 4 \times 10^{-4}, \qquad \textbf{(19b)}$$

where we have used equation 16 and[11] $m_b' \simeq 2m_t'$. Equation 19a is not a strong constraint, though equation 19b is quite limiting. Alternately, equations 19 imply $|\tan\theta_3| \lesssim 3 \times 10^{-2}$ and $|\sin\theta_5| \lesssim 6 \times 10^{-2}$.

SUMMARY AND CONCLUSIONS

It is possible to relate proton decay rates in $N = 1$ supergravity models with the rare K-meson decays $K_L \rightarrow \mu\bar{\mu}$ and $K^+ \rightarrow \pi^+\nu\bar{\nu}$. The requirement that the triplet Higgs mass obey $M_H \lesssim M_{\mathrm{GUT}}$ puts strong constraints on the KM matrix elements and generally allows one to distinguish three generations from four generations. For three generation, the branching ratio obeys the constraint $\mathrm{BR}(K^+ \rightarrow \pi^+\nu\bar{\nu}) \lesssim 8 \times 10^{-10}$ for any m_t (and $\lesssim 3 \times 10^{-10}$ for $m_t < 70$ GeV), while for four generations $\mathrm{BR}(K^+ \rightarrow \pi^+\nu\bar{\nu}) \approx 1 \times 10^{-9}$, and the parameters $A_i = V_{is}^* V_{id}$ are constrained to obey equation 18. The matrix element V_{ub} and V_{ub}' are constrained by equations 19.

REFERENCES

1. TOTSUKA, Y. 1985. *In* Proceedings of the 1985 Int. Symp. on Lepton and Photon Interactions. N. Konuma and K. Takahashi, Eds.
2. WEINBERG, S. 1982. Phys. Rev. **D26:** 287; SAKAI, N. & T. YANAGIDA. 1982. Nucl. Phys. **B197:** 533.
3. NATH, P., A. H. CHAMSEDDINE & R. ARNOWITT. 1985. Phys. Rev. **D32:** 2348.
4. INAMI, T. & C. S. LIM. 1981. Prog. Theor. Phys. **65:** 297.

5. ELLIS, J. & J. HAGELIN. 1983. Nucl. Phys. **B217:** 189.
6. SAKAI, N. 1984. Nucl. Phys. **B229:** 317; CAMPBELL, B. A., J. ELLIS, & D. V. NANOPOULOS. 1984. Phys. Lett. **141B:** 229.
7. ENQVIST, K., A. MASIERO, & D. V. NANOPOULOS. 1985. Phys. Lett. **156B:** 209.
8. YUAN, T. C. 1986. Phys. Rev. **D33:** 1894.
9. ARNOWITT, R. & P. NATH. 1987. Phys. Rev. **D36:**3423.
10. CLINE, D. B. 1987. Missing transverse energy events and a search for additional neutrino families and a fourth-generation lepton. This volume.
11. GOLDBERG, H. 1985. Phys. Lett. **165B:** 292; BAGGER J., S. DIMOPOULOS & E. MASSO. 1985. Phys. Rev. Lett. **55:** 920.

Strongly Coupled Generation in the Standard Model

R. JOHNSON[a] AND J. SCHECHTER

Physics Department
Syracuse University
Syracuse, New York 13244-1130

INTRODUCTION

For sufficiently strong Yukawa couplings the Standard Model goes over to the linear sigma model. We point out that the associated heavy quarks may behave like strongly interacting low-energy u- and d-quarks, so one cannot argue that very heavy quark systems will provide an easy test of quantum chromodynamics (QCD). An upper quark mass bound of about 1.5 TeV is found. (This should also apply to techniquarks.) Experimental confirmation might come from the observation of very tightly bound baryons made from three heavy quarks.

DISCUSSION

The possible existence of a still heavier fourth generation of quarks and leptons is a question of great interest and importance for understanding physics at the next level of accelerator energies. In addition to the many phenomenological constraints on further generations discussed at this conference, there exist some interesting theoretical mass bounds[1-5] on heavy fermions. These bounds are all in the range of several hundred gigaelectronvolts to a few teraelectronvolts, and are derived by demanding that various aspects of the perturbative Standard Model not break down. Interestingly, a simple extrapolation of the up-quark mass spectrum yields a value of $m_{t'} \simeq m_t/m_c \cdot m_t \simeq 1$ TeV for m_t about 45 GeV. In this paper we discuss yet another way of deriving a mass bound for heavy quarks that, unlike other approaches, attempts to describe the model after a transition to a new "strongly coupled" phase is reached.

Let us consider the minimal Standard Model with one Higgs doublet and one heavy quark doublet,

$$\mathcal{L} = \mathcal{L}_{\text{gauge}} - \tfrac{1}{2}|D\mu\Phi|^2 - \bar{q}\gamma_\mu D_\mu q - c(\Phi^\dagger\Phi - v^2)^2$$
$$- (g_1\bar{q}_L\Phi b'_R + ig_2\bar{q}_L\sigma_2\Phi^* t'_R + \text{h.c.}), \qquad \textbf{(1)}$$

where

$$\Phi = \begin{bmatrix} H^+ \\ H^\circ \end{bmatrix} \quad \text{and} \quad q_L = \begin{bmatrix} t' \\ b' \end{bmatrix}_L .$$

[a]Present address: Iowa State University, Ames, Iowa 50010.

Normally, one does perturbation theory about the classical solution

$$\Phi = \begin{bmatrix} 0 \\ v \end{bmatrix}$$

with quarks aquiring masses,

$$m_{b'} = g_1 v \qquad m_{t'} = g_2 v \qquad v = 174 \text{ GeV}. \tag{2}$$

One can easily see from equation 2 that a mass of 1 TeV would correspond to a Yukawa coupling as large as $g = 6$. In order to treat this large coupling problem we draw on an analogy with calculations done for the nucleon–pion system.[6] If we use the notation

$$\Phi = \begin{bmatrix} \sqrt{2} i\pi^+ \\ \sigma - i\pi^0 \end{bmatrix}$$

equation 1 can be rewritten,

$$\mathcal{L} = \mathcal{L}_{\text{gauge}} + \mathcal{L}_\sigma + \mathcal{L}'$$

$$\mathcal{L}' = \frac{(g_1 - g_2)}{2} [i\pi^0 \bar{q}(1 - \gamma_5)q + \bar{q}\tau_3(\sigma - i\tau \cdot \pi)q]$$

$$\mathcal{L}_\sigma = -\frac{1}{2}(\partial_\mu \sigma)^2 - \frac{1}{2}(\partial_\mu \pi)^2 - \bar{q}\not{\partial} q - c(\sigma^2 + \pi^2 - v^2)^2 - g\bar{q}(\sigma - i\gamma_5 \tau \cdot \pi)q$$

$$g \equiv \frac{g_1 + g_2}{2}. \tag{3}$$

It seems reasonable as a first approximation to neglect the gauge degrees of freedom when the Yukawa couplings are large. As for the strength of $\mathcal{L}'$, the ρ-parameter constraint[1] yields roughly $|m_{t'} - m_{b'}| < 300$ GeV, or for our case, $|g_2 - g_1|v/|g_2 + g_1|v \lesssim 300 \text{ GeV}/2 \text{ TeV} = 15\%$. Thus it would seem reasonable to neglect $\mathcal{L}'$ (as well as possible mixing to other generations) in this first approximation. We are left with just the Gell-Mann–Levy linear sigma model,[7] $\mathcal{L}_\sigma$, which has recently been studied in the context of strong coupling for the nucleon–pion system.[6] We now show that the classical solution (equation 2) does not always have the lowest energy in the fermion sector. Following Soni,[8] let us compare its energy with the ground-state energy estimated in a semiclassical manner using a hedgehog ansatz for the scalar field configuration,

$$\sigma = v \cos \theta(r) \qquad \pi_i = v\hat{x}_i \sin \theta(r)$$

$$q = \begin{bmatrix} \dfrac{iF(r)}{r} \chi \\ \sigma \cdot \hat{x} \dfrac{G(r)}{r} \chi \end{bmatrix} \qquad \chi = \frac{\alpha_1\beta_2 - \alpha_2\beta_1}{\sqrt{2}}, \tag{4}$$

where α_i and β_i are two component spin and isospin vectors, respectively. Thus q has positive parity and $\mathbf{K} = \mathbf{I} + \mathbf{J} = 0$. The total energy of the system of N fermions in a single orbital[b] is

$$E_{\text{tot}}[\theta, F, G] = N\epsilon[\theta, F, G] + E_H[\theta]$$

$$E_H[\theta] = 4\pi v^2 \int_0^\infty r^2\, dr\left[\frac{1}{2}\theta'^2 + \frac{\sin^2\theta}{r^2}\right]$$

$$\epsilon[\theta, F, G] = \int_0^\infty dr\left[gv(F^2 - G^2)\cos\theta - 2FG\left(gv\sin\theta + \frac{1}{r}\right) + GF' - FG'\right] \quad (5)$$

and its self-consistent minimization is to be compared with Ngv, the energy of N fermions with Yukawa masses gv. We take for the consistent boundary conditions on $\theta(r)$

$$\theta(\infty) = 0 \qquad \text{and} \qquad \theta(0) = -n\pi \qquad n = 0, \pm 1, \pm 2. \quad (6)$$

Before presenting the exact numerical work, we illustrate the qualitative features of the solution in a simplified model with $\theta(r)$ specified by

$$\theta(r) = -\pi\left(1 - \frac{r}{R}\right) \qquad r < R$$

$$\theta(r) = 0 \qquad r > R. \quad (7)$$

Kahana *et al.*[6] found that the energy of the fermions is well approximated by $\epsilon \simeq (\pi/R - gv)$ and the scalar field energy is given by $E_H = 2\pi v^2 R(1 + \pi^2/3)$. Thus the total energy can be minimized with respect to the size parameter R, and one finds,

$$R \simeq \frac{\sqrt{N}}{3v}$$

$$E \simeq 6\pi v\sqrt{N} - Ngv. \quad (8)$$

Note that the energy decreases as g increases. Compared to the energy of N free fermions whose masses rise as gv we see that the hedgehog ansatz must yield a lower energy for g greater than some critical g_c. The exact numerical results verify this and are displayed in FIGURE 1 for the cases $N = 1$ and $N = 3$ and the boundary condition for $\theta(0)$ being chosen $n = 1$.

For the case $N = 1$ (baryon number $B = 1/3$) the self-consistent numerical calculation predicts $g_c \simeq 8$ and $E_{\text{tot}} < 1.4$ TeV. However, this state is not yet an eigenstate of I and J. One can introduce collective coordinates for the scalar field analogous to the treatment for the nucleon.[9] Then one finds a spectrum of particles with $I = J$ and

[b] We use color purely as an index here. Obviously a one-quark state will be a color triplet. However, for two quarks in the same space–spin–isospin state they must be in an antitriplet state to obey the Pauli principle. Also three quarks must be in a color singlet state.

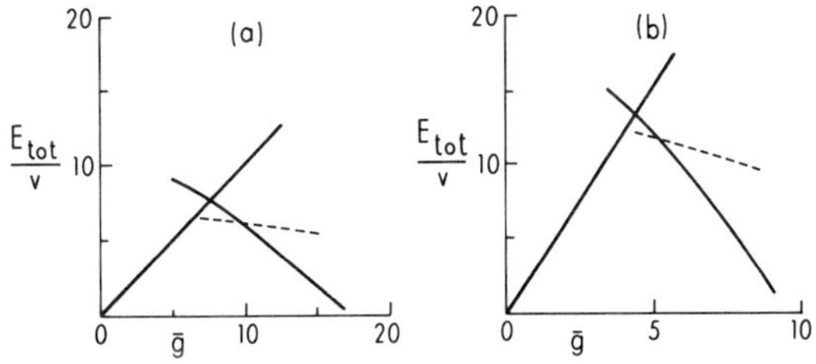

FIGURE 1. The self-consistently determined energy E_{tot}/v plotted against g for the (a) $N = 1$ and (b) $N = 3$ cases. The straight lines with positive slope represent the energy in the uniform Higgs phase.

masses $m(I = J) = E_{tot} + J(J + 1)/2\lambda$, where $\lambda = \frac{8}{3}\pi v^2 \int_0^\infty \sin^2\theta(r)r^2\, dr$ is the "moment of inertia." We see that for $g > g_c$ this model has the exciting possibility of exotic quarks with $J = I = \frac{3}{2}$ and $Q = I_3 + B/2 = \frac{5}{3}, \frac{2}{3}, -\frac{1}{3}, -\frac{4}{3}$, although we expect these states are rather unstable. For the standard $J = I = \frac{1}{2}$, $B = \frac{1}{3}$ quark state there is a bound on its mass,

$$m_{q'} < 1.8 \text{ TeV}. \tag{9}$$

Thus an $e^+ e^-$ collider with center-of-mass energy 3–4 TeV should be able to see all generations!

For the $N = 3$ $(B = 1)$ case, FIGURE 1 shows that $g_c \simeq 4.5$ and $E_{tot} < 2.2$ TeV. Adding collective coordinates now gives "heavy" nucleon and delta states $\tilde{N}$ and $\tilde{\Delta}$. If we consider, for example, $g = 6$, FIGURE 1 tells us that a 2-TeV baryon is to be composed of three quarks each of mass 1 TeV. This indicates that above $g_c \simeq 4.5$ we expect very strongly bound baryons—more analogous to the situation of ordinary QCD for light quarks rather than for heavy quarks. One might be able to look for $e^+e^- \rightarrow \tilde{N}\bar{\tilde{N}}$ + W's and Z's at center-of-mass energy around 5 TeV.

A few remarks should be made about the reliability of these semiclassical calculations. First we note that changing the boundary condition on n from $n = 1$ to $n = 0$ or $n = 2$ raises the energy slightly. The overall shape for $\theta(r)$ remained unchanged, as shown in FIGURE 2. Thus we are not artificially stabilizing a solution

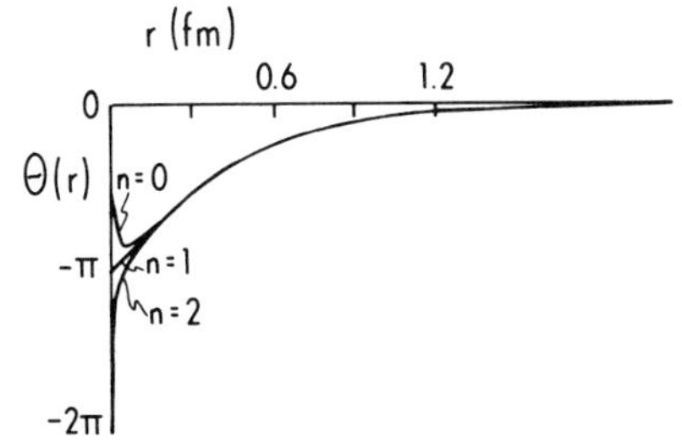

FIGURE 2. Plot of $\theta(r)$ vs. r for the choice of boundary conditions $[\theta(0) = -n\pi]$; $n = 0, 1$, and 2. One quark is present in the valence level and g has been set to 10.

with a topological number that, of course, is not conserved when the gauge interactions are present. We also note that the inclusion of $\mathcal{L}'$ in first-order perturbation theory still gives no contribution to the masses. Secondly, one can generalize the ansatz (equation 4)

$$\sigma = H(r)\cos\theta(r)$$

$$\pi_i = H(r)\hat{x}_i \sin\theta(r) \qquad \text{with} \quad H(\infty) = v. \tag{10}$$

The effect of the new degree of freedom, H, is to lower the energy slightly, depending on the Higgs mass, and make the mass bounds even stronger. Similarly one can include gauge field degrees of freedom using an ansatz like that given in reference 10. We expect that the mass bound will remain and perhaps be lowered. Finally, one can see from FIGURE 1 that as g gets large, the energy eventually becomes negative. Kahana and Ripka[11] claim that this is an artifact of the approximation that neglects the variation of the "Dirac Sea" levels. Taking this "Casimir energy" into account also has the effect of lowering the mass bound.

CONCLUSION

To conclude, we find that a semiclassical nonperturbative treatment of the Yukawa sector of the minimal Standard Model[12] seems to imply the existence of a mass bound on heavy quarks

$$m_{q'} < 1.8 \text{ TeV}$$

and a possible new phase of strongly bound baryons with mass

$$M_{B'} < 2.2 \text{ TeV}.$$

Furthermore, if one interprets $\mathcal{L}_\sigma$ as arising from a Technicolor theory, these same bounds should apply for techniquarks and technibaryons. It is not immediately obvious how the addition of nonminimal Higgs would affect these results, but one might suspect that the additional Higgs fields would not be excited unless they contributed to a lower energy for the fermion system. Thus the mass bound should only be lowered. Evidence[13] for a nonminimal Higgs sector may have already been seen in the $\tau \longrightarrow \eta\pi\nu$ decays reported by J. Repond in this volume.

REFERENCES

1. (ρ parameter) VELTMAN, M. 1977. Nucl. Phys. **B123:** 89; EINHORN, M. B. *et al.* 1981. Nucl. Phys. **B191:** 146; COHEN, A. *et al.* 1984. Nucl. Phys. **B232:** 61.
2. (Partial wave unitarity) LEE, B. W. *et al.* 1977. Phys. Rev. Lett. **38:** 883; LEE, B. W. *et al.* 1977. Phys. Rev. **D16:** 1519; CHANOWITZ, M. S. *et al.* 1978. Phys. Lett. **78B:** 285.
3. (Higgs potential) HUNG, P. Q. 1979. Phys. Rev. Lett. **42:** 873; POLITZER, H. D. & S. WOLFRAM. 1979. Phys. Lett. **82B:** 242; DUNCAN, M. J. *et al.* 1985. Phys. Lett. **153B:** 165.
4. (Fixed point) HILL, C. T. 1981. Phys. Rev. **D24:** 691; PENDLETON, M. & G. G. ROSS. 1981. Phys. Lett. **98B:** 291.
5. ($1/N$ expansion) EINHORN, M. B. & G. J. GOLDBERG. 1986. Phys. Rev. Lett. **57:** 2115.

6. Kahana, S., G. Ripka & V. Soni. 1984. Nucl. Phys. **A415:** 351; Birse, M. C. & M. K. Banerjee. 1984. Phys. Lett. **136:** 284.
7. Gell-Mann, M. & M. Levy. 1960. Nuovo Cimento **16:** 705.
8. Soni, V. 1985. *In* Proceedings of the Second Hellenic School on Elementary Particle Physics. Corfu, Greece.
9. Adkins, G., C. Nappi & E. Witten. 1983. Nucl. Phys. **B228:** 552.
10. Eilam, G. *et al.* 1986. Phys. Rev. Lett. **56:** 1331.
11. Kahana, S. & G. Ripka. 1985. Phys. Lett. **155B:** 327.
12. For more details, see Johnson, R. & J. Schechter. 1986. Phys. Rev. **D36:**1484.
13. Johnson, R., N. W. Park & J. Schechter. 1987. Syracuse Univ. preprint SU-4228-350. Syracuse, N.Y.

Beyond Truth and Beauty[a]

HARALD FRITZSCH[b]
CERN
Geneva, Switzerland

This meeting started three days after receiving the exciting news that a supernova had exploded in our cosmic neighborhood. Last Monday (February 23, 1987) the first photons from the new supernova, which exploded about 150,000 years ago in the Great Magellanic Cloud, reached the earth and alerted the astronomers. From a particle physics point of view the interesting question arises whether anything else has reached the earth at about the same time or slightly before, for example, neutrinos or gravitons. The theoretical models of supernova explosions suggest that a large part of the energy in a supernova explosion is emitted in terms of neutrino radiation. Those neutrinos should be detectable here on earth, and a careful investigation of the neutrino pulse might even reveal details, such as information about the neutrino mass spectrum. The times of flight of the various neutrinos arriving here on earth depend on the masses. For mass splittings in the electronvolt-range differences in the time of flight of the order of a few seconds are expected, and they may be measurable, since the initial neutrino burst of a supernova explosion is expected to be very sharply peaked. Probably the supernova of February 1987 is still too far away and too weak to allow such an investigation. However, the next and probably more powerful supernova, this time not in the backyard of our galaxy, but right in its basement, may be just around the corner. Let us hope that all neutrino and gravity wave detectors are ready then to receive the signal.

The title of this summary talk reflects the theme of this conference. Everyone attending this meeting is interested in the possibility that the spectrum of leptons and quarks may extend beyond the heavy *b*- and *t*-quarks and the τ-lepton discovered (or in the case of *t*, anticipated firmly) thus far. And everyone will agree that such an insight into the lepton–quark spectrum would have immediate consequences with respect to our theoretical understanding of the particle physics universe. Thus far all particle physics phenomena have been found to be in accordance with the Standard Model—the well-known superposition of quantum chromodynamics (QCD) and the electroweak theory. No doubt, the Standard Model exhibits signs of eternal truth. Our understanding of, in particular, the basic interaction of photons with electrons and quarks [quantum electrodynamics (QED)] and of the gluons with themselves or with quarks (QCD), and to a somewhat lesser extent the interaction of the weak bosons with the leptons and quarks show the simplicity and beauty that one expects from a basic and true theory. The old saying of Plato, "Beauty is the Splendor of Truth," finds here its well-justified application.

All the complexities of atomic physics can be explained in terms of the fundamen-

[a]This work was supported in part by DFG Fr. 412/7-1.
[b]Present address: Physics Department, University of Munich, Munich, Federal Republic of Germany.

tal interaction among charged objects (electrons, protons, nuclei) and photons, dictated by the principle of the electromagnetic gauge invariance. The coupling strength, given in terms of the coupling constant α, and the various masses of the particles serve as independent and free parameters.

Likewise the huge body of complex nuclear and hadronic phenomena is uniquely determined by the QCD interaction among quarks and gluons, which is determined by requiring the non-Abelian gauge invariance in the color space. Unlike in atomic physics, in nuclear and hadronic physics there is no strength parameter like α that can be changed arbitrarily, since QCD strength parameter $\alpha_s(\mu)$, which depends on the associated energy scale μ, determines not only the basic strength of the quark–gluon or gluon–gluon interaction, but at the same time the hadronic mass parameters like the nucleon mass. A change of $\alpha_s(\mu)$ would simply result in a rescaling of all mass parameters, for example, the QCD scale Λ_c (we disregard all quark masses for the moment). But the absolute scale of the latter is a free parameter; the ratio Λ_c/m_e, for example, is a parameter that cannot be determined by the QCD interaction, but only within an extended theory of QCD and QED, possibly including the weak interactions. Of course, the various quark masses serve as independent and free parameters in QCD.

It is surprising that, on the one hand, a dramatic breakthrough has been made in understanding the structure of the basic interactions, but on the other hand, little progress has been made in order to understand the rather bizarre looking tower of mass scales in the Standard Model, especially those of the mass spectrum of the leptons and quarks. The "explanation" of these mass scales given in the Standard Model by relating them to a set of Yukawa coupling constants describing the strengths of the couplings of the leptons and quarks to the scalar Higgs field is nothing but a convenient parametrization. It offers no insight into the dynamical details of the mass generation. A deeper understanding of the problem can only be reached by going beyond the Standard Model, which probably means also going beyond some of the apparent beauty and truth of the Standard Model and replacing some of its features by something else.

If we look with scrutiny at the physics displayed by the Standard Model, including gravity, we observe that three of the basic interactions are mediated by massless gauge quanta and are intimately related to exact conservation laws, namely gravity, the QED, and the QCD interaction. On the other hand, the fourth interaction, the weak interaction, violates many symmetries. But the mass generation and the breaking of symmetries seem to be related to each other. One may suspect that some, if not all the clues to understanding the complex tower of quark and lepton masses are hidden in the weak interaction sector, and a careful study of the dynamics of the weak bosons, for example, at LEP II, could reveal enough secrets to enable us to find these clues.

At present the theorists contemplating how physics might look beyond the Standard Model can be divided into two groups:

(1) The large number of theorists who have adopted the "string dogma." For them, the real world is simply a superficial and nearly irrelevant shadow of the complex underlying ten-dimensional world of superstrings. The theory stands out as a consistent theory of gravity and other interactions. But thus far this is the only achievement of the theory. Whether some or all of the outstanding puzzles of the Standard Model can

eventually be explained by the underlying dynamics of the superstring world remains to be seen. Personally I am sceptical that such a step will be made soon. It may well be that the superstring theories remain in the future what they are today—new approaches to generalize the quantum field theory. Direct contact with the Standard Model might come on a different level, as assumed by most researchers today, for example, one at which a nontrivial substructure of the quarks and leptons plays a significant role.

The superstring theory starts with the dynamics at energies of the order of the Planck energy (10^{19} GeV). The physics of the Standard Model is considered as a low-energy phenomenon eventually explained as the low-energy tail of the superstring dynamics. Thus one starts from high up on the energy scale and works oneself down, from a speculative 10^{19}-GeV utopia to the 10^{2}-GeV reality.

(2) The second group of theorists tries to go the other way, the inductive way. They start from the observed dynamics of the Standard Model at energies of about 0.1 TeV and try to extrapolate to the next level up in the energy scale, taking into account both theoretical insights and experimental facts. Physics has always developed that way. But this aspect is largely ignored by the theorists of the first group. Maybe, this time the direction of research is indeed reversed, and those that start from the speculative heights of the superstring mountain gain eventually enough sight to overlook the Standard Model. But thus far the mountain is surrounded by clouds, which may never disappear. Most theorists in this audience, however, seem like myself to belong to the second group that slowly climbs upward, keeping in constant touch with the Standard Model physics, that is, with the real world.

The center of the discussion at this meeting was a possibly existing fourth family of leptons and quarks. This is, of course, a legitimate topic, since thus far no good explanation exists why there should be just the three families observed today. We do not even know whether the number of lepton–quark families is a well-defined number. It might well be that leptons and quarks are composite systems, in which case an infinite number of families is expected to exist, just as there exists an infinite number of baryon octet or decuplet states. Thus many opportunities would arise to organize meetings like this one.

But one could also have a situation in which the number of families plays a significant role as an important algebraic number of the underlying theory, as in the Standard Model the number of space dimensions or the number of QCD colors. Since at present a large number of theorists work on superstring theories, I would like to remind the audience that the space we live in has three dimensions, not nine or ten. It is rather suspicious that the number three enters three times in the present form of the Standard Model: there are three space dimensions, three QCD colors, and three families. Could there be a connection between these apparently unrelated different kinds of "threenesses"?

Of course, everyone knows that we should extend the three-dimensional space to the four-dimensional or rather 3 + 1-dimensional space–time manifold. Besides the three QCD colors another conserved quantum number is important, the electric charge. The group of exactly conserved gauged charges in the Standard Model is $SU(3)^c \times U(1)$, that is, again a 3 + 1-pattern appears. Following these analogies, one might suspect that another fourth family exists, or that in particular the physics of the family space exhibits an underlying 3 + 1 pattern as follows:

$$\begin{pmatrix} \nu_e & u \\ e^- & d \end{pmatrix} \begin{pmatrix} \nu_\mu & c \\ \mu^- & s \end{pmatrix} \begin{pmatrix} \nu_\tau & t \\ \tau^- & b \end{pmatrix}$$

$$\boxed{\begin{pmatrix} \nu_\sigma & h \\ \sigma^- & l \end{pmatrix}.}$$

The fourth family consists of a new lepton called σ, its neutral partner ν_σ, a new quark h of charge $2/3$, and a new quark 1 of charge $-1/3$. In naming these objects, I followed the tradition of denoting new charged leptons (all, except the electron) by Greek letters, and the neutral partner by the symbol ν with the corresponding subscript, although the ν_σ-particle might not be a traditional neutrino with a vanishing or very small mass, but rather a very heavy particle. The letters denoting the new quarks h and l are derived from "high" and "low," in analogy to "top" and "bottom." (People who are less pragmatically oriented might prefer "hate" and "love," in analogy to "truth" and "beauty.")

As alluded to previously, there might be a 3 + 1-structure in the family space, not a simpleminded generalization from three to four, in the sense that the fourth family is perhaps nearly decoupled from the first three families, or that it plays a significant role in the generation of the masses for the members of the first three families.

Of course, we currently do not know whether a fourth family exists, not to mention possible values for the masses of its members. Possible guesses can be obtained by looking at the masses of the members of the first three families.

The observed mass spectrum shows an interesting feature, which is not understood: the mass ratios m_i/m_{i+1} (m_i: mass of a quark or a charged lepton belonging to family i; m_{i+1}: mass of the corresponding quark or lepton of family $i + 1$) are all small and, very roughly speaking, of similar order of magnitude such that one is tempted to connect the mass values by straight lines on a logarithmic scale. This works well for the quarks of charge $-1/3$, since $m_d/m_s \approx m_s/m_b \approx 1/20$. If this trend continues, the fourth quark of charge $-1/3$, the l-quark, should have a mass of about $20 \cdot m_b \approx 105$ GeV. But this is the mass value, obtained at an energy scale of 1 GeV, where the other quark mass values are defined. The physical l-mass, relevant for the spectroscopy of $\bar{l}l$-onia or l-flavored mesons, is reduced by a factor of about 0.6 such that $m_l \approx 60, \ldots, 65$ GeV.

An important parameter to extrapolate to the fourth family is the t-quark mass. But one might expect that the ratios m_h/m_l, m_t/m_b, and m_c/m_s are of similar order of magnitude, in which case we would have $m_h \approx 5 \ldots 10 \cdot m_l \approx \approx 250 \ldots 600$ GeV.

The ratio m_b (1 GeV)$/m_\tau$ is about three, in accordance with the expectation within the minimal $SU(5)$ theory, and a reasonable guess would be to take a similar value for the ratio m_l (1 GeV)$/m_\sigma$, in which case we have $m_\sigma \approx m_l(1\text{ GeV})/3 \approx 35$ GeV.

These values are, of course, nothing but rough guesses, with little theoretical foundation, but they might provide some preliminary guidance for the experimentalists searching for phenomena due to the presence of a fourth family.

The observed lepton–quark spectrum is of an exponential type, that is, the mass values increase roughly like exp (const. $\cdot$ i) [i: family quantum number]. Thus this spectrum is unlike any other spectrum of states observed in physics before, for example, the spectrum of atoms, nuclei, or hadrons. Atomic levels are most suitably displayed on a linear energy scale; likewise, nuclear levels. The spectrum of the light hadrons, on the other hand, should be displayed on a scale quadratic in the energy; this reflects the relativistic dynamics of the light quarks.

It looks as if the masses of the lighter families are generated by a perturbative mechanism involving the masses of the members of the heavier families, a mechanism that at the same time could lead to the phenomenon of weak interaction mixing.

Suppose there exists a fourth generation, and suppose that the masses of the h and l quarks serve as driving terms for the generation of the lighter quark masses by a mixing phenomenon that at the same time leads to the observed flavor mixing. In this case, the mass matrix for the up- and down-type quarks could be

$$M = m_{h,l} \cdot \begin{pmatrix} 0 & a & 0 & 0 \\ a & 0 & b & 0 \\ 0 & b & 0 & c \\ 0 & 0 & c & 1 \end{pmatrix} \qquad (1)$$

It is well known that such an ansatz, in the case of three generations, describes well the observed flavor mixing angles in terms of quark mass ratios. In the case of four generations the masses m_h and m_l would be the dominating and most important mass terms. The ratio m_h/m_l might be of special relevance, since this ratio would presumably not be influenced by the flavor mixing phenomenon, due to the large mass values. It could be close to four, that is, given by the ratio of the squares of the quark charges, as suggested by simple composite models in which the lepton and quark masses are of electromagnetic origin. (If the ratio were four, the h-mass would be $m_h = 4m_l \approx 250$ GeV, using $m_l \approx 60, \ldots, 65$ GeV, as mentioned previously.)

Independent of any particular model for the mass matrices, the idea of connecting the mass eigenvalues and the mixing angles leads one to the possibility of estimating the order of magnitude of the t-quark mass. Since the u and d masses are very small compared to m_t or m_b, we can neglect the first family and look solely at the interplay between the second and the third families. The flavor mixing between these two families is observed to be rather small, the relevant mixing angle being of the order of 0.04. Thus it would be a good approximation to start from the limit in which there is no flavor mixing. In this case, one might expect that the various masses are proportional: $m_t/m_b = m_c/m_s$, that is, $m_t(1\text{ GeV}) \approx m_b \cdot (1\text{ GeV}) \cdot (m_c/m_s)$. The ratio m_c/m_s is not well determined, due to the rather uncertain value of m_s, which may vary between about 120 MeV and 220 MeV. The ratio m_c/m_s correspondingly varies between 5 and 10; thus, m_t (1 GeV) $\approx$ 26 GeV . . . 55 GeV. Taking into account the QCD rescaling factor (about 0.6) to obtain the physical t mass, one obtains $m_t \approx$ 15 GeV . . . 34 GeV. The lower part of this spectrum is already ruled out by the e^+e^--annihilation experiments, while the upper part may still be consistent with the present experimental constraints.

Taking into account the flavor mixing between the second and third family, a larger range of t-quark values is obtained. The exact range depends on the specific ansatz, which determines the interplay between the masses and the mixing angles. In general, we expect

$$\frac{m_c}{m_t} \cong \frac{m_s}{m_b} - c_1 \cdot \theta - c_2 \cdot \theta^2 \ldots . \tag{2}$$

Such an expansion makes sense, since $\theta = \theta_{cb}$ is observed to be rather small: $\theta \approx 0.045$, and the second-order term $c_2\theta^2$ can already be neglected. If we start from a mass matrix of the type of equation 1 (with a specific choice of the phase parameters), one finds, for example,

$$\frac{m_c}{m_t} \cong \frac{m_s}{m_b} - 2\theta \sqrt{\frac{m_s}{m_b}} . \tag{3}$$

Given the uncertainty of the various quark mass values (especially the uncertainty of m_s), one obtains in this case a rather broad range for the physical t-quark mass: 30 GeV $< m_t <$ 80 GeV. The uncertainty of the strange-quark mass comes essentially from our inability to extract the expectation value of the s-quark bilinear $s^+\gamma_{0s}$ in a hyperon state. The naive quark model suggests that this expectation value is close to the expectation value of s^+s, which counts the number of strange quarks in the wave function. Personally I believe that in reality one is close to that situation, in which case the strange-quark mass is most likely in the range between 150 and 180 MeV. In this case the physical t-quark mass is restricted to the range 40 GeV $< m_t <$ 60 GeV. Notice that rather little room is left for the possibility of producing $\bar{t}t$-pairs at the Stanford Linear Center (SLC) (this would require $m_t \lesssim$ 47 GeV).

Searching for the fourth family of leptons and quarks means for the experimentalists searching for the effects of the production of the new heavy lepton or of the new quarks. Useful information in this respect was given in the talks of Cline and Branson. The UAI-group has searched for a standard new charged lepton, produced in the W-decay and decaying into a massless or light neutrino under the emission of a virtual W disintegrating into a $\bar{q}q$-pair. The signal would be quite clear: A large amount of missing energy, accompanied by two quark jets. Masses up to 41 GeV for the new lepton are excluded. This limit is very useful. It implies, that the new lepton, if it exists, cannot be lighter than about twenty times the τ-mass. This already poses a slight problem for our guess of the masses described in the introduction. In any case, it is demonstrated that a hadron-collider is able to yield useful information even about purely leptonic phenomena—a field that used to be thus far the topic of e^+e^--colliders. It is not excluded that the new lepton is eventually found that way—it may be just around the corner. If our guess made previously contains some truth, the mass of the fourth charged lepton should not be much heavier than 50 GeV.

The mass constraint is not valid if the new charged lepton is accompanied by a heavy, perhaps even stable neutrino such that the mass difference is relatively small (of the order of a few GeV or less). Such a possibility, employing a stable heavy neutrino, can be utilized, as described by West, to solve the solar neutrino problem and the missing matter problem of cosmology at the same time. In order to take part actively in the solar dynamics, the new heavy neutrino must have a rather large magnetic moment—an aspect that I find less attractive, since special new Higgs fields have to be designed in order to generate it. In composite models, however, the situation may be less serious, since here a magnetic moment may appear due to the internal structure, like the neutron magnetic moment in hadron physics.

How the UAI bound on the heavy lepton mass can be evaded has been discussed in detail by Barger. Especially the case $m_{\nu_L} > M_W - m_L$ is worth being considered. In this case, no bound exists. In the case $m_L < m_{\nu_L} < M_W - m_L$, the bound is reduced to about 22 GeV (i.e., the limit set by PETRA).

Thus far the search for the t-quark at the CERN collider has not led to a positive result, as described by Eggert. However, the prospects are still there to find the t-quark at a hadron collider, either at CERN or at FNAL (as emphasized by Yoh), and a hopeful method employing a sophisticated jet analysis for the UA2-detector at CERN was described by Minkowski.

The new quarks were searched for in the e^+e^--annihilation at DESY, as reported by Branson. The appearance of the t-quark throughout the energy region searched at DESY (up to about 47 GeV) can be ruled out safely, posing a limit on the t mass of the order of 24 GeV. As noted especially by the Mark J and Jade groups, one seems to find a signal in low thrust hadronic events with isolated muons for $\sqrt{s} > 46.3$ GeV. It is completely unclear how this signal should be interpreted. It may come from the production of a new quark of charge $-1/3$, the b' or l quark, but due to the planarity of the events and the lack of the corresponding inclusive electron signal, this interpretation begins to look unlikely. Perhaps new, exotic objects, for example, leptoquark bosons, are produced. In any case, one may have a repetition of the ADONE–Doris syndrome. The ADONE storage ring was operating for a long time just below the $\bar{c}c$-systems, and Doris likewise just below the $\bar{b}b$-systems. Thus PETRA and PEP are perhaps unlucky by not quite reaching the energy for the abundant production of the new objects, and what Branson has reported was just a short look through the peephole before the slide came down and PETRA was closed. In any case, the matter will be settled soon, since Tristan just started to operate.

Atwood presented the news from the construction site of the SLC, which soon will be ready as the first Z^0 factory of the century, followed more than one year later by LEP. If new quarks and/or leptons exist with a mass less than $M_{Z^0}/2$, they will be discovered first at the SLC, and presumably investigated in detail at LEP. The event topology of the new quarks and leptons is such that 10,000 Z^0 events would be enough to obtain a clear signal. However, the prospects of the SLC finding the t, b' (l)-, or t' (h)-quark or the $L^-(\sigma^-)$-lepton are somewhat reduced by the UAI limit on the heavy lepton mass and by the naive guesses for the mass scales made previously (of course, these guesses could be entirely wrong, and the UAI limit would not be applicable if the new lepton is accompanied by a sufficiently heavy neutrino).

Let me try to assess the probabilities of finding new quarks or leptons at the SLC, taking for granted (100% chance) that a fourth family exists. On the basis of the mass ranges discussed previously, the guess is:

t-quark:	20%
l-quark:	5%
h-quark:	1%
σ-lepton:	30%
ν_σ-neutrino:	95%.

The SLC will also offer for the first time the possibility of counting the number of neutrinos at $E > M_Z$, by looking for the reaction $e^+e^- \longrightarrow \gamma Z^0 \longrightarrow \gamma(\bar{\nu}\nu)$. All neutrinos with masses $m_\nu < M_{Z^0}/2$ will be counted that way. That way one will find out whether the neutrino counting of high-energy physics and the neutrino counting of the astrophysicists or cosmologists converge to the same number. The latter way was described in the talk by Dr. Schramm, who during his talk was in a more excited state than usual, as a result of the supernova explosion 150,000 years ago.

In cosmology one finds constraints on neutrinos (and possibly other particles) provided they have a mass of less than 10 MeV. Originally the constraints on the number of (left-handed) neutrinos came from the observed abundance of ^{4}He. As emphasized by Schramm, other constraints can be found by considering the D and ^{3}He abundances, as well as the ^{7}Li abundance. All constraints converge to the same number, allowing 3, at most 4, neutrinos. As he said, the possibility of having four neutrinos begins to look unlikely but, in view of the systematic errors affecting most astrophysical observations, cannot be ruled out. Thus cosmology and the fourth family are not yet married, but are only dating each other occasionally. It is safe to predict that the marriage will happen quickly, once the first member of the fourth family is discovered.

Whether a fourth family exists at relatively low energies depends on the details of the mass generation mechanism, which in the Standard Model is provided by the Higgs mechanism. If the Higgs particle(s) exist(s), it can be found at LEP, as described by S. Wu. LEP I and SLC could find the H^0 particle, if it has a mass of less than about 40 GeV. If it is heavier, we have to wait a few years for the completion of the LEP 200 stage. As Wu explained in detail, LEP 200 has good prospects of finding the H^0, if its mass is in the range 40 GeV . . . ~80 GeV. The range could perhaps be slightly extended, up to ~110 GeV. This shows that it may take many years until the details of the "Higgs sector" are understood. If the H^0 is heavier than 100 GeV, only the experimentalists in the next millennium, still in high school at present, may be able to discover it. But in those days the theorists may have come up with another idea!

D. Nanopoulos, the conductor of the Madison superstring orchestra, has discussed the problems in the superstring theory that suddenly arise if a fourth family is admitted. The problem is not one connected with the basic foundations of the superstring theories, but is rather related to the extrapolations of the various mass scales entering the theory. But even for three families related problems exist, although less serious. In any case, we predict that the problems will evaporate once the fourth family is discovered.

The spectrum of views regarding the fourth family displayed at this meeting can be appreciated by quoting the following three statements from three different talks:

WEST: The fourth family is the most important one, providing more than 90% of all matter in the universe.

SCHRAMM: The fourth family is almost ruled out.

NANOPOULOS: The fourth family and superstrings do not mix.

I should like to mention an interesting aspect of the superstring theory as far as the lepton–quark mass spectrum is concerned. Since the details of the mass generation are linked to specific algebraic properties of the underlying Calabi–Yau or orbifold compactification, one has a realistic chance to obtain specific algebraic patterns for the mass matrices, such as described in equation 1. More work needs to be done in this direction. The observed pattern of the flavor mixing matrix and the possible underlying pattern of the quark and lepton mass matrix, discussed also in the introduction, was described by Gronau, supplemented by a specific analysis of the fourth family by Shin. Mohapatra analyzed these patterns with respect to their stability under radiative corrections.

The mass matrices were studied by C. Jarlskog from a different perspective based on invariant functions. To underline the beauty of her approach, she started with a beautiful transparency with Gothic letters painted by her daughter. She emphasized correctly that the mass matrices are not directly observable, but rather are theoretical constructions, which are by no means unique. Especially noteworthy is her approach to the flavor mixing problem based on invariants. For a dynamical theory of flavor mixing to be developed in the future, this may be very important, if it turns out that these invariants enter in the theory in a crucial manner. A good example in this respect is General Relativity. The metric tensor of space–time, which can be considered as the analog of the mass matrix, is not the relevant dynamical quantity of General Relativity, but invariant functions like the space–time curvature R, which enters directly in the Einstein field equations.

Arguments speaking for the existence of a fourth family were presented by C. Hill, who noted that the masses of the members of a fourth family are expected to be such that their couplings to the Higgs field are about the same as the $SU(2)$ gauge couplings.

Repond described the new experimental τ-decay result from PEP (HRS detector). The observation of the decay $\tau^- \longrightarrow \nu_\tau \eta \pi^-$ (BR 5.1 ± 1.0 ± 1.2%) by the HRS group would constitute a major blow for the standard electroweak model, if it is confirmed by other experiments. This decay is forbidden in the Standard Model, since the W^- couples to the quark current $\bar{u}\gamma_\mu d_L$ (L: left-handed), which acts like a superposition of ρ^- and A_1^-. G-parity forbids the transition $\rho^- \longrightarrow \eta\pi^-$, while parity conservation of the strong interaction does not allow the transition $A_1^- \longrightarrow \eta\pi^-$. Typical estimates of the violation of G-parity via isospin violations due to electromagnetism or the u-d mass difference give a branching ratio of at most 10^{-3} for the decay $\tau \longrightarrow \nu_\tau \eta\pi$. If the effect is indeed confirmed, new weak currents (e.g., second-class currents, scalar contributions to the weak current) would have to be introduced. I am sceptical as far as such modifications are concerned, since new weak currents imply new contributions to the total decay rate, and it would be a miracle if the τ-lifetime is consistent with estimates based on the standard electroweak model. I am reminded of the quotation of Sir Arthur Eddington: "It is also a good rule not to put too much confidence in experimental results, until they have been confirmed by theory."

No such problems exist with the theory as far as neutrinos from supernovas are concerned. Nussinov discussed in his talk the neutrino flux emitted by a supernova. He estimated that the Kamioka detector should observe about a dozen neutrino events in a short time interval (~10 s). (His prediction came quite close to the observation of the Kamioka group reported one week after this meeting.)

A number of talks were concerned with possible signatures of a fourth family in the physics observed today, that is, below the energy for exciting the new flavors directly. Gilman discussed the constraints on the weak interaction parameters imposed by the low- and high-energy experiments. By now the various weak interaction mixing parameters are determined quite well, such that due to the observed small size of the transition $b \rightarrow u$, a possible problem exists as far as the measured value of the *CP* parameter ϵ is concerned. Since the contribution from short-distance physics to ϵ is proportional to m_t^2, it might be that one needs a fairly large value of m_t ($m_t > 60$ GeV) in order to get agreement with the experiments. However, the uncertainty in the relevant hadronic matrix elements is still large (a factor that could be as large as 5).

Still a lot of fog surrounds the ratio ϵ'/ϵ. Assuming that the $|\Delta I| = 1/2$ weak transitions are dominated by the penguin contributions, one may connect the latter and the ratio ϵ'/ϵ, predicting $\epsilon'/\epsilon \sim 10^{-2}$. Such a relatively large value seems unlikely now, in view of the new experimental results, reported here by Bock (experiment E731 at FNAL). As discussed by Holder, the new experiments, for example, the one under way at CERN, will either establish a nonzero value of ϵ'/ϵ or set a new limit at the 1‰ level.

Gilman discussed possible ways to accommodate such a small value of ϵ'/ϵ, especially by raising the value of m_t, or the *B*-parameter. In this context, I should like to emphasize that a well-established direct link doesn't yet exist between the $|\Delta I| = 1/2$ enhancement of the nonleptonic *K* decays, the ϵ-parameters, and the penguin terms. Furthermore, as mentioned by both Gronau and Paschos, the ϵ-parameters are rather sensitive to possibly existing contributions of further families.

Very interesting new results were presented in the talk of Albrecht (Argus group, DESY), which established the existence of $B_d^0 - \bar{B}_d^0$ mixing. In the standard three-generation model, taking into account the observed flavor mixing pattern, such a mixing must exist, since it can always proceed via the box diagram that likewise in the two-generation model causes the $K^0 - \bar{K}^0$ mixing. The short-distance contribution to the $B_d^0 - \bar{B}_d^0$ transition mass $|M_{12}| = (\Delta M/2)$ is proportional to $\eta f_B^2 B_B |V_{td}|^2 m_t^2$, where η is a QCD parameter estimated to be of order 1, f_B is the *B*-meson decay constant, and B_B is the parameter relating the relevant four-quark matrix element to the one derived from the vacuum insertion approximation. The mixing is only observable if the B^0-$\bar{B}^0$ transition takes place fast enough, that is, before the decay takes place. The relevant experimental quantity is $r = (\Delta M/\Gamma)^2/[2 + (\Delta M/\Gamma)^2]$.

The Argus result implies $r_d \sim 20\%$, which gives $(\Delta M/\Gamma) \sim 0.7$. It is very surprising that this value of $\Delta M/\Gamma$ is of the same order of magnitude as $\Delta M_K/\Gamma_{K_s}$. In view of the vastly different decay mechanisms for *B*- and *K*-mesons—the *B*-mesons decay dominantly by the well-understood β-decay of the *b*-quark, while the large value of Γ_{K_s} is still not understood—the coincidence of $\Delta M_K \sim \Gamma_{K_s}$ and $\Delta M_B \sim \Gamma_B$ looks like a miraculous coincidence, and that's what it probably is. It seems a useless undertaking to find a dynamical explanation for this coincidence.

As emphasized by Paschos (see also the talk of Gilman), the large value of the B_d^0 mixing observed at DESY did not come as a complete surprise for all theorists. In fact, using $f_B = f_K$, $B_B = 1$, and $V_{td} \approx 0.02$, one finds agreement with the observation for $m_t \approx$ 60 GeV, as mentioned by Gilman. In view of the uncertainties involved, it is remarkable that one obtains at least the correct order of magnitude. I should like to mention that the *B* decay constant f_B might be slightly larger than f_K (note that $f_K/f_\pi \sim$ 1.25; thus one might have $f_B/f_K > 1$). Nevertheless it seems, as emphasized by Bigi, that in view of the observation of the B_d^0 – mixing at the level seen at DESY, one should be prepared for a rather large value of m_t (probably larger than 50 GeV), and the prospects of seeing the $t\bar{t}$-production at the SLC or LEP I dwindle. Furthermore, it is remarkable that a large value of m_t, say $m_t > 60$ GeV, would solve three potential problems at the same time:

(a) The observed magnitude of the B_d^0-mixing does not cause problems.
(b) The observed value of the ϵ-parameter of *CP*-violation is easily reproduced.
(c) The smallness of the ratio ϵ'/ϵ is understood.

The evidence for *B*-meson mixing found some time ago by the UAI group, and reported here by Eggert, fits nicely together with the new DESY result, as in particular as emphasized by Ali. One observes about twice as many like-sign dimuon events as expected in the case of no mixing. A sizable mixing of the B^0-mesons is the only explanation of the observed effect within the Standard Model. However, the like-sign dimuons will, in general, originate both from an original $B_d^0 - \bar{B}_d^0$ or $B_s^0 - \bar{B}_s^0$ production. Theoretically one expects $\Gamma_{B_s^0}/\Gamma_{B_d^0} \approx 1, |\Delta M_{B_s}/\Delta M_{B_d}| \approx (f_{B_s}/f_{B_d})^2\ (V_{ts}/V_{td})^2$. The latter ratio is expected to be much larger than 1; typical estimates range between 5 and 50. Therefore, it is likely that $\Delta M_{B_s} \gg \Gamma_{B_s}$, that is, the B_s-mesons are fully mixed.

The UAI results are consistent with a 100% mixing of the B_s-mesons. In this case, essentially all the excess of the like-sign dimuons observed in the experiment would be due to $B_s - \bar{B}_s$ production.

It is quite possible that *CP*-violation is larger for the *B*-mesons than for the *K*-mesons. Presumably such effects are largest for the B_s-mesons. If the $B_s - \bar{B}_s$ transitions are indeed as rapid as previously discussed, in accordance with the UAI results, it will be extremely difficult, if not impossible, to measure the *CP*-violating decays directly by sophisticated heavy flavor detectors of the type described by Bjorken.

As mentioned by Eggert, a large part of the J/ψ particles observed in the collider experiments are due to $\bar{b}b$-production, where one of the *b*-quarks decays by emitting a J/ψ. This decay may play an important role in the future, since at collider energies above $\sqrt{s} \sim 1$ TeV (Tevatron, SSC, LHC), essentially all J/ψ particles produced in the collisions are expected to be due to the *b* decay. Thus the J/ψ's that can easily be identified by their leptonic decay modes could be used as a *b*-flavor tagging device. Since *b*-quarks are always produced in pairs, a hadronic collider could be converted into a relatively cheap *B*-factory, identifying one of the *b*-quark jets via its J/ψ decay mode and studying the decay of the other *b*-quark, emitted in the opposite direction.

The experimental status of the *B* decays was reviewed by Horwitz. He concentrated on rare decays, especially on those that can be caused by penguin diagrams, for example, $B \to K\pi$, $B \to Kl^+l^-$ (l: e, μ), or $B \to K\bar{\nu}\nu$. These decays proceed via a loop involving the *t*-quark (and possibly heavier quarks of charge 2/3).

The theoretical aspects of such decays were discussed by Soni. In the absence of any contributions due to the fourth or higher families, one expects the leptonic decays like $B \to Ke^+e^-$ to have a rather small branching ratio of the order of 10^{-6}. However, these decays are very sensitive to the presence of higher families. If a fourth family is present, one may have branching ratios of the order of 0.1%.

No rare decays of this type were found. The BR limit from CLEO for the decay $B \to Ke^+e^-$ is $2 \cdot 10^{-4}$, ruling out the upper part of the theoretical estimates for this decay. Likewise no signal has been found in the channels $B \to K^+\pi^-$ or $B \to \phi K^+$.

Of particular interest is the decay $b \to s + \gamma$. In the Standard Model with three generations this is a flavor-changing radiative decay that is expected to have a sufficiently large branching ratio in order to be seen in the forthcoming experiments, as mentioned by Soni and in the contributed paper of Trampetic. QCD radiative corrections cause the amplitude for this decay not to be multiplied by the suppression factor $(m_t^2 - m_c^2)/M_w^2$, but by $\alpha_s \cdot \ln(m_t^2/m_c^2)$, which is much larger than the previous factor such that BR($b \rightarrow s\gamma$) for $m_t \sim 50$ GeV is of the order of 0.1%. As mentioned by Trampetic, about 10–20% of all decays $b \rightarrow s + \gamma$ are expected to lead to the final state $K^*\gamma$, that is, a monoenergetic γ-ray.

The decay rate for $b \rightarrow s\gamma$ is, of course, dependent on possibly existing further generations. Thus it provides another window to look for effects of the fourth family.

Horwitz presented the status of the so-called "charm deficit," observed in B decays. In the Standard Model one expects to see in the average 1.15 charm particles per B-decay (one from the $b \rightarrow c$ transition, 0.15 from the transition $W \rightarrow \bar{c}s$). However, the measurements (from CLEO and Argus) give 0.86 ± 0.09, that is, the B decays seem to be less charming than expected by theorists. As Horwitz stated, the effect may not be as serious as indicated by the number given previously, since there may be errors in the measurements of the various branching ratios involved, in particular those for the decays of the charmed particles in the channels detected in the B decays. Nevertheless, it seems to me that the issue is important and needs to be clarified. If it turns out that a large fraction (say, 15–30%) of the B decays lead to final states without charmed particles, a serious difficulty for the Standard Model arises. As mentioned by Soni, the decay $b \rightarrow s$ + gluon would lead to a final state with nonzero strangeness, but no charm. In the Standard Model this decay contributes at most to a 1% level of all B decays. However, we must keep in mind that the weak interaction coupling properties of the t-b system are not yet fully known experimentally. It could be that the W-boson coupling to the t-b system has a nonvanishing but small right-handed component with strength ϵ, in which case the gluonic b decay would proceed via a chromomagnetic transition, and one finds

$$\frac{b \rightarrow s + \text{gluon}}{b \rightarrow s + X} \sim \epsilon^2 \alpha_s \cdot \left(\frac{m_t}{m_b}\right)^2 .$$

That is, the branching ratio for the $b \rightarrow s$ + gluon decay could easily be of the order of 10–20%. A small right-handed current would also enhance the $B - \bar{B}$ transitions, and might be the reason for the large effect observed by the Argus group.

Of course, in view of the great success of the Standard Model, new right-handed currents are not exactly what everybody would like to see. The present status of the electroweak sector of the Standard Model was reviewed by Marciano. It is impressive to see that the measurements of $\sin^2\theta_w$ are slowly reaching a precision at the 1% level. Furthermore, the neutral current strength parameter ρ is settling down on a value very close to 1 ($\rho = 1.001 \pm 0.004$), not allowing mass splittings among members of weak isospin doublets of more than about 200 GeV. In view of this constraint, it is expected that the masses of the members of a fourth family (if it exists) are not larger than a few hundred gigaelectronvolts.

The status of the theory of the weak decays of heavy mesons and baryons was the subject of the talk of L. Chau. Of course, the problems associated with these decays represent a challenge, since one is studying here a rather subtle interplay not only between the weak and the strong interactions, but also between the short-distance QCD dynamics, which is accessible by perturbative methods, and the long-distance dynamics plagued by confinement effects. L. Chau's Quark Diagram Scheme takes both into account. During her talk and Hitlin's talk on the Mark III results, it became clear that the decays of charm and b particles will remain a rich field of study for many years to come.

Bjorken's talk set the scene for the future. As he emphasized, a truly rich data sample for b decay physics can be obtained in hadron–hadron machines (fixed-target machines or colliders). In order to be able to investigate CP-violation in b decays, one needs about $10^9 \bar{b}b$/experiment, a number that cannot be reached in e^+e^- collisions. Of course, a very sophisticated and expensive detector would be needed to select and investigate the B particles produced in hadronic collisions, such as Bjorken's $6\pi \times 2$ heavy flavor "taster."

The role of high-energy e^+e^--colliders in the search for new quarks and leptons was the topic of the talk of Dorfan. Such machines will be available in the future. LEP II will reach an energy E_{cm} of about 200 GeV. New linear collider projects under study—especially at SLAC—should be able to reach energies above 1 TeV. Since the annihilation cross section behaves at high energies like $1/s$, a high luminosity ($\mathcal{L} \sim 10^{33}$ cm^{-2} sec^{-1}) is a must for these machines.

As Dorfan pointed out, it is not possible to find new quarks and leptons by looking for heavy quarkonia production or by measuring the increase of R. Instead it looks promising to search for distinctive topologies of the final states, especially for those involving leptons, for example, by observing a hadronic decay of the h-quark on the one side and a semileptonic decay of the $\bar{h}$-quark on the other side, leading to a quark jet, an electron or a muon, and a neutrino ("missing energy"). It is curious to note that at energies above the LEP II range, the W^+W^- production, that is, the process that will be the major topic of investigation at LEP II, is the most serious background for finding and investigating new flavors. In general one can say that linear e^+e^- colliders in the teraelectronvolt range with a luminosity at about 10^{33} cm^{-2} sec^{-1} seem to be the best machines, albeit not the ones that can be built today, to study new phenomena, especially new lepton and quark flavors, at energies above the weak interaction energy scale.

The search for new leptons and quarks at the superconducting super collider (SSC) was discussed by Wojcicki, who also described the status of the project today shortly after the decision of the president to approve the project. Of course, the final decision to build the SSC has to be made by Congress. I am impressed by the general enthusiasm toward the SSC prevailing in the physics community of this country—an important factor that Congress must take into account in its final decision and that will set constraints for the future projects of high-energy physics in Europe. But I should like to stress that in view of the unprecedented scale and costs of the SSC project, the particle physics community should aim at developing a coherent worldwide program of research in the future. Such a program should include the study of hadronic collisions using the SSC and/or the LHC-project (a hadron ring in the LEP tunnel at CERN), of e^+e^--collisions using a linear collider in the teraelectronvolt energy range, and of electron–proton collisions in the teraelectronvolt energy range. Carrying out such a program will require a worldwide collaboration, implying, for example, the participation of Europe in U.S. projects and vice versa. High-energy physics as a fundamental experimental field of research will only survive in the future if the available resources provided by the taxpayers are shared wisely and used in the most efficient way.

Note added in proof: A neutrino signal that is most likely related to the supernova in the Great Magellanic Cloud has been observed by several groups (Kamioka, IMB detector, Mont Blanc detector), as reported two weeks after this meeting at the Moriond Conference in Les Arcs (March 1987).

ACKNOWLEDGMENT

It is a pleasure to thank the organizers of this meeting, especially the scientific organizers N. Byers, D. Cline, J. Cornwall, and A. Soni, for all the effort made so that this meeting on a very speculative subject turned out to be as lively and stimulating as if one were dealing with observed facts. I am sure that as a result nature cannot resist but to provide us soon with the first hints that the spectrum of leptons and quarks does not end abruptly at the *t*-quark mass.

Index of Contributors